Advances in Manufacturing and Characterization of Functional Polyes

Advances in Manufacturing and Characterization of Functional Polyesters

Editors

Rafael Balart
Sergio Torres-Giner
Octavio Fenollar
Nestor Montanes
Teodomiro Boronat

MDPI • Basel • Beijing • Wuhan • Barcelona • Belgrade • Manchester • Tokyo • Cluj • Tianjin

Editors

Rafael Balart
Universitat Politècnica de
València (UPV)
Spain

Sergio Torres-Giner
Universitat Politècnica de
València (UPV)
Spain

Octavio Fenollar
Universitat Politècnica de
València (UPV)
Spain

Nestor Montanes
Universitat Politècnica de
València (UPV)
Spain

Teodomiro Boronat
Universitat Politècnica de
València (UPV)
Spain

Editorial Office
MDPI
St. Alban-Anlage 66
4052 Basel, Switzerland

This is a reprint of articles from the Special Issue published online in the open access journal *Polymers* (ISSN 2073-4360) (available at: https://www.mdpi.com/journal/polymers/special_issues/functional_polyesters).

For citation purposes, cite each article independently as indicated on the article page online and as indicated below:

LastName, A.A.; LastName, B.B.; LastName, C.C. Article Title. *Journal Name* **Year**, *Volume Number*, Page Range.

ISBN 978-3-0365-0280-9 (Hbk)
ISBN 978-3-0365-0281-6 (PDF)

Cover image courtesy of Alejandro J. Müller.

© 2021 by the authors. Articles in this book are Open Access and distributed under the Creative Commons Attribution (CC BY) license, which allows users to download, copy and build upon published articles, as long as the author and publisher are properly credited, which ensures maximum dissemination and a wider impact of our publications.

The book as a whole is distributed by MDPI under the terms and conditions of the Creative Commons license CC BY-NC-ND.

Contents

About the Editors . . . . **vii**

Preface to "Advances in Manufacturing and Characterization of Functional Polyesters" . . . . **ix**

Rafael Balart, Nestor Montanes, Octavio Fenollar, Teodomiro Boronat and Sergio Torres-Giner
Advances in Manufacturing and Characterization of Functional Polyesters
Reprinted from: *Polymers* **2020**, *12*, 2839, doi:10.3390/polym12122839 . . . . **1**

Jaemin Jeong, Fiaz Hussain, Sangwon Park, Soo-Jung Kang and Jinhwan Kim
High Thermal Stability, High Tensile Strength, and Good Water Barrier Property of Terpolyester Containing Biobased Monomer for Next-Generation Smart Film Application: Synthesis and Characterization
Reprinted from: *Polymers* **2020**, *12*, 2458, doi:10.3390/polym12112458 . . . . **7**

Maryam Safari, Agurtzane Mugica, Manuela Zubitur, Antxon Martínez de Ilarduya, Sebastián Muñoz-Guerra and Alejandro J. Müller
Controlling the Isothermal Crystallization of Isodimorphic PBS-*ran*-PCL Random Copolymers by Varying Composition and Supercooling
Reprinted from: *Polymers* **2020**, *12*, 17, doi:10.3390/polym12010017 . . . . **23**

Juan Ivorra-Martinez, Luis Quiles-Carrillo, Teodomiro Boronat, Sergio Torres-Giner and José A. Covas
Assessment of the Mechanical and Thermal Properties of Injection-Molded Poly(3-hydroxybutyrate-*co*-3- hydroxyhexanoate)/Hydroxyapatite Nanoparticles Parts for Use in Bone Tissue Engineering
Reprinted from: *Polymers* **2020**, *12*, 1389, doi:10.3390/polym12061389 . . . . **45**

Elena Torres, Ivan Dominguez-Candela, Sergio Castello-Palacios, Anna Vallés-Lluch and Vicent Fombuena
Development and Characterization of Polyester and Acrylate-Based Composites with Hydroxyapatite and Halloysite Nanotubes for Medical Applications
Reprinted from: *Polymers* **2020**, *12*, 1703, doi:10.3390/polym12081703 . . . . **67**

Yun Zhao, Hui Liang, Shiqiang Zhang, Shengwei Qu, Yue Jiang and Minfang Chen
Effects of Magnesium Oxide (MgO) Shapes on In Vitro and In Vivo Degradation Behaviors of PLA/MgO Composites in Long Term
Reprinted from: *Polymers* **2020**, *12*, 1074, doi:10.3390/polym12051074 . . . . **81**

Sergi Montava-Jorda, Victor Chacon, Diego Lascano, Lourdes Sanchez-Nacher and Nestor Montanes
Manufacturing and Characterization of Functionalized Aliphatic Polyester from Poly(lactic acid) with Halloysite Nanotubes
Reprinted from: *Polymers* **2019**, *11*, 1314, doi:10.3390/polym11081314 . . . . **95**

Tao Yang, Ferina Saati, Jean-Philippe Groby, Xiaoman Xiong, Michal Petrů, Rajesh Mishra, Jiří Militký and Steffen Marburg
Characterization on Polyester Fibrous Panels and Their Homogeneity Assessment
Reprinted from: *Polymers* **2020**, *12*, 2098, doi:10.3390/polym12092098 . . . . **117**

Ángel Agüero, David Garcia-Sanoguera, Diego Lascano, Sandra Rojas-Lema, Juan Ivorra-Martinez, Octavio Fenollar and Sergio Torres-Giner
Evaluation of Different Compatibilization Strategies to Improve the Performance of Injection-Molded Green Composite Pieces Made of Polylactide Reinforced with Short Flaxseed Fibers
Reprinted from: *Polymers* **2020**, *12*, 821, doi:10.3390/polym12040821 . . . . **131**

Franco Dominici, Fabrizio Sarasini, Francesca Luzi, Luigi Torre and Debora Puglia
Thermomechanical and Morphological Properties of Poly(ethylene terephthalate)/Anhydrous Calcium Terephthalate Nanocomposites
Reprinted from: *Polymers* **2020**, *12*, 276, doi:10.3390/polym12020276 . . . . **153**

Diego Lascano, Giovanni Moraga, Juan Ivorra-Martinez, Sandra Rojas-Lema, Sergio Torres-Giner, Rafael Balart, Teodomiro Boronat and Luis Quiles-Carrillo
Development of Injection-Molded Polylactide Pieces with High Toughness by the Addition of Lactic Acid Oligomer and Characterization of Their Shape Memory Behavior
Reprinted from: *Polymers* **2019**, *11*, 2099, doi:10.3390/polym11122099 . . . . **169**

Marina Ramos, Franco Dominici, Francesca Luzi, Alfonso Jiménez, Maria Carmen Garrigós, Luigi Torre and Debora Puglia
Effect of Almond Shell Waste on Physicochemical Properties of Polyester-Based Biocomposites
Reprinted from: *Polymers* **2020**, *12*, 835, doi:10.3390/polym12040835 . . . . **189**

Sergi Montava-Jorda, Diego Lascano, Luis Quiles-Carrillo, Nestor Montanes, Teodomiro Boronat, Antonio Vicente Martinez-Sanz, Santiago Ferrandiz-Bou and Sergio Torres-Giner
Mechanical Recycling of Partially Bio-Based and Recycled Polyethylene Terephthalate Blends by Reactive Extrusion with Poly(styrene-*co*-glycidyl methacrylate)
Reprinted from: *Polymers* **2020**, *12*, 174, doi:10.3390/polym12010174 . . . . **207**

Maria Jorda, Sergi Montava-Jorda, Rafael Balart, Diego Lascano, Nestor Montanes and Luis Quiles-Carrillo
Functionalization of Partially Bio-Based Poly(Ethylene Terephthalate) by Blending with Fully Bio-Based Poly(Amide) 10,10 and a Glycidyl Methacrylate-Based Compatibilizer
Reprinted from: *Polymers* **2019**, *11*, 1331, doi:10.3390/polym11081331 . . . . **227**

About the Editors

Rafael Balart received his Ph.D. from the Polytechnic University of Valencia (UPV) in 2003. At present, he is Full Professor of Materials Science and Engineering at the Department of Mechanical and Materials Engineering. With more than 25 years of experience in the field of polymer and composite materials, he currently leads the Research Group on Environmentally Friendly Polymers and Composites (GiPCEco), which focuses its activity on the development and formulation of sustainable polymers and the manufacture of composite materials for industrial applications. In recent years, he has focused his research on upgrading industrial, agricultural, and agroforestry wastes with a biorefinery approach to obtain high added-value products with applications in polymers and composites.

Sergio Torres-Giner received a Dipl-Ing in Chemical Engineering in 2003 from the Polytechnic University of Valencia (UPV), Spain. In 2004, he achieved an MSc in Process Systems Technology at Cranfield University, England, followed by an MBA in Industrial Management in 2005 at the Catholic University of Valencia 'San Vincente Martir', Spain. He was able to complete his Ph.D. in 2010 in Food Science at the University of Valencia, Spain. He currently works as a scientist at the Research Institute of Food Engineering for Development (IIAD) in the field of macromolecular science of application interest in food packaging technology. He has more than 15 years of experience in both public research agencies and industrial R&D organizations. He has published over 80 peer-reviewed scientific papers indexed in JCR, 10 book and encyclopedia chapters, and 4 patents. His research activity has strongly contributed to advancing the knowledge on biopolymers and transferring it to applications and products for food-related applications.

Octavio Fenollar received his Ph.D. in Engineering from the Polytechnic University of Valencia (UPV) in 2011 and currently, he is Associate Professor of Materials Science and Engineering at the Department of Mechanical and Materials Engineering (UPV Campus d'Alcoi). He has expertise in the field of polymer processing, industrial formulations, and upgrading plastic wastes. With more than 14 years of experience at the university, he leads the research group of Environmentally Friendly Polymers and Composites (GIPCEco). He joined the Institute of Materials Technology (ITM) in 2006 and, since then, he has specialized in polymer processing and advanced characterization as well as recycling and upgrading plastic and other industrial wastes. He has wide experience in the manufacturing, formulation, optimization, and characterization of both thermoplastic and thermosetting polyesters. He is also a member of the Design Factory group at EPSA-UPV, working in additive manufacturing with polymers, mainly polyesters.

Nestor Montanes received his Ph.D. in Engineering from the Polytechnic University of Valencia (UPV) in 2017 and currently, he is Associate Professor of Materials Science and Engineering at the Department of Mechanical and Materials Engineering (UPV Campus d'Alcoi). With more than 10 years of experience in the university and over 10 years of experience in the packaging industry, he has actively participated in the implementation of new teaching methodologies in the classroom: reverse teaching, gamification, and so on. He joined the Technological Institute of Materials (ITM) in 2014 and specializes in polymer analysis and characterization, mainly of biopolymers and polymer-based composites. His areas of interest include polymer manufacturing, compounding, formulation, additives, polymer blends, and upgrading industrial wastes into the polymer industry.

Teodomiro Boronat is a Mechanical Engineer and received his Ph.D. in Engineering from the Polytechnic University of Valencia (UPV) in 2009. Currently, he is Associate Professor of Manufacturing Engineering at the Department of Mechanical and Materials Engineering (UPV Campus d'Alcoi). He is also a researcher at the Technological Institute of Materials (ITM) and specializes in polymer manufacturing processes and additive manufacturing. His main research topic is to study the relationship between processing parameters and their obtained final properties. Other areas of interest include thermoplastic manufacturing processes, thermosetting systems, rheology, and additive manufacturing.

Preface to "Advances in Manufacturing and Characterization of Functional Polyesters"

In the field of polymer science, bio-based and biodegradable polymers are rapidly developing due to the rapid depletion of fossil fuels and environmental pollution deriving from traditional plastics. This trend includes aromatic polyesters derived totally or partially from natural resources as well as recycled polyesters, such as bio-based polyethylene terephthalate (bio-PET) and recycled polyethylene terephthalate (r-PET), respectively, and biodegradable aliphatic polyesters. In some cases, despite being obtained from petroleum, these polyesters can undergo biodisintegration under controlled compost soil. Examples of biodegradable petroleum derived polyesters include poly(ε-caprolactone) (PCL) and poly(butylene succinate) (PBS). Furthermore, some other biodegradable polyesters are also obtained from biomass, such as polylactide (PLA), or even microorganisms, such as polyhydroxyalkanoates (PHAs). Blends and composites based on these polyesters also represent a recurrent solution due to their good balance in terms of easy manufacturing, improved sustainability profile, and tailor-made performance. Therefore, all the above polyesters contribute to sustainable development. Furthermore, they can be functionalized by copolymerization, the use of micro- and nanoparticles or reactive compatibilizers to tailor the desired thermal and mechanical properties, barrier performance, biodegradation, and biocompatibility. Functionalization can widen their potential use in both commodity and technical areas such as packaging, textiles, automotive, building and construction, and also in specialized fields such as tissue engineering and the controlled release of drugs, electronics, or shape-memory devices.

This book is divided into thirteen chapters that gather a series of research articles focused on the manufacturing and characterization of functional polyesters. The book starts with two chapters devoted to the synthesis of functional copolyesters with adjustable properties depending on the polymerization conditions, their monomer contents, or thermal and mechanical post-treatments. Thereafter, the book follows with four chapters covering the development of functional polyesters as resorbable materials in bone tissue engineering. This part of the book is based on research articles reporting the use of blends and nanocomposites based on PCL, PLA, and PHA that can satisfy most of the current requirements for bone fixation of load-bearing devices in terms of both mechanical performance and degradation rates. The next two chapters describe the potential application of functional polyesters in the development of sound absorbing materials and sustainable advanced composites. It then continues with two more chapters focused on the packaging field, which show that PLA and its green composites with food waste derived fillers can be used for compostable food containers and disposable articles with high performance. Finally, the book ends with three chapters that discuss strategies to valorize the waste of polyester-based materials such as those found in bottles for water and beverages, food trays, or fibers for textiles. All these research articles describe the use of nanostructured functional fillers and/or reactive additives to increase the physical–mechanical properties of r-PET and their blends with virgin polyester such as bio-PET by reinforcement or a process of the chain-extension mechanism of the polyester chains. All the studies gathered in this section indicate that secondary recycling of bio-based but non-biodegradable polyesters can be feasible from environmental and economic points of view.

The present book can be of interest to a diverse audience of researchers and professionals working in any area of polymer science and engineering and in academia, industry, or government, including but not limited to scientists, engineers, consultants, etc. This book also intends to make original and cutting-edge research works in the field of synthesis, characterization, manufacturing, and applications of functional polyesters accessible to a broad readership.

Rafael Balart, Sergio Torres-Giner, Octavio Fenollar, Nestor Montanes, Teodomiro Boronat
Editors

Editorial

Advances in Manufacturing and Characterization of Functional Polyesters

Rafael Balart [1,*], Nestor Montanes [1], Octavio Fenollar [1], Teodomiro Boronat [1] and Sergio Torres-Giner [2,*]

[1] Technological Institute of Materials (ITM), Universitat Politècnica de València (UPV), Plaza Ferrándiz y Carbonell 1, 03801 Alcoy, Spain; nesmonmu@upvnet.upv.es (N.M.); ocfegi@epsa.upv.es (O.F.); tboronat@dimm.upv.es (T.B.)
[2] Research Institute of Food Engineering for Development (IIAD), Universitat Politècnica de València (UPV), Camino de Vera s/n, 46022 Valencia, Spain
* Correspondence: rbalart@mcm.upv.es (R.B.); storresginer@upv.es (S.T.-G.)

Received: 13 November 2020; Accepted: 26 November 2020; Published: 29 November 2020

In the last few years, a remarkable growth in the use of functional polyesters has been observed. This trend comprises the development of aromatic polyesters derived either from renewable resources or recycling processes on account of the rapid depletion of fossil fuels and the cost of extracting polymers from petroleum. Furthermore, biodegradable aliphatic polyesters are also being rapidly developed due to pollution deriving from traditional plastics. The latter group includes both bio-based and petroleum derived polyesters that are biodegradable, that is, they can undergo biodisintegration under controlled compost soil or natural conditions. Moreover, blends and composites based on these novel polyesters represent a recurrent and cost-effective solution due to their good balance in terms of easy manufacturing, improved sustainability profile, and tailor-made performance. All these polyesters do not only contribute to sustainable development but they can also be functionalized to tailor their desired properties in terms of thermal and mechanical properties, barrier performance, biodegradation, and biocompatibility. For instance, copolymerization and the use of micro- and nanoparticles or reactive compatibilizers can widen the potential use of polyesters in both commodity and technical areas such as packaging, textiles, automotive, building and construction, and also in specialized fields such as tissue engineering and controlled release of drugs, electronics or shape-memory devices. The present Special Issue gathers a series of thirteen articles focused on the manufacturing and characterization of functional polyesters.

In terms of technical applications, the synthesis of functional materials is increasingly important for optical devices and electronics. In this regard, Jeong et al. [1] developed a novel copolyester for next-generation flexible devices in optics. To this end, authors synthesized by melt polymerization and subsequent solid-state polycondensation (SSP) a copolyester, named PCITN, based on 2,6-naphthalene dicarboxylic acid (NDA), terephthalic acid (TPA), 1,4-cyclohexanedimethanol (CHDM), and isosorbide (ISB). Polymerization carried out in two steps led to a PCITN copolyester having a high molecular weight (Mw = 68,900 g/mol), which was successfully shaped thereafter into films by extrusion at 290 °C. The resultant randomly-oriented PCITN films were preheated for 25 min and then tensile stretched and uniaxially cold drawn at 150 °C in machine direction (MD) at different draw ratios (λ = 1~4) to increase molecular chain orientation. The uniaxially oriented films showed high glass transition temperature (T_g of up to 140 °C) and Young's modulus (E = 2.6 GPa), which was ascribed to the use of the low segmental mobility and high thermal stability of the building block ISB. The PCITN films also presented lower water absorption (0.54 wt.%) and birefringence (Δn = 0.09) when compared with other conventional substrate polymers used for flexible electronics such as polyimide (PI), poly(ethylene 2,6-naphthalate) (PEN), and polyethylene terephthalate (PET). Copolymerization in two stages was also explored by Safari et al. [2] to develop random copolyesters with adjustable properties depending on their crystallinity. Authors synthesized poly(butylene succinate)-*ran*-poly(ε-caprolactone) (PBS-*ran*-PCL) copolyesters by transesterification/ring opening polymerization (ROP) reaction of 1,4-butanediol (BD), dimethyl succinate (DMS), and ε-caprolactone (CL) followed by polycondensation at reduced pressure. The PBS-*ran*-PCL copolyesters showed isodimorphic behavior with a controllable balance between comonomer inclusion and exclusion, where at some intermediate compositions the crystal lattice of each one of the components partially tolerate the presence of the other. In particular, the pseudo-eutectic point was found at 55 mol.% CL, where both PBS- and PCL-rich phases crystallized, whereas single PBS- and PCL-rich crystals were respectively formed at compositions based on lower and higher CL-unit molar contents. Furthermore, the phase crystallization of PBS-*ran*-PCL with pseudo-eutectic composition was successfully controlled by varying the isothermal crystallization temperature.

Only specifically developed functional polyesters can fulfill the specific challenges of the constantly developing biomedical field. At present, biodegradable aliphatic polyesters are being intensively studied as resorbable materials in bone tissue engineering, particularly for low-stress parts such as small orthopedic plates, rods or bone screws. Accordingly, these polyesters can develop scaffolds and biomaterial devices with desired geometries and special functionalities to attain osteoconductive properties. However, polyesters cannot satisfy most of the current mechanical requirements for bone fixation of load-bearing devices. In this regard, the use of micro- and nano-scale mineral particles with high hardness and different bioactives is very promising. For instance, nanoparticles of hydroxyapatite (nHAs) were used by Ivorra-Martinez et al. [3] to improve the mechanical performance of the poly(3-hydroxybutyrate-*co*-3-hydroxyhexanoate) [P(3HB-*co*-3HHx)] copolyester. It was observed that the incorporation of the osteoconductive nanofiller yielded balanced properties to the microbial copolyester in terms of strength and ductility, showing values of tensile modulus (E_t) and elongation at break (ε_b) ranging from approximately 1 up to 1.7 GPa and 6.5 to 19.4%, respectively. As a result, the P(3HB-*co*-3HHx)/nHA nanocomposites presented a closer mechanical performance to that found in the natural bone when compared to titanium (Ti) and alloys of metals such as stainless steel and cobalt-chrome (Co–Cr) alloys.

Another challenge that polyesters is facing in the biomedical area is related to their inadequate degradation rates, which would increase the risk of requiring additional chirurgical interventions, eventually causing adverse tissue reactions or even infections. In this context, Torres et al. [4] also employed nHAs and halloysite nanotubes (HNTs) to increase the surface wettability of hydrophobic and hydrophilic sets of poly(ε-caprolactone) (PCL), polylactide (PLA), and their blends as well as poly(2-hydroxyethyl methacrylate) (PHEMA) and its copolymer with ethyl methacrylate (EMA), that is, poly(2-hydroxyethyl methacrylate-*co*-ethyl methacrylate) P(HEMA-*co*-EMA). Authors demonstrated that both the blending of PCL with PLA and the incorporation of nHAs and HNTs provided hydrophilic units and decreased the crystallinity of PCL, favoring the accessibility of water molecules to the ester linkages. Therefore, in comparison with the unfilled PCL/PLA blend, mass loss increased up to 48% after incubation for 12 weeks in phosphate buffered saline (PBS) during the evaluation of their degradation rate. Consequently, the mechanical properties of the polyesters decreased above 60% after the incubation time due to the higher hydrolytic cleavage achieved. Similarly, Zhao et al. [5] developed nanocomposites of PLA with 1 wt.% of nanoparticles and whiskers of magnesium oxide (MgO) that were chemically modified with stearic acid. The in vitro degradation of the nanocomposites was analyzed by PBS soaking for up to 12 months, showing that the addition of the stearic acid-modified MgO accelerated the PLA's water uptake rate, especially for the whiskers. Furthermore, the dissolution of MgO through the neutralization of the acidic product of the PLA degradation contributed to regulate the pH value of PBS. The in vivo results of the histological morphologies attained with the nanocomposites further suggested that the presence of MgO can improve bone repair since, when dissolving, it releases magnesium ions (Mg^{2+}) that can activate a variety of enzymes that promote the synthesis of proteins. In addition, PLA-based materials can also serve as a drug carrier for controlled release applications. Micro- and nanoencapsulation using HNTs is promising to deliver different active and bioactive compounds such as antioxidants, antimicrobials, antibiotics, or growth factors

(GFs). In this regard, Montava-Jorda et al. [6] developed PLA composites with HNTs loadings of 3, 6, and 9 wt.%. Authors found that water uptake of PLA increased due to the hydrophilic nature of the nanotubes, which offer a high surface area with hydroxyl (–OH) groups, and thus can favor the degradation rate of the resultant wound dressings.

One of the most exciting areas of functional polyesters is the development of sound absorbing materials. It has been demonstrated that, for instance, nonwoven structures of polyester fibers are potential candidates in noise reducing panels. The research article of Yang et al. [7] estimated the non-acoustic parameters of high-loft nonwoven panels made of a commercial polyester blend composed of 45 wt.% staple polyester, 30 wt.% hollow polyester, and 25 wt.% bicomponent polyester provided by the Technical University of Liberec (Liberec, Czech Republic). The panels, with densities ranging from 16.93 to 45.56 kg/m^3, were manufactured by perpendicular laying technology and were tested by means of the Bayesian reconstruction procedure, implementing the Johnson-Champoux-Allard-Lafarge model and Markov chain Monte Carlo optimization technique. The inversed method showed that the polyester-based panels were homogeneous along with the panel thickness, presenting the same inferred tortuosity. Mean relative differences of airflow resistivity and porosity of 0.019 and 0.004 were respectively attained, which mostly affected the thermal characteristic length. In the textile area, linen (*Linum usitatissimum*) is among the most usable and profitable plants. One of its processing by-products is flax fiber (FF), which can also be applied to replace glass fiber (GF) as a suitable reinforcement in advanced composites of polyesters for several engineering applications such as panel boards and insulation panels. However, lignocellulosic fibers habitually show a poor interfacial adhesion with hydrophobic polyester matrices such as PLA. In this context, Agüero et al. [8] evaluated the influence of different compatibilization strategies on the performance of PLA/FF composites. In particular, the compatibilization routes consisted of silanization with (3-glycidyloxypropyl) trimethoxysilane (GPTMS) and reactive extrusion (REX) with a commercial random copolymer of poly(styrene-*co*-glycidyl methacrylate) (PS-*co*-GMA, Xibond™ 920, Polyscope, Geleen, The Netherlands), a multi-functional epoxy-based styrene-acrylic oligomer (ESAO, Joncryl ADR 4368®, BASF S.A., Barcelona, Spain), and maleinized linseed oil (MLO). Among the tested routes, the petroleum derived ESAO yielded the highest mechanical resistance and toughness improvement and also the highest thermal stability due to the chain-extension or cross-linking effect on PLA, whereas the most ductile green composites were attained with MLO by plasticization.

Focusing on the packaging field, the PLA biopolyester is widely used for food containers (e.g., food films and trays) or disposable articles (e.g., straws and cutlery). However, it habitually results in extremely brittle materials with low ductility and toughness. To improve the impact properties of PLA, Lascano et al. [9] incorporated an oligomer of lactic acid (OLA) in the contents of 5–20 wt.%. At a loading of 15 wt.% OLA, it was observed a percentage increase of nearly 171% in the impact strength of PLA. Furthermore, for low deformation angles, the OLA-containing PLA samples successfully recovered over 95% of their original shape, while for the highest angles they still reached a recovery of approximately 70%. In packaging applications, the development of eco-friendly or green composites by the valorization of agrofood waste also shows several benefits including the enhancement of the environmental profile, improved biodegradability, lightweight, or cost reduction. Among the wide variety of agricultural wastes and food processing by-products, almond (*Prunus amygdalus* L.) shell powder (ASP) was used by Ramos et al. [10] as the filler for a commercial biopolyester blend (INZEA F2®, Nurel, Zaragoza, Spain) at 10 and 25 wt.%. In this study, the lignocellulosic filler was first subjected to two grinding levels, yielding sizes of 125–250 and 500–1000 µm, and MLO and low-functionality ESAO (Joncryl ADR 4400®, BASF S.A.) were added as compatibilizers. Authors reported that ASP successfully improved the biodisintegration rate of PLA under composting conditions, while full disintegration was obtained after 90 and 28 days for the green composites containing 10 and 25 wt.% ASP, respectively. Authors concluded that the presence of ASP at high contents produced a significant discontinuity in the polyester matrix, facilitating water penetration and favoring the growth of microorganisms responsible for biodisintegration.

Finally, disposal of waste polyester-based materials has become an urgent environmental problem in the last decades. For instance, PET, which is widely used worldwide in bottles for water and beverages, food trays or fibers for textiles, produces a plastic waste that is neither biodegradable nor compostable, and recycling currently represents the only solution. Although the post-consumer uses of recycled polyethylene terephthalate (r-PET) streams has increased, it is currently limited to low contents in mixed formulations with virgin PET due to the weakening of its physical-mechanical properties as a result of the M_W reduction during processing derived from the cleavage of the polyester chains. Thus, the use of nanostructured additives and/or reactive additives has emerged as a novel route to deal with this technical issue. In the study of Dominici et al. [11], anhydrous calcium terephthalate anhydrous salts (CATAS), a nanometric metal organic framework consisting of calcium ions (Ca^{2+}) as metal clusters coordinated to TPA as organic ligand, were developed and incorporated by melt processing into rPET at different two levels of loadings, that is, 0.1–1 wt.% and 2–30 wt.%. Authors found that 0.4 wt.% of CATAS led to the formation of a rigid amorphous fraction due to the aromatic interactions (π–π conjugation) between the rPET matrix and Ca-based salts, which was located at the rPET/CATAS interface. In contrast, tangible changes were observed below this threshold, whereas a restriction of rPET/CATAS molecular chains mobility was detected above 0.4 wt.% CATAS due to the formation of percolation networks with hybrid mechanical characteristics. In another study, Montava-Jorda et al. [12] melt-mixed at 15–45 wt.% partially bio-based polyethylene terephthalate (bio-PET) with r-PET flakes that were obtained from pre-consumer waste streams of the food-use bottle industry and PS-*co*-GMA at 1–5 parts per hundred resin (phr) of polyester blend. For the polyester blend containing 45 wt.% of r-PET, the addition of 5 phr of the reactive compatibilizer led to an enhancement in the elongation-at-break value from 10.8 to 378.8%, whereas impact strength also increased from 1.84 to 2.52 kJ/m^2. It was concluded that, due to a chain-extension mechanism based on the reaction of the–OH and carboxyl (–COOH) terminal groups of both bio-PET and r-PET chains with the multiple groups of glycidyl methacrylate (GMA) present in PS-*co*-GMA, branched and larger interconnected macromolecules were formed and contributed to improve the ductile performance of the polyester blends. Additionally, in terms of improving the mechanical or secondary recycling of polyesters with other polymers, Jorda et al. [13] reported binary blends of bio-PET with polyamide 1010 (PA1010), a fully bio-based polyamide (bio-PA), compatibilized with PS-*co*-GMA. Authors reported that the addition of 30 wt.% of PA1010, which provides a final renewable content of nearly 50 wt.% in the polymer blend, yielded an immiscible droplet-like structure in which PA1010 droplets are embedded in the bio-PET matrix. However, the intrinsic stiffness of the biopolyester was improved by the bio-PA also the addition of PS-*co*-GMA at 3 phr, which induced a remarkable reduction of the droplet size from approximately 4 to 1 mm. According to these studies, secondary recycling of bio-based but non-biodegradable polyesters can be feasible from environmental and economic points of view.

From the above, the Special Issue *Advances in Manufacturing and Characterization of Functional Polyesters* published in *Polymers*, brings together a broad range of research works dealing with functional polyesters to update the "state-of-the-art" knowledge in this field. Functionalization of polyesters was successfully achieved by means of copolymerization, the use of additives at micro- and nano-scale or reactive compatibilizers. The resultant materials can find potential uses in technical applications, such as optical devices and electronics, in the biomedical field for tissue engineering and controlled release of drugs, or as sound absorbing panels, textiles and apparel, advanced composites, and sustainable food packaging.

Author Contributions: All the guest editors wrote and reviewed this editorial letter. All authors have read and agreed to the published version of the manuscript.

Funding: This research work was funded by the Spanish Ministry of Science and Innovation (MICI) project number MAT2017-84909-C2-2-R.

Acknowledgments: S.T.-G. acknowledges MICI for his Ramón y Cajal contract (RYC2019-027784-I). The guest editors thank all the authors for submitting their work to this Special Issue and for its successful completion. We also acknowledge all the reviewers participating in the peer-review process of the submitted manuscripts for enhancing their quality and impact. We are also grateful to Chris Chen and the editorial assistants of Polymers who made the entire Special Issue creation a smooth and efficient process.

Conflicts of Interest: The authors declare no conflict of interest.

References

1. Jeong, J.; Hussain, F.; Park, S.; Kang, S.J.; Kim, J. High thermal stability, high tensile strength, and good water barrier property of terpolyester containing biobased monomer for next-generation smart film application: Synthesis and characterization. *Polymers* **2020**, *12*, 2458. [CrossRef] [PubMed]
2. Safari, M.; Mugica, A.; Zubitur, M.; Martínez de Ilarduya, A.; Muñoz-Guerra, S.; Müller, A.J. Controlling the isothermal crystallization of isodimorphic PBS-ran-PCL random copolymers by varying composition and supercooling. *Polymers* **2020**, *12*, 17. [CrossRef] [PubMed]
3. Ivorra-Martinez, J.; Quiles-Carrillo, L.; Boronat, T.; Torres-Giner, S.; Covas, J.A. Assessment of the mechanical and thermal properties of injection-molded poly(3-hydroxybutyrate-co-3-hydroxyhexanoate)/hydroxyapatite nanoparticles parts for use in bone tissue engineering. *Polymers* **2020**, *12*, 1389. [CrossRef] [PubMed]
4. Torres, E.; Dominguez-Candela, I.; Castello-Palacios, S.; Vallés-Lluch, A.; Fombuena, V. Development and characterization of polyester and acrylate-based composites with hydroxyapatite and halloysite nanotubes for medical applications. *Polymers* **2020**, *12*, 1703. [CrossRef] [PubMed]
5. Zhao, Y.; Liang, H.; Zhang, S.; Qu, S.; Jiang, Y.; Chen, M. Effects of magnesium oxide (MgO) shapes on in vitro and in vivo degradation behaviors of PLA/MgO composites in long term. *Polymers* **2020**, *12*, 1074. [CrossRef] [PubMed]
6. Montava-Jorda, S.; Chacon, V.; Lascano, D.; Sanchez-Nacher, L.; Montanes, N. Manufacturing and characterization of functionalized aliphatic polyester from poly(lactic acid) with halloysite nanotubes. *Polymers* **2020**, *11*, 1314. [CrossRef] [PubMed]
7. Yang, T.; Saati, F.; Groby, J.P.; Xiong, X.; Petrů, M.; Mishra, R.; Militký, J.; Marburg, S. Characterization on polyester fibrous panels and their homogeneity assessment. *Polymers* **2020**, *12*, 2098. [CrossRef]
8. Agüero, Á.; Garcia-Sanoguera, D.; Lascano, D.; Rojas-Lema, S.; Ivorra-Martinez, J.; Fenollar, O.; Torres-Giner, S. Evaluation of different compatibilization strategies to improve the performance of Injection-molded green composite pieces made of polylactide reinforced with short flaxseed fibers. *Polymers* **2020**, *12*, 821. [CrossRef]
9. Lascano, D.; Moraga, G.; Ivorra-Martinez, J.; Rojas-Lema, S.; Torres-Giner, S.; Balart, R.; Boronat, T.; Quiles-Carrillo, L. Development of injection-molded polylactide pieces with high toughness by the addition of lactic acid oligomer and characterization of their shape memory behavior. *Polymers* **2019**, *11*, 2099. [CrossRef] [PubMed]
10. Ramos, M.; Dominici, F.; Luzi, F.; Jiménez, A.; Garrigós, M.C.; Torre, L.; Puglia, D. Effect of almond shell waste on physicochemical properties of polyester-based biocomposites. *Polymers* **2020**, *12*, 835. [CrossRef] [PubMed]
11. Dominici, F.; Sarasini, F.; Luzi, F.; Torre, L.; Puglia, D. Thermomechanical and morphological properties of poly(ethylene terephthalate)/anhydrous calcium terephthalate nanocomposites. *Polymers* **2020**, *12*, 276. [CrossRef] [PubMed]
12. Montava-Jorda, S.; Lascano, D.; Quiles-Carrillo, L.; Montanes, N.; Boronat, T.; Martinez-Sanz, A.V.; Ferrandiz-Bou, S.; Torres-Giner, S. Mechanical recycling of partially bio-based and recycled polyethylene terephthalate blends by reactive extrusion with poly(styrene-co-glycidyl methacrylate). *Polymers* **2020**, *12*, 174. [CrossRef] [PubMed]
13. Jorda, M.; Montava-Jorda, S.; Balart, R.; Lascano, D.; Montanes, N.; Quiles-Carrillo, L. Functionalization of partially bio-based poly(ethylene terephthalate) by blending with fully bio-based poly(amide) 10,10 and a glycidyl methacrylate-based compatibilizer. *Polymers* **2019**, *11*, 1331. [CrossRef] [PubMed]

Publisher's Note: MDPI stays neutral with regard to jurisdictional claims in published maps and institutional affiliations.

© 2020 by the authors. Licensee MDPI, Basel, Switzerland. This article is an open access article distributed under the terms and conditions of the Creative Commons Attribution (CC BY) license (http://creativecommons.org/licenses/by/4.0/).

Article

High Thermal Stability, High Tensile Strength, and Good Water Barrier Property of Terpolyester Containing Biobased Monomer for Next-Generation Smart Film Application: Synthesis and Characterization

Jaemin Jeong [1,†], Fiaz Hussain [1,2,†], Sangwon Park [1], Soo-Jung Kang [1] and Jinhwan Kim [1,*]

1. Department of Polymer Science and Engineering, Sungkyunkwan University, 2066, Seobu-ro, Jangan-gu, Suwon, Gyeonggi 16419, Korea; jjm123kr@gmail.com (J.J.); fiazravian@gmail.com (F.H.); bluswon@gmail.com (S.P.); soojung@skku.edu (S.-J.K.)
2. Institute of Advanced Materials, Bahauddin Zakariya University, Multan 60800, Pakistan
* Correspondence: jhkim@skku.edu
† These authors contributed equally as first author to this work.

Received: 24 September 2020; Accepted: 19 October 2020; Published: 23 October 2020

Abstract: This research synthesizes novel copolyester (PCITN) containing biobased isosorbide, 1,4-cyclohexandimethanol, terephthalic acid, and 2,6-naphthalene dicarboxylic acid and characterize its properties. The PCITN copolyester was extruded into film, and its performance properties including: tensile strength, Young's modulus, thermal, dimensional stability, barrier (water barrier), and optical (birefringence and transmittance) were analyzed after uniaxial stretching. The films have higher T_g, T_m, dimensional stability, and mechanical properties than other polyester-type polymers, and these performance properties are significantly increased with increasing stretching. This is due to the increased orientation of molecular chains inside the films, which was confirmed by differential scanning calorimetry (DSC), X-ray diffraction (XRD), and birefringence results. Good water barrier (0.54%) and lower birefringence (Δn: 0.09) of PCITN film compared to poly(ethylene terephthalate) (PET), poly(ethylene 2,6-naphthalate) (PEN), and polyimide (PI) films, used as conventional substrate materials for optical devices, make it an ideal candidate as performance material for next-generation flexible devices.

Keywords: bio-based; copolyester; dimensional stability; flexible optical devices; uniaxial stretching; birefringence; and barrier properties

1. Introduction

Biobased polymers have gained the attention of scientists due to the rapid depletion of fossil fuels and increasing environmental pollution [1–4]. With the advancement and innovation of biotechnology and chemical industry, a wide range of biobased monomers such as isosorbide (ISB) [5], lactic acid [6], tannic acid [7], 2,5-furandicarboxylic acid [8], succinic acid [9], etc. have been developed and are being used widely for the development of biobased polymers. However, their poor mechanical and thermal properties limit their industrial applications for engineering polymers.

Among various biomass monomers, isosorbide (1,4:3,6-dianhydrohexitol; ISB) is well known due to its superior performance properties. It is derived from glucose, and its bifunctional hydroxyl groups facilitate the condensation or addition polymerization. Its unique rigid structure and chirality significantly improve the glass transition temperature (T_g) and transparency of the resulting polymers. Many efforts have been done to develop the ISB-based polyesters [10], epoxies [11], polyurethane [12], and polycarbonates [13,14], which are not only environmentally friendly but also have superior thermal and optical properties. However, it is very difficult to get a high molecular weight product using ISB alone due to its low reactivity and degradability during the polymerization process.

Poly(ethylene terephthalate) (PET) and poly(ethylene naphthalate) (PEN) are well known conventional homopolyesters. High crystallizability, low T_g, and inferior barrier properties of PET act as obstacles in its commercial applications, especially at high temperatures (>100 °C). PEN has good thermal stability, however, expensive raw materials used for its synthesis, high birefringence, and necking of the film during stretching limit its commercial applications. To replace the conventional polymeric substrate for flexible electronics, materials need to offer ease of processing, high T_g, good dimensional and thermal stability, good optical, and moisture barrier properties. The properties of the polymers can be significantly improved by blending [15]. The performance properties of the polyester-based composites are significantly higher than copolyesters [16]. However, polyester-based composites are not suitable for the application where transparency and flexibility are required. So, it would be of great interest to develop a biobased transparent and flexible copolyester for versatile industrial applications.

This research focuses on the synthesis of terephthalic acid (TPA), 2,6-naphthalene dicarboxylic acid (NDA), 1,4-cyclohexanedimethanol (CHDM), and isosorbide (ISB)-based copolyester that has high T_g and good mechanical, thermal, and barrier properties, even with low ISB content compared to conventional polyesters. It is important to note that the low reactivity of ISB prevents the obtaining of high molecular weight polymer products, which limits the commercial application of polymers as engineering plastics [17–19]. To overcome this problem, CHDM was incorporated to facilitate between TPA, NDA, and ISB. Finally, PCTIN copolyesters containing 25 mole% of ISB (feed ratio) with high weight average molecular weight (Mw: 68,900 g/mol) and T_g (121 °C) were successfully synthesized by two-step melt polymerization. Novel PCITN copolyester film was fabricated using a melt extrusion process to extend their commercial applications and analyze their performance properties. We significantly improved the thermal, mechanical, dimensional stability, and barrier properties, which is the weak point of bio-based polymers, by utilizing additional monomers: CHDM, TPA, and NDA.

In this study, the fabrication and characterization of biobased novel PCITN film are being reported to explore its versatile industrial applications such as a smart film. The fabricated semi-crystalline film was uniaxially stretched (λ1–λ4) in machining direction (MD) at 150 °C. The influence of stretching on thermal, mechanical, dimensional stability, optical, and barrier properties was studied. Analyzed performances were found to increase linearly with increasing stretching, which is due to increased molecular orientation in the stretched films. It is obvious from the results that based on the unique performance properties of PCITN compared to PET, PEN, and Polyimide (PI) [20,21], this novel material has the potentional to be used as a performance material in next-generation flexible electronic devices.

2. Materials and Methods

2.1. Materials

TPA (99.9%), ISB (99.8%), and CHDM (99.8%) with 70 mole% trans-isomers were purchased from SK Chemicals (SKC; Suwon, Korea). NDA (99.8%) was purchased from BASF (Ludwigshafen, Germany). The ultra-pure (99%) catalyst (titanium-N-butoxide (TNBT) and thermal stabilizer (phosphorous acid) were used as received from Sigma Aldrich (Soul, South Korea). The high-purity monomers were used as received. High purity (99.9%) solvents, like deuterated chloroform ($CHCl_3$-d) (Sigma Aldrich, Soul, South Korea), deuterated trifluoroacetic acid (TFA-d) (Sigma Aldrich, Soul, South Korea), and o-chlorophenol (Sigma Aldrich, Soul, South Korea) used for the characterization of specimens, were used as received without any further purification.

2.2. Synthesis of Copolyester Followed by Solid-State Polycondensation (SSP)

2.2.1. Copolyester Synthesis

PCITN copolyester was melt polymerized using two different 5 L batch reactors of the pilot-scale melt polymerization reactor [22]. For esterification, the residual oxygen was removed from the reactor by purging pure nitrogen (N_2). All the chemicals: TPA (249 g, 1.5 moles), NDA (973 g, 4.5 moles), CHDM (1089 g, 6.8 moles), ISB (331 g, 2.3 moles) titanium butoxide catalyst (3.906 g; 300 ppm relative to the total expected weight of the product), and phosphorous acid stabilizer (0.4850 g; 100 ppm relative to the total expected weight of the product) were charged into the esterification reactor at a molar ratio of diacid:diol = 1.0:1.5 for PCITN. The temperature of the reaction mixture was increased slowly up to 270 °C and held at this temperature until the completion of the esterification reaction. At this stage, water (by-product) generated during esterification reaction has been distilled out. After, this product was transferred into the polycondensation reactor. For transesterification, the temperature was gradually increased to 300 °C, the vacuum was reduced to 0.5 Torr, and these conditions were maintained until the completion of the reaction. Torque meter value (Nm) was used to determine the end-point of the reaction. The stirring speed was reduced gradually as the degree of polymerization increased. The polycondensation reaction continued isothermally at 90 RPM, 60 RPM, and 30 RPM until the torque value reached 1.6 Nm at each RPM. Finally, the reactor pressure was returned to atmospheric pressure by purging N_2 gas to avoid the oxidative degradation of the polymer. The resulting PCITN copolyester was extruded from the reactor, quenched, and pelletized. The schematic diagram for the melt polymerization of PCITN is shown in Scheme 1.

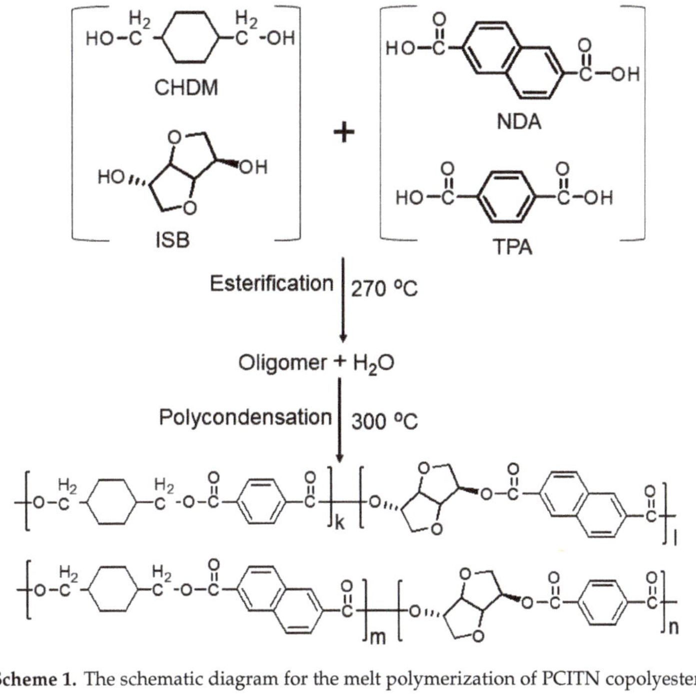

Scheme 1. The schematic diagram for the melt polymerization of PCITN copolyester.

Solid-state polycondensation (SSP) is an important step of the polymer industry as it significantly increases the molecular weight of synthesized polymers making them suitable for a wide range of commercial applications as an engineering plastic [23–25]. SSP of synthesized PCITN copolyester was also carried out at optimized conditions (Temperature and time) under the nitrogen atmosphere. The synthesized PCITN copolyester was thoroughly characterized for its chemical structure, thermal, and degradation behavior after SSP at 250 °C for 20 h.

2.2.2. Film Fabrication and Uniaxial Stretching

The extrusion grade PCITN copolyester having a high molecular weight (M_w: 68,900) and intrinsic viscosity (IV: 0.77) was successfully fabricated into a film using the melt extrusion process at 290 °C. The randomly-oriented film was successfully fabricated (thickness; 550 µm and width; 18 cm) and wound up on a cold stainless-steel winder roll (25 °C).

As synthesized, randomly-oriented film specimens were preheated for 25 min before tensile stretching and then uniaxially cold drawn (at 150 °C) in MD using a tensile stretching machine with a speed of 1.3 mm/sec. To freeze the orientation of aligned polymeric chains obtained during drawing, the stretched films were cooled down to room temperature using cold air without removing the stress. The films with different draw ratios (λ) (1~4) were thus obtained and they were characterized for their thermal, mechanical and thermal degradation behavior; dimensional stability; and optical and barrier properties.

2.3. Characterization

The IV (dL/g) of PCITN copolyester was measured in the o-chloroform (OCP) solvent (15 g/15 mL) using an automated Ubbelohde viscometer (no. 1C) at 30 °C. The number average molecular weight (M_n), weight average molecular weight (M_w), and polymer dispersity index (PDI) were measured by gel permeation chromatography (GPC, Agilent, Santa Clara, California, United States) using m-cresol as a mobile phase at a velocity of 0.7 mL/min. The separation was performed with two Shodex LF804 columns at 100 °C equipped with a Malvern TDA 305 refractive index detector. The sample solutions were filtered using a polytetrafluoroethylene microporous membrane (Merck Millipore, 0.45 µm pore size) before injection. M_n, M_w, and PDI (M_w/M_n) were determined from universal calibration with polystyrene standards.

For the actual chemical composition of synthesized copolyester, proton nuclear magnetic resonance (^{1}H-NMR, Bruker Cooperation, Billerica, Massachusetts, United States) spectra were obtained at 25 °C by a Unity Inova 500NB High Resolution 500 MHz NMR Console. Deuterated chloroform ($CHCl_3$-d) and deuterated trifluoroacetic acid (TFA-d) (1:1) were used as solvent and tetramethylsilane (TMS) as an internal standard and as reference for chemical shifts (parts per million, ppm). Chemical shifts were expressed in ppm (parts per million) relative to an internal standard (TMS).

Thermal properties of polymer pallets were analyzed by differential scanning calorimeter (DSC Q20, TA instruments, New Castle, Delaware, USA) thermograms. About 8–10 mg of the sample was melted at 300 °C and then quenched to 40 °C at a cooling rate of 200 °C/min. Then, the temperature was raised from 40 to 300 °C at a heating rate of 10 °C/min under N_2 atmosphere. Thermal decomposition behavior was analyzed by thermo-gravimetric analyzer (TGA Q50, TA instruments, New Castle, Delaware, USA) thermograms. For this, the sample placed in the alumina pan of TGA was heated from 40–600 °C at a scan rate of 10 °C/min under a constant N_2 flow of 50 mL/min. The temperature at which 5% weight loss is observed ($T_{id5\%}$), the onset temperature of main degradation ($T_{d50\%}$), and residue at 600 °C (%) were recorded from TGA analysis.

2.4. Characterization of Film Properties

Fabricated film specimens were thoroughly characterized using DSC for their T_g, T_m, cold crystallization temperature (T_{cc}), cold crystallization enthalpy (ΔH_c in J/g), and the melting enthalpy (ΔH_m in J/g). Only, the first heating thermogram (40 to 300 °C) of DSC was recorded to study the effect of stretching on the thermal properties of the films. The degree of crystallinity (X_c) of all the film specimens was also calculated from XRD analysis by the integration of corresponding amorphous and crystalline peaks. For accurate results, fabricated films were analyzed four times. T_g was estimated as the onset point in heat capacity associated with a transition.

The dimensional stability of the film specimens was determined using a thermo-mechanical analyzer (TMA6100, Seiko Exstar 6000, Seiko Instrument Inc., Chiba, Japan). For this, the values for the coefficient of linear thermal expansion (CTE) of the film specimens (60 × 5 × 0.1 mm^3) were recorded from 30 to 120 °C with a heating rate of 5 °C/min and analyzed.

The influence of the stretching on the crystalline structure of the films was analyzed by high-power wide-angle X-ray diffraction (XRD, D8 Advance, Bruker Cooperation, Billerica, Massachusetts, United States) spectroscopy. For this, spectrograms of film specimens were obtained using CuKα radiations (λ = 0.154 nm) in the 2θ range of 20 to 50° with a scanning speed of 2θ/min. Tensile testing of the developed samples was carried out using a universal testing machine (UTM, E3000, E3000, Instron, Norwood, Massachusetts, USA), with a crosshead speed of 50 mm/min. Samples were prepared (50 mm × 5 mm) according to the standard test method, ASTM D882 [26], before the testing. For the reliability and accuracy of the mechanical data, the film specimens with different draw ratios (λ) were analyzed four times. The tensile strength was measured as the ultimate tensile strength at the fracture point while Young's modulus was measured from the initial slope of the S-S curve (strain < 0.1%). The water barrier (%) property was determined by the cup method by following the standard test method ASTM D570-98 [27]. Before the testing, the film samples were conditioned for 1 week at 23 ± 2 °C and 65% RH. The film specimens were placed in the in DI water at 23 ± 2 °C and continued to record the values for the wet weight until they reached a steady-state, usually after 11–12 h.

The water absorption (%) of samples was determined using the wet weight (g) and dry weight (g) data as follows:

Water absorption (%) = [Wet weight (g) − Dry weight (g)] ÷ Dry weight (g) × 100%

Optical properties: transmittance and birefringence (Δn) of fabricated films were determined using a UV–Vis spectrophotometer (S-400, Scinco, Seoul, Korea) and a wide range 2-D birefringence analyzer system (WPA-100-L, Photonic Lattice Inc., Sendai, Japan), respectively. The percentage transmittance (%) for wavelengths from 300–800 nm was recorded and analyzed. For Δn measurement, retardation of white light in the absence of any sample between the light source and detector was used as a reference. The retardation of the film sample was measured against the reference and analyzed. The Δn values of film specimens were determined from their retardation (R) and thickness (d) values using the formula as follow:

$$\Delta n = R/d$$

3. Results and Discussion

3.1. Actual Composition of PCITN Copolyester

The composition of monomers: ISB, CHDM, TPA, and NDA was estimated from the integration of peak intensities of each proton of each monomer. The actual chemical composition of synthesized PCITN copolyester was determined by ^{1}H-NMR spectroscopy. ^{1}H-NMR and assignments of characteristic chemical shifts (δ) (ppm) are shown in Figure 1. Characteristics δ (ppm) peaks at 4.60, 2.12, 1.95, and 1.35 are assigned to trans-CHDM. While δ (ppm) peaks at 4.70, 2.25, 1.83, and 1.70 ppm are assigned to cis-CHDM isomers. δ (ppm) peaks observed at 4.52, 5.84, 5.54, 5.10, 5.80, and 4.36 ppm are assigned to hydrogen atoms 1, 2, 3, 4, 5, and 6 of the ISB linked with terephthalic units. The peaks for 1 and 6 overlap with the hydrogens (11) of the CHDM moiety. δ (ppm) peaks observed at 4.50, 5.73, 5.50, 5.06, 5.78, and 4.34 ppm are assigned to hydrogen atoms 1′, 2′, 3′, 4′, 5′, and 6′ of ISB between the naphthalic unit and the terephthalic unit. The peaks for 1′ and 6′ overlap with the hydrogens (11) of the CHDM moiety.

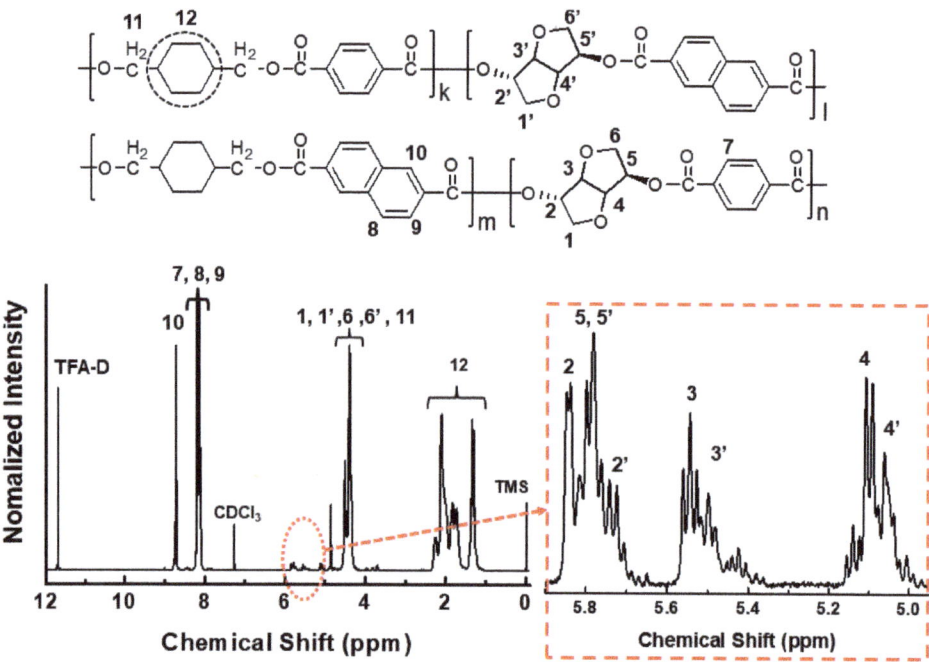

Figure 1. ^{1}H-NMR spectra of PCITN copolyester.

The δ peaks at 3.40~3.81 ppm are assigned to protons of CH$_2$OH at the chain end of the polymer chain. The δ peaks of diacid moieties: TPA (7) and NDA (8 and 9) appear at 8.11~8.21 ppm. However, the δ peak at 8.73 ppm represented by 10 in the figure is assigned to the hydrogen atom of NDA, which enables us to calculate the actual chemical composition of diacid moieties present in PCITN copolyester. The amount of each monomer in PCITN copolyester was determined by the integration of their corresponding δ peaks. Analyzed results (Table 1) indicate synthesized PCITN has a higher amount of CHDM and TPA than feeding monomers, which indicates that CHDM and TPA have higher reactivity than isosorbide and NDA, respectively. Our findings are also supported by the previous research works [28–30].

Table 1. Characteristics of synthesized PCITN copolyester pallets

T_g (°C) [a]	123.3	M_n [b]	25,400
T_m (°C) [a]	287.4	M_w [b]	68,900
T_{cc} (°C) [a]	222.9	PDI (M_n/M_w) [b]	2.70
ΔH_m (J/g) [a]	46.99	IV [c] (dL/g)	0.77
CHDM/ISB Feed Ratio	75/25	CHDM/ISB Composition Ratio [d]	93.92/6.08
TPA/NDA Feed Ratio	25/75	TPA/NDA Composition Ratio [d]	30.95/69.05

[a] Glass transition temperature (T_g), cold crystallization temperature (T_{cc}), melting temperature (T_m), and melting enthalpy ΔH_m (J/g) were determined by differential scanning calorimeter (DSC). [b] Number average molecular weight (M_n, g/mol), weight average molecular weight (M_w, g/mol), and polymer dispersity index (PDI) were determined by gel permeation chromatography (GPC). [c] The intrinsic viscosity (IV) was determined in o-chloroform (OCP) solvent by using automated Ubbelohde viscometer (no. 1C) at 30 °C. [d] Actual chemical composition of each monomer in PCITN was determined by proton nuclear magnetic resonance (^{1}H-NMR) spectroscopy.

3.2. Mechanical Behavior of PCITN Films

Tensile testing of the film specimens with different draw ratios (λ) was conducted and the representative stress-strain (S-S) curves are shown in Figure 2. PCITN films showed high tensile strength, as can be seen in Figure 2a. The mechanical properties of as-synthesized PCITN film (λ1) show the ultimate tensile strength as 50.5 MPa and strain as 35.8%. In the case of films having different stretching (λ) such as λ2, λ3, λ3.5, and λ4, the values for the tensile strength are 71.2, 101.5, 132.8, and 154.5 MPa, respectively. It is obvious from the results that as the λ is increased, both tensile strength (MPa) and Young's modulus (GPa) are also increased (Figure 2b). Initially, tensile strength increased linearly with increasing stretching (λ1–λ3) due to an increment in the molecular orientation of the polymeric chain along the stretching direction. Then, a rapid increment in tensile was observed at higher stretching (λ3–λ4) due to the stress-induced crystallization (SIC). Additionally, the film with the highest stretching (λ4) exhibited the lowest strain at break. It can also be seen from the S-S curve that a decrement in the strain at break is observed with increasing λ. These findings can be attributed to the higher degree of molecular orientation during the stretching of the films. Lee et al. and Hoik et al. also found that mechanical properties of the polymeric materials are directly influenced by the molecular orientation [31,32]. Moreover, the unique V-shaped structure of ISB with a 120° angle between the rings causes the hindrance of polymerization of polymer chains [33]. So, mechanical performance can be affected by the orientation of polymeric chains.

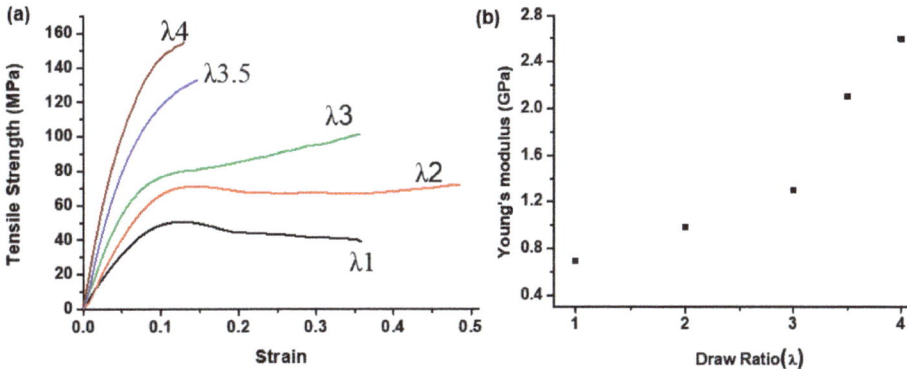

Figure 2. Mechanical properties of cold-drawn PCITN films: (**a**) Tensile strength and (**b**) Young's modulus as a function of draw ratio (λ).

3.3. Direct Evidence of SIC by High Power X-ray Diffraction (XRD) Analysis

The XRD spectrum of the developed PCITN film specimens at the ambient room temperature from 5 to 50 is shown in Figure 3. Three diffraction peaks for the randomly oriented PCTIN film with λ1 that appeared at 7.84°, 18.72° and 42.90° are corresponding to the Miller indices of (001), ($1\bar{1}2$) and ($\bar{1}05$), respectively. The peaks observed at 18.70° and 42.92° are the characteristic peaks of amorphous regions, while the peaks observed at 7.84° are corresponding to the crystalline regions of undrawn (DR1) film. This indicates that as-synthesized PCITN film is semi-crystalline, and it has relatively low crystalline regions compared to stretched films. However, based on the XRD analysis, the reduction in amorphousness of PCITN film can be seen as the intensity of the amorphous peaks decreases with increasing λ. The new peaks are observed at 15.76°, 17.68°, 20.12°, and 23.08° at higher stretching (λ > 3), indicating that at the microscopic level the regular arrangement of molecular chains of the films increases with increasing stretching. These peaks are attributed to the new crystallites that appeared due to stretching and they are attributed to the Miller indices of (010), ($1\bar{1}1$), ($1\bar{1}0$), and (100), respectively [34,35]. It indicates the appearance of the new crystalline regions at the expense of amorphous regions due to the SIC at higher stretching (λ > 3).

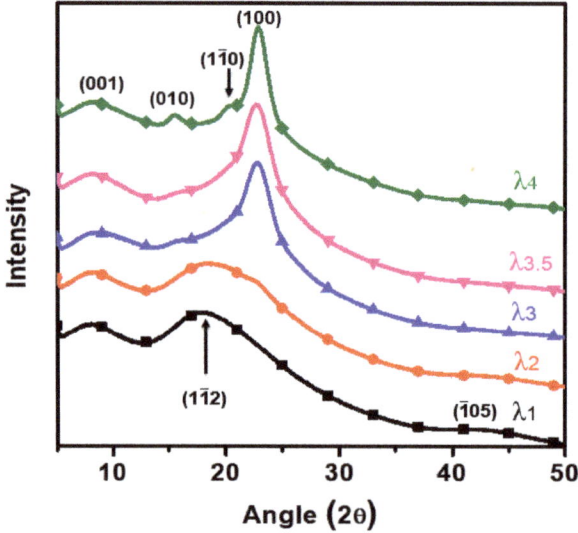

Figure 3. X-ray diffraction pattern of PCITN films stretched at different draw ratios (λ).

3.4. Thermal Properties and SIC Analysis of PCITN Film

The unique chemical structure of ISB with high thermal stability and low segmental mobility makes it an interesting building block for a polymer having high thermal properties, especially the T_g. Thermal properties of the developed uniaxially stretched PCITN films with different λ were analyzed (Figure 4), and the results are summarized in Table 2. It is obvious from DSC analysis that the thermal properties and X_c are directly influenced by the λ of the film. T_g was improved linearly while both T_{cc} and ΔH_{cc} were reduced with increasing λ. The gradual reduction in ΔH_{cc} indicates that thermally induced crystallites are reduced, and stress-induced crystallites are increased with increasing λ. ΔH_m (J/g) and X_c (%) also confirm the appearance of SIC at higher stretching (λ >3). Initially, the thermal properties and crystallinity of the stretched films increase linearly with λ due to the gradual growth of molecular chains along the stretching direction. Then, a rapid increment is observed at higher stretching, which can be attributed to new crystallites due to SIC. The effect of the stretching (λ) on thermal and X_c (%) behavior observed in this study can be due to increased symmetry of polymeric chains, which was improved significantly by stretching. The beauty of this analysis is that its findings are compatible with the previously reported research [36–38].

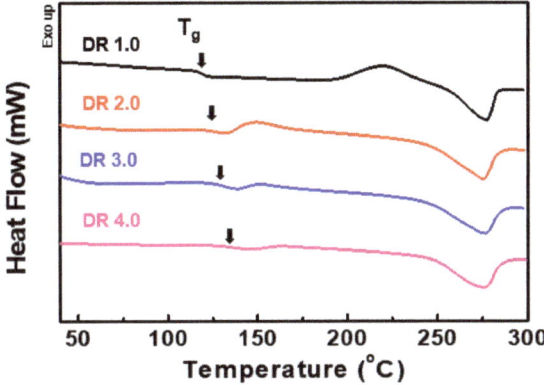

Figure 4. DSC thermograms of the fabricated PCITN films.

Table 2. Influence of stretching on the thermal properties and X_c of PCITN film.

Draw Ratio	T_g (°C)	T_{cc} (°C)	ΔH_{cc} (J/g)	T_m (°C)	ΔH_m (J/g)	X_c (%)	$T_{id5\%}$ (°C)	Residue at 600 °C (%)
1	123.9	220.5	18.20	277.5	20.49	8.65	394.7	1.9
2	125.1	149.2	10.61	275.1	21.86	43.43	398.3	4.3
3	131.9	-	-	276.1	25.32	75.92	398.4	8.7
4	134.9	-	-	275.4	26.66	97.65	396.4	13.4

3.5. Thermal Stability of the Fabricated PCITN Films

Thermal degradation behavior of the fabricated PCITN films stretched at different λ were determined using TGA and the corresponding thermograms, indicating the high thermal stability, are shown in Figure 5. The temperature corresponding to 5% wt loss of the initial weight and residue (%) at 600 °C were recorded from TGA thermograms, and the results are summarized in Table 2. It is evident from the TGA thermograms that all the analyzed samples have a one-stage decomposition, indicating that they have a random copolymer structure. The residue (%) at 600 °C is increased with increasing λ due to increased crystallinity (%) of the resultant stretched polymeric films. The same finding was reported by Um et al. [39]. The high thermal stability of the fabricated film makes it suitable for applications where high thermal stability is required.

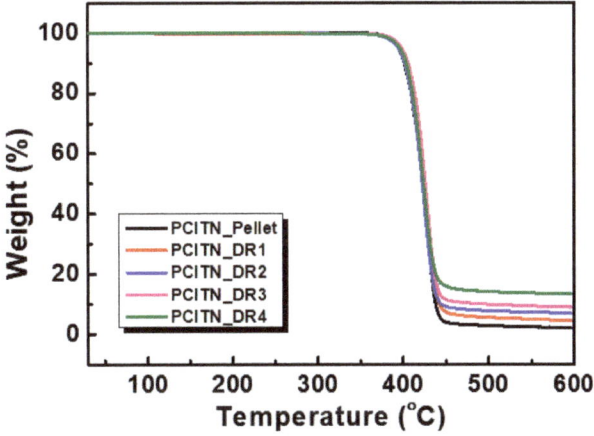

Figure 5. TGA thermograms of the fabricated PCITN films.

3.6. Improving the Dimensional Stability of Uniaxially Oriented Films

Dimensional stability is one of the most important parameters of polymer films for their various industrial applications, especially in the field of flexible electronic devices. The dimensional stability of the fabricated films was determined as their CTE behavior, and the results are shown in Figure 6. It is found that the thermal stability of the PCITN film improves with increasing λ. The least thermal expansion of PCITN film with λ4 indicates that it has the highest dimensional stability, which is due to the higher degree of molecular orientation in this film than other films. Just like conventional plastic films, novel PCITN films also show an undesirable dimensional change around T_g. This phenomenon can be attributed to the facts: (a) the molecular relaxation due to the increased mobility of polymer chains and (b) a rapid shrinkage or expansion due to the residual frozen stress within the stretched film structure. Compared to stretched films, the unstretched film expands abruptly around its T_g due to increased mobility of molecular chains. CTE behavior of the PCITN films was improved after uniaxial stretching. It can be attributed to the fact that the orientation of molecular chains is significantly improved with increasing λ. It is also reported by the previous researches that that increased molecular orientation improves the dimensional stability of the resultant films [40,41].

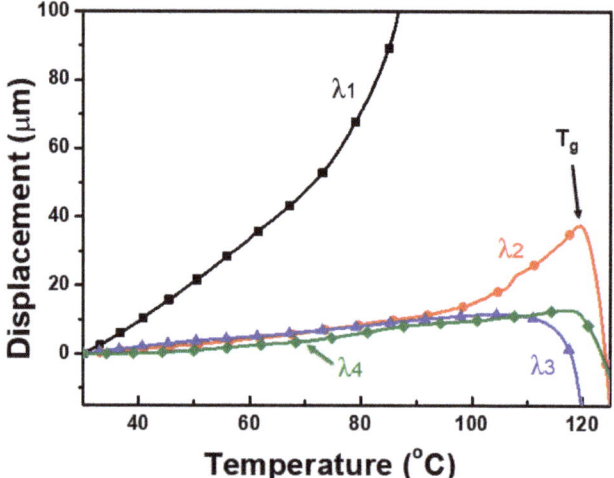

Figure 6. The coefficient of linear thermal expansion (CLTE) plots between 30–120 °C for the uniaxially stretched PCITN films.

3.7. Water Barrier Property

It is well known that good barrier properties of smart polymeric films are critical for their industrial applications especially for next-generation flexible electronic applications. The results for the water barrier of fabricated PCITN in comparison with commercial PET and PEN films are shown in Figure 7. It shows that PCITN film has a better water barrier compared to conventional homopolyester films. This superior water barrier is due to the presence of rigid ISB and cycloaliphatic CHDM units present in the synthesized PCITN films compared to PET and PEN films. This superior water barrier is due to the presence of rigid ISB and cycloaliphatic CHDM units present in the synthesized PCITN films compared to PET and PEN films. It is also obvious from Figure 7 that water barrier property is also improved with increasing λ. The increased water barrier of PCITN film with increasing λ is due to the well-known facts of the improved orientation of polymeric chains and the appearance of new crystallites due to SIC at higher stretching. It is important to note that our findings of water barrier properties are also supported by XRD and DSC results (Sections 3.3 and 3.4). It is important to note that our findings of water barrier properties are also supported by XRD and DSC results (Sections 3.3 and 3.4).

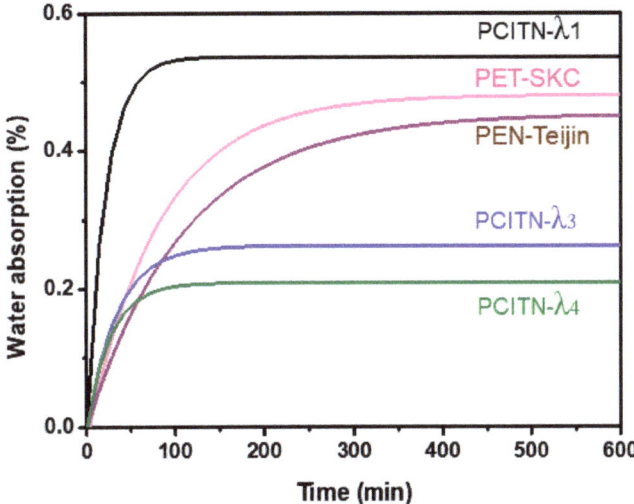

Figure 7. Comparison of the water absorption (%) of developed PCITN films and commercial PET and PEN homopolyester films.

3.8. Birefringence of As-Synthesized and Uniaxially Stretched PCITN Films

The birefringence can be used to evaluate the molecular orientation and relaxation phenomena of the polymer films. To analyze the orientation development during uniaxial cold-drawing of PCITN films, birefringence has been measured for drawn PCITN films, and the results in comparison with uniaxially stretched PET film [37] are shown in Figure 8. It can be seen from the results (Figure 8) that like PET film, PCITN film also exhibits three regimes during stretching. In the first regime, for λ values up to 1.0~2.5, birefringence increases linearly with increasing λ. In regime II, for λ values up to 3.0~4.0, a rapid increment in the birefringence is observed. In regime III (λ > 4), birefringence reaches its saturation where further stretching cannot cause a significant variation in birefringence. The birefringence results indicate how the molecular orientations are developed during the uniaxial stretching of PCITN. In regime I, the birefringence of isotropic and randomly-oriented films increases linearly with increasing λ due to the steady growth of molecular chains along the stretching direction. An abrupt increment of birefringence in regime II is attributed to the development of SIC phases and the reduction of amorphousness of the stretched films. In regime III, birefringence reaches its saturation due to the achievement of maximum alignment polymer chains. Our results for the birefringence are also supported by the previous research works [37,42]. Our research indicates that novel PCITN film with lower birefringence can be used as an alternative performance material to conventional PET and PEN films for optical applications (PEN > PET > PCITN) [43].

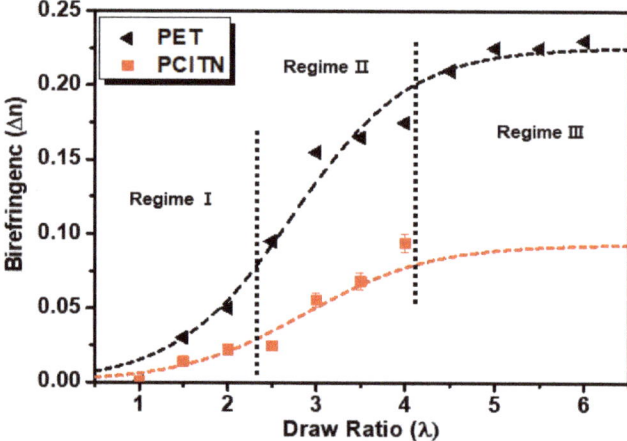

Figure 8. Influence of uniaxial stretching on the birefringence of PCITN films in comparison with uniaxially stretched PET film.

It is worthy to note that the analyzed properties of the developed PCITN film are superior or comparable to the conventional polymeric substrates for flexible electronic devices (Table 3) [20,37,44–46]. Uniaxially stretched PCTIN has better optical (Yellow color vs transparent) and water barrier properties (water absorption; 1.8% vs 0.20%) than PI film [47]. Compared to conventional polymer (PET, PEN, and PI), fabricated semi-crystalline PCITN film has unique performance characteristics such as high T_g, wide processing window, good thermal degradation behavior, low CTE, good water barrier, and low birefringence, which make this novel smart film an ideal substrate for next-generation flexible electronics.

Table 3. Characteristics of uniaxially stretched PCITN film in comparison with PET, PEN, and PI films; conventional polymeric substrate for flexible electronics

Property	PET (Melinex)	PEN (Teonex)	PI (Kapton)	PCITN (Uniaxially Stretched)
T_g (°C)	78	120	360	140
T_m (°C)	260	268	-	275
Commercial Availability	Yes	Yes	Yes	No
Transmission (400–700 nm), %	89	87	Yellow	86
CTE (−55 to 85 °C) (ppm °C^{-1})	15	13	30–60	5.8
Young's modulus (GPa)	5.3	6.1	2.5	2.6
Birefringence (Δn) (*Uniaxially Oriented*)	0.24	0.48	-	0.09
Water absorption (%) (*Randomly Oriented*)	1.08	0.98	-	0.54

The analyzed performance properties of PCITN can be further improved by heat setting and biaxial stretching. The availability of raw materials, ease of synthesis, thermal stability, and wide processing window of PCITN make it suitable for scale-up at the industrial scale. So, it is very easy to develop PCITN film on a commercial scale by adopting the reported methods.

4. Conclusions

A novel PCITN copolyester containing a biobased monomer, isosorbide, with a high T_g and wide processing window was synthesized by melt polymerization. The unique structure of ISB increased the rigidity and T_g of synthesized copolyester. PCTIN films were successfully fabricated and thoroughly characterized for their thermal, mechanical, optical, and barrier properties after the uniaxial cold drawing (λ = 1~4) at optimized conditions. The performance properties of PCITN were significantly improved due to the increase in molecular chain orientation with increasing λ. The good thermal behavior, high strength, good dimensional stability, good barrier, and optical properties are very useful for polymers as a smart film for industrial applications. The unique characteristics of PCITN compared to conventional polymers make it at futuristic performance material for flexible devices.

Author Contributions: Conceptualization, J.J. and F.H.; methodology, J.J. and F.H.; validation, S.P.; formal analysis, J.J. and F.H.; investigation, J.J. and F.H.; writing—original draft preparation, J.J. and F.H.; writing—review and editing, J.K.; supervision, S.-J.K. and J.K.; All authors have read and agreed to the published version of the manuscript.

Funding: This research received no external funding.

Acknowledgments: The author would like to express appreciation to BASF for financial support.

Conflicts of Interest: The authors declare no conflict of interest.

References

1. Behera, K.; Chang, Y.H.; Chiu, F.C.; Yang, J.C. Characterization of poly(lactic acid)s with reduced molecular weight fabricated through an autoclave process. *Polym. Test.* **2017**, *60*, 132–139. [CrossRef]
2. Iwata, T. Biodegradable and bio-based polymers: Future prospects of eco-friendly plastics. *Angew. Chem. Int. Ed.* **2015**, *54*, 3210–3215. [CrossRef] [PubMed]
3. Bond, J.Q.; Alonso, D.M.; Wang, D.; West, R.M.; Dumesic, J.A. Integrated Catalytic Conversion of -Valerolactone to Liquid Alkenes for Transportation Fuels. *Science* **2010**, *327*, 1110–1114. [CrossRef] [PubMed]
4. Hussain, F.; Khurshid, M.F.; Masood, R.; Ibrahim, W. Developing antimicrobial calcium alginate fibres from neem and papaya leaves extract. *J. Wound Care* **2017**, *26*, 778–783. [CrossRef] [PubMed]
5. Aricò, F. Isosorbide as biobased platform chemical: Recent advances. *Curr. Opin. Green Sustain. Chem.* **2020**, *21*, 82–88. [CrossRef]
6. Hamad, K.; Kaseem, M.; Ayyoob, M.; Joo, J.; Deri, F. Polylactic acid blends: The future of green, light and tough. *Prog. Polym. Sci.* **2018**, *85*, 83–127. [CrossRef]
7. Girard, M.; Bee, G. Invited review: Tannins as a potential alternative to antibiotics to prevent coliform diarrhea in weaned pigs. *Animal* **2020**, *14*, 95–107. [CrossRef]
8. Wang, J.; Liu, X.; Jia, Z.; Sun, L.; Zhang, Y.; Zhu, J. Modification of poly(ethylene 2,5-furandicarboxylate) (PEF) with 1, 4-cyclohexanedimethanol: Influence of stereochemistry of 1,4-cyclohexylene units. *Polymer* **2018**, *137*, 173–185. [CrossRef]
9. Amulya, K.; Kopperi, H.; Venkata Mohan, S. Tunable production of succinic acid at elevated pressures of CO_2 in a high pressure gas fermentation reactor. *Bioresour. Technol.* **2020**, *309*, 123327–123335. [CrossRef]
10. Wang, X.; Wang, Q.; Liu, S.; Sun, T.; Wang, G. Synthesis and properties of poly(isosorbide 2,5-furandicarboxylate-co-ε-caprolactone) copolyesters. *Polym. Test.* **2020**, *81*, 106284–106290. [CrossRef]
11. Musa, C.; Kervoëlen, A.; Danjou, P.E.; Bourmaud, A.; Delattre, F. Bio-based unidirectional composite made of flax fibre and isosorbide-based epoxy resin. *Mater. Lett.* **2020**, *258*, 126818–126821. [CrossRef]
12. Zenner, M.D.; Xia, Y.; Chen, J.S.; Kessler, M.R. Polyurethanes from isosorbide-based diisocyanates. *ChemSusChem* **2013**, *6*, 1182–1185. [CrossRef] [PubMed]
13. Choi, Y.H.; Lyu, M.Y. Comparison of Rheological Characteristics and Mechanical Properties of Fossil-Based and Bio-Based Polycarbonate. *Macromol. Res.* **2020**, *28*, 299–309. [CrossRef]
14. Eo, Y.S.; Rhee, H.W.; Shin, S. Catalyst screening for the melt polymerization of isosorbide-based polycarbonate. *J. Ind. Eng. Chem.* **2016**, *37*, 42–46. [CrossRef]
15. Behera, K.; Chang, Y.H.; Yadav, M.; Chiu, F.C. Enhanced thermal stability, toughness, and electrical conductivity of carbon nanotube-reinforced biodegradable poly(lactic acid)/poly(ethylene oxide) blend-based nanocomposites. *Polymer* **2020**, *186*, 122002–122014. [CrossRef]
16. Hussain, F.; Jeong, J.; Park, S.; Jeong, E.; Kang, S.-J.; Yoon, K.; Kim, J. Fabrication and characterization of a novel terpolyester film: An alternative substrate polymer for flexible electronic devices. *Polymer* **2020**, *210*, 123019–123026. [CrossRef]
17. Kricheldorf, H.R.; Gomourachvili, Z. Polyanhydrides 10: Aliphatic polyesters and poly(ester-anhydride)s by polycondensation of silylated aliphatic diols. *Macromol. Chem. Phys.* **1997**, *198*, 3149–3160. [CrossRef]

18. Bersot, J.C.; Jacquel, N.; Saint-Loup, R.; Fuertes, P.; Rousseau, A.; Pascault, J.P.; Spitz, R.; Fenouillot, F.; Monteil, V. Efficiency increase of poly(ethylene terephthalate-co-isosorbide terephthalate) synthesis using bimetallic catalytic systems. *Macromol. Chem. Phys.* **2011**, *212*, 2114–2120. [CrossRef]
19. Yoon, W.J.; Hwang, S.Y.; Koo, J.M.; Lee, Y.J.; Lee, S.U.; Im, S.S. Synthesis and characteristics of a biobased high-Tg terpolyester of isosorbide, ethylene glycol, and 1,4-cyclohexane dimethanol: Effect of ethylene glycol as a chain linker on polymerization. *Macromolecules* **2013**, *46*, 7219–7231. [CrossRef]
20. MacDonald, W.A. Latest advances in substrates for flexible electronics. In *Large Area and Flexible Electronics*; Caironi, M., Noh, Y., Eds.; Wiley: Weinhim, Germany, 2015; pp. 306–307. ISBN 9783527329786.
21. Lewis, J.S.; Weaver, M.S. Thin-film permeation-barrier technology for flexible organic light-emitting devices. *IEEE J. Sel. Top. Quantum Electron.* **2004**, *10*, 45–57. [CrossRef]
22. Hussain, F.; Park, S.; Jeong, J.; Kang, S.J.; Kim, J. Structure–property relationship of poly(cyclohexane 1,4-dimethylene terephthalate) modified with high trans-1,4-cyclohexanedimethanol and 2,6-naphthalene dicarboxylicacid. *J. Appl. Polym. Sci.* **2020**, *137*, 48350–48959. [CrossRef]
23. Steinborn-Rogulska, I.; Rokicki, G. Solid-state polycondensation (SSP) as a method to obtain high molecular weight polymers: Part II. Synthesis of polylactide and polyglycolide via SSP. *Polimery* **2013**, *58*, 85–92. [CrossRef]
24. Kasmi, N.; Papageorgiou, G.Z.; Achilias, D.S.; Bikiaris, D.N. Solid-State polymerization of poly(Ethylene Furanoate) biobased Polyester, II: An efficient and facile method to synthesize high molecular weight polyester appropriate for food packaging applications. *Polymers* **2018**, *10*, 471. [CrossRef] [PubMed]
25. Park, S.; Hussain, F.; Kang, S.; Jeong, J.; Kim, J. Synthesis and properties of copolyesters derived from 1,4-cyclohexanedimethanol, terephthalic acid, and 2,6-naphthalenedicarboxylic acid with enhanced thermal and barrier properties. *Polymers* **2018**, *42*, 662–669. [CrossRef]
26. *ASTM (D882-02) Standard Test Method for Tensile Properties of Thin Plastic Sheeting*; ASTM International: West Conshohocken, PA, USA, 2002; Volume 14, pp. 1–11.
27. *ASTM D570-98 Standard Test Method for Water Absorption of Plastics*; ASTM International: West Conshohocken, PA, USA, 1998; Volume 16, pp. 1–4.
28. Hussain, F.; Park, S.; Jeong, J.; Jeong, E.; Kang, S.; Yoon, K.; Kim, J. Fabrication and characterization of poly(1,4-cyclohexanedimthylene terephthalate-co-1,4-cyclohxylenedimethylene 2,6-naphthalenedicarboxylate) (PCTN) copolyester film: A novel copolyester film with exceptional performances for next generation flexible elec. *J. Appl. Polym. Sci.* **2020**, *138*, 49840–49850. [CrossRef]
29. Hussain, F.; Jeong, J.; Park, S.; Kang, S.-J.; Kim, J. Single-step solution polymerization and thermal properties of copolyesters based on high trans-1,4-cyclohexanedimethanol, terephthaloyl dichloride, and 2,6-naphthalene dicarboxylic chloride. *Polymer* **2019**, *43*, 475–484. [CrossRef]
30. Koo, J.M.; Hwang, S.Y.; Yoon, W.J.; Lee, Y.G.; Kim, S.H.; Im, S.S. Structural and thermal properties of poly(1,4-cyclohexane dimethylene terephthalate) containing isosorbide. *Polym. Chem.* **2015**, *6*, 6973–6986. [CrossRef]
31. Lee, H.; Koo, J.M.; Sohn, D.; Kim, I.S.; Im, S.S. High thermal stability and high tensile strength terpolyester nanofibers containing biobased monomer: Fabrication and characterization. *RSC Adv.* **2016**, *6*, 40383–40388. [CrossRef]
32. Hwang, K.Y.; Kim, S.D.; Kim, Y.W.; Yu, W.R. Mechanical characterization of nanofibers using a nanomanipulator and atomic force microscope cantilever in a scanning electron microscope. *Polym. Test.* **2010**, *29*, 375–380. [CrossRef]
33. Yoon, W.J.; Oh, K.S.; Koo, J.M.; Kim, J.R.; Lee, K.J.; Im, S.S. Advanced polymerization and properties of biobased high Tg polyester of isosorbide and 1,4-cyclohexanedicarboxylic acid through in situ acetylation. *Macromolecules* **2013**, *46*, 2930–2940. [CrossRef]
34. Jeong, Y.G.; Jo, W.H.; Lee, S.C. Synthesis and isodimorphic cocrystallization behavior of poly(1,4-cyclohexylenedimethylene terephthalate-*co*-1,4-cyclohexylenedimethylene 2,6-naphthalate) copolymers. *J. Polym. Sci. Part B* **2004**, *42*, 177–187. [CrossRef]
35. Jeong, Y.G.; Jo, W.H.; Lee, S.C. Crystal structure determination of poly(1,4-trans-cylcohexylenedimethylene 2,6-naphthalate) by X-ray diffraction and molecular modeling. *Macromolecules* **2003**, *36*, 5201–5207. [CrossRef]
36. Özen, İ.; Bozoklu, G.; Dalgıçdir, C.; Yücel, O.; Ünsal, E.; Çakmak, M.; Menceloğlu, Y.Z. Improvement in gas permeability of biaxially stretched PET films blended with high barrier polymers: The role of chemistry and processing conditions. *Eur. Polym. J.* **2010**, *46*, 226–237. [CrossRef]

37. Ben Doudou, B.; Dargent, E.; Grenet, J. Relationship between draw ratio and strain-induced crystallinity in uniaxially hot-drawn PET-MXD6 films. *J. Plast. Film Sheeting* **2005**, *21*, 233–251. [CrossRef]
38. Ajji, A.; Guèvremont, J.; Cole, K.C.; Dumoulin, M.M. Orientation and structure of drawn poly(ethylene terephthalate). *Polymer* **1996**, *37*, 3707–3714. [CrossRef]
39. Um, I.C.; Ki, C.S.; Kweon, H.Y.; Lee, K.G.; Ihm, D.W.; Park, Y.H. Wet spinning of silk polymer: II. Effect of drawing on the structural characteristics and properties of filament. *Int. J. Biol. Macromol.* **2004**, *34*, 107–119. [CrossRef]
40. MacDonald, B.A.; Rollins, K.; Eveson, R.; Rakos, K.; Rustin, B.A.; Handa, M. New developments in polyester film for flexible electronics. *MRS Proc.* **2003**, *769*, 1–8. [CrossRef]
41. Blumentritt, B.F. Anisotropy and dimensional stability of biaxially oriented poly(ethylene terephthalate) films. *J. Appl. Polym. Sci.* **1979**, *23*, 3205–3217. [CrossRef]
42. Nobukawa, S.; Nakao, A.; Songsurang, K.; Pulkerd, P.; Shimada, H.; Kondo, M.; Yamaguchi, M. Birefringence and strain-induced crystallization of stretched cellulose acetate propionate films. *Polymer* **2017**, *111*, 53–60. [CrossRef]
43. Huijts, R.A.; Peterst, S.M. The relation between molecular orientation and birefringence in PET and PEN fibres. *Polymer* **1994**, *35*, 3119–3121. [CrossRef]
44. MacDonald, W.A. Engineered films for display technologies. *J. Mater. Chem.* **2004**, *14*, 4–10. [CrossRef]
45. MacDonald, W.A.; Looney, M.K.; MacKerron, D.; Eveson, R.; Adam, R.; Hashimoto, K.; Rakos, K. Latest advances in substrates for flexible electronics. *J. Soc. Inf. Disp.* **2007**, *15*, 1075–1081. [CrossRef]
46. Cakmak, M.; Wang, Y.D.; Simhambhatla, M. Processing characteristics, structure development, and properties of uni and biaxially stretched poly(ethylene 2,6 naphthalate) (PEN) films. *Polym. Eng. Sci.* **1990**, *30*, 721–733. [CrossRef]
47. Cheng, I.-C.; Wagner, S. Overview of flexible electronics. In *Flexible Electronics: Materials and Applications Electronic Materials: Science and Technology*; Wong, W.S., Salleo, A., Eds.; Springer: New York, NY, USA, 2009; pp. 1–28.

Publisher's Note: MDPI stays neutral with regard to jurisdictional claims in published maps and institutional affiliations.

© 2020 by the authors. Licensee MDPI, Basel, Switzerland. This article is an open access article distributed under the terms and conditions of the Creative Commons Attribution (CC BY) license (http://creativecommons.org/licenses/by/4.0/).

Article

Controlling the Isothermal Crystallization of Isodimorphic PBS-*ran*-PCL Random Copolymers by Varying Composition and Supercooling

Maryam Safari [1], Agurtzane Mugica [1], Manuela Zubitur [2], Antxon Martínez de Ilarduya [3], Sebastián Muñoz-Guerra [3] and Alejandro J. Müller [1,4,*]

1. POLYMAT and Polymer Science and Technology Department, Faculty of Chemistry, University of the Basque Country UPV/EHU, Paseo Manuel de Lardizabal 3, 20018 Donostia-San Sebastián, Spain; maryam.safari@polymat.eu (M.S.); agurtzane.mugica@ehu.es (A.M.)
2. Chemical and Environmental Engineering Department, Polytechnic School, University of the Basque Country UPV/EHU, Plaza Europa 1, 20018 Donostia-San Sebastián, Spain; manuela.zubitur@ehu.eus
3. Departament d'Enginyeria Química, Universitat Politècnica de Catalunya, ETSEIB, Diagonal 647, 08028 Barcelona, Spain; antxon.martinez.de.ilarduia@upc.edu (A.M.d.I.); sebastian.munoz@upc.edu (S.M.-G.)
4. IKERBASQUE, Basque Foundation for Science, María Díaz Haroko Kalea, 3, 48013 Bilbao, Spain
* Correspondence: alejandrojesus.muller@ehu.es; Tel.: +34-943018191

Received: 1 December 2019; Accepted: 18 December 2019; Published: 20 December 2019

Abstract: In this work, we study for the first time, the isothermal crystallization behavior of isodimorphic random poly(butylene succinate)-*ran*-poly(ε-caprolactone) copolyesters, PBS-*ran*-PCL, previously synthesized by us. We perform nucleation and spherulitic growth kinetics by polarized light optical microscopy (PLOM) and overall isothermal crystallization kinetics by differential scanning calorimetry (DSC). Selected samples were also studied by real-time wide angle X-ray diffraction (WAXS). Under isothermal conditions, only the PBS-rich phase or the PCL-rich phase could crystallize as long as the composition was away from the pseudo-eutectic point. In comparison with the parent homopolymers, as comonomer content increased, both PBS-rich and PCL-rich phases nucleated much faster, but their spherulitic growth rates were much slower. Therefore, the overall crystallization kinetics was a strong function of composition and supercooling. The only copolymer with the eutectic composition exhibited a remarkable behavior. By tuning the crystallization temperature, this copolyester could form either a single crystalline phase or both phases, with remarkably different thermal properties.

Keywords: isodimorphism; random copolymers; crystallization; nucleation; growth rate

1. Introduction

Biocompatible and biodegradable polymers are being developed for a wide range of applications due to their potential to solve the environmental concerns caused by traditional nondegradable plastics [1–4]. Among the biodegradable polymers that have been most intensively studied are aliphatic polyesters such as Poly(glycolide) (PGA), Poly(L-lactide) (PLLA), Poly(ethylene succinate) (PES), Poly(butylene succinate) (PBS), and Poly(ε-caprolactone) (PCL) [5,6]. Although aliphatic polyesters have been used for many years in industrial, biomedical, agricultural, and pharmaceutical applications, there is still room for many improvements [7,8]. The synthesis of random copolyesters, using biobased comonomers, can overcome some of the drawbacks of biodegradable polyesters, such as slow biodegradation rate (due to high crystallinity degrees) and undesirable mechanical properties [9–11].

The properties of crystallizable random copolymers constituted by two semicrystalline parent components have been recently reviewed [12]. Depending on their ability to share crystal lattices, three different cases have been reported [12–14]: (a) total comonomer exclusion occurs when the chemical repeat units are very different and the crystal lattice of each one of the components cannot tolerate the presence of the other; (b) total comonomer inclusion or isomorphic behavior can only be obtained in cases where the components can cocrystallize in the entire composition range (as their chemical structures are very similar), forming a single crystal structure [15,16]; (c) an intermediate and complex case, where a balance between comonomer inclusion and exclusion occurs, leading to isodimorphic copolymers.

In isodimorphic random copolymers, at least one of the two crystalline phases includes some repeat units of the minor component in its crystal lattice. When the melting point is plotted as a function of composition a pseudo-eutectic behavior is commonly observed, where, on each side of the pseudo-eutectic point, only the crystalline phase of the major component is formed, which may contain a limited amount of the minor comonomer chains included in the crystal lattice [12].

In our previous works [10,17], we have synthesized and studied the morphology and crystallinity of poly (butylene succinate-*ran*-caprolactone) (PBS-*ran*-PCL) copolyesters. In situ wide angle X-ray scattering (WAXS) indicated that changes were produced in the crystalline unit cell dimensions of the dominant crystalline phase. In addition, differential scanning calorimetry (DSC) measurements showed that all copolymers could crystallize, regardless of composition, and their thermal transitions temperatures (i.e., T_c and T_m) went through a pseudo-eutectic point when plotted as a function of composition. Therefore, all this evidence demonstrated an isodimorphic behavior. At the pseudo-eutectic composition, both PBS-rich and PCL-rich phases can crystallize [10,17].

In the current work, we perform a detailed isothermal crystallization study of PBS-*ran*-PCL copolymers to determine the nucleation and crystallization kinetics of the copolyesters and study the influence of composition on the crystallization kinetics. This information is very important as it allows tailoring the properties of random copolymers as well as their applications. The analysis of the isothermal crystallization kinetics of PBS-*ran*-PCL was performed using differential scanning calorimetry (DSC), polarized light optical microscopy (PLOM), and in situ wide angle X-ray scattering (WAXS).

2. Materials and Methods

2.1. Synthesis

The PBS-*ran*-PCL copolymers were synthesized by a two-stage melt-polycondensation reaction. First, transesterification/ROP reaction of dimethyl succinate (DMS), 1,4-butanediol (BD), and ε-caprolactone (CL), and then polycondensation at reduced pressure, as reported in detail previously [17]. Samples are denoted in an abbreviated form, e.g., BS_xCL_y, indicating the molar ratio of each component determined by ^{1}H-NMR, as subscripts (x and y). Table 1 shows molar composition, number- and weight-average molar mass and thermal transitions of the isodimorphic random copolyesters under study in this work.

Table 1. Molar composition determined by ^{1}H-NMR, number- and weight-average molar mass determined by gel permeation chromatography (GPC), and thermal transitions determined by differential scanning calorimetry (DSC) (at 10 °C/min) of the materials employed in this work.

	Copolyester	M_n	M_w	T_g (°C)	T_c (°C)	T_m (°C)
1	PBS	7500	21,470	−33	76	114
2	$BS_{91}CL_9$	8790	21,640	−36	61	104
3	$BS_{78}CL_{22}$	6580	18,000	−40	46	93
4	$BS_{66}CL_{34}$	7830	19,700	−44	14	73
5	$BS_{62}CL_{38}$	9750	27,300	−45	12	70
6	$BS_{55}CL_{45}$	8970	24,700	−46	7	62
7	$BS_{51}CL_{49}$	7400	23,500	−47	4	58
8	$BS_{45}CL_{55}$	8000	17,300	−48	−5/−22	13/49
9	$BS_{38}CL_{62}$	11,000	24,300	−50	−13	17
10	$BS_{34}CL_{66}$	10,000	29,900	−51	−10	18
11	$BS_{27}CL_{73}$	11,540	28,700	−53	−8	19
12	$BS_{11}CL_{89}$	6300	19,500	−56	10	36
13	PCL	5400	17,400	−60	34	55

2.2. Polarized Light Optical Microscopy (PLOM)

A polarized light optical microscope, Olympus BX51 (Olympus, Tokyo, Japan), equipped with an Olympus SC50 digital camera and with a Linkam-15 TP-91 hot stage (Linkam, Tadworth, UK) (coupled to a liquid nitrogen cooling system) was used to observe spherulites nucleation and growth. Films with around 100 μm thickness were prepared by melting the samples in between two glass slides. For the isothermal experiments, the conditions were very similar to those employed during DSC measurements. The samples were heated to 30 °C above their melting point to erase their thermal history, and then were rapidly cooled from the melt at 60 °C/min to the selected isothermal crystallization temperature, T_c. Then, the sample was kept at T_c while the spherulites appeared and grew. To better compare the compositions in a fixed common crystallization temperature, T_c values were chosen at the same supercooling degree for PBS-rich compositions at ΔT = 40, 38, 36, 34, and 32 °C and for PCL-rich compositions at ΔT = 40, 39, 38, 37, and 36 °C. The supercooling was calculated from the equilibrium melting temperatures determined by Hoffman–Weeks extrapolations after isothermal crystallization in the DSC, see below.

2.3. Differential Scanning Calorimetry (DSC)

Isothermal differential scanning calorimetry experiments were performed using a Perkin Elmer 8500 calorimeter equipped with a refrigerated cooling system Intracooler 2P, under a nitrogen atmosphere (with a flow of 20 mL/min) and calibrated with high purity indium and tin standards. The weight of the samples was about 5 mg and samples were hermetically sealed in standard aluminum pans.

To investigate the overall crystallization kinetics, an isothermal protocol was applied. First, the minimum isothermal crystallization temperature $T_{c,min}$ was determined by trial and error following Müller et al. [18,19]. Samples were quenched to T_c values (estimated from the nonisothermal DSC runs) at 60 °C/min and then immediately reheated at 20 °C/min up to temperatures above the melting point of the crystalline phase involved. If any latent melting enthalpy is detected, this means that the sample was able to crystallize during the cooling to T_c, therefore, this T_c value cannot be used as $T_{c,min}$ and a higher T_c value is explored.

After the T_c range was determined, the isothermal crystallization experiments were performed, closely following the procedure suggested by Lorenzo et al. [19]: (I) heating from room temperature to 30 °C above their melting point at 10 °C/min; (II) holding the sample for 3 min at that temperature to erase thermal history; (III) quenching the sample to a predetermined crystallization temperature (T_c) at 60 °C/min. T_c was in the range between −3 and 90 °C depending on composition; (IV) isothermal crystallization until maximum saturation; (V) heating from T_c to 30 °C above the melting point of the sample at 10 °C/min, to record the melting behavior after the isothermal crystallization. This final melting run provided the values of apparent melting points that were employed to perform the Hoffman–Weeks extrapolation to calculate the equilibrium melting temperature of each material.

2.4. Simultaneous WAXS Synchrotron Measurements

To study the crystal structure during isothermal crystallization at the pseudo-eutectic point, the $BS_{45}CL_{55}$ sample was examined by in situ WAXS performed at beamline BL11-NCD at the ALBA Synchrotron radiation facility, Cerdanyola del Vallés, Barcelona, Spain. The samples in DSC aluminum pans were placed in a Linkam THMS-600 stage coupled to a liquid nitrogen cooling system. WAXS scans were taken periodically every 30 s during the isothermal crystallization. The energy of the X-ray source was 12.4 keV (λ = 1.0 Å). In the WAXS configuration, the sample-detector, Rayonix LX255-HS with an active area of 230.4 × 76.8 mm (pixel size: 44 µm^2) distance employed was 15.5 mm with a tilt angle of 27.3°. The scattering vector was calibrated using chromium (III) oxide for WAXS experiments.

3. Results

We have studied previously [17] the nonisothermal crystallization behavior of the same PBS-ran-PCL random copolymers employed in this work. The results demonstrated that these copolymers exhibit an isodimorphic behavior.

Figure 1 presents a phase diagram for the PBS-ran-PCL system. These random copolymers exhibit a single-phase melt and a single glass transition temperature, as expected for random copolymers. Upon cooling from the melt, the materials are capable of crystallizing in the entire composition range, in spite of being random, as demonstrated by NMR studies [17]. The copolymers display a pseudo-eutectic point at the composition $BS_{45}CL_{55}$. This $BS_{45}CL_{55}$ copolymer is the only one in the series that can form two crystalline phases upon cooling from the melt, i.e., a PBS-rich phase and a PCL-rich phase (as evidenced earlier by WAXS and DSC [17]), hence the two melting point values reported in Figure 1 for this composition. To each side of the pseudo-eutectic point, a single crystalline phase is formed, either a PBS-rich phase (i.e., left-hand side of the eutectic) or a PCL-rich phase (i.e., right-hand side of the eutectic), with crystalline unit cells resembling those of PBS and PCL respectively.

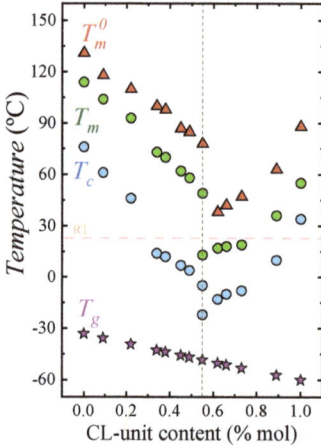

Figure 1. Phase diagram based on data published in [17] on the nonisothermal crystallization of the PBS-*ran*-PCL copolymers under study. Additionally, equilibrium melting temperatures obtained in the present work by Hoffman–Weeks analysis of isothermally obtained data, are also included. The dashed vertical line indicates the pseudo-eutectic point. The dashed horizontal line indicates an arbitrary room temperature value.

In the present work, we performed isothermal crystallization studies and calculated the equilibrium melting temperatures (T_m^0) of homopolymers and copolymers by employing the Hoffman–Weeks extrapolation. Examples of Hoffman–Weeks plots can be found in Figure SI-3, while Figure 1 reports the variation of the equilibrium melting temperatures obtained with composition. The T_m^0 values show a similar trend with composition as the apparent melting peak temperatures determined by DSC during nonisothermal experiments, and they also display a pseudo-eutectic point. These T_m^0 values will be employed throughout this paper, as they are needed to fit the Lauritzen and Hoffman nucleation and crystallization theory to analyze the experimental data.

The phase diagram shown in Figure 1 illustrates the versatility of isodimorphic copolymers. It is well known that the optimal mechanical properties in terms of ductility and toughness of thermoplastic semicrystalline materials are generally observed at temperatures in between T_g and T_m. Thanks to random copolymerization, the copolymers exhibit a single T_g value that is independent of the melting point of the phase (or phases in the case of the composition at the pseudo-eutectic point) that is able to crystallize. This remarkable behavior provides a separate control of T_g and T_m which cannot be obtained in homopolymers. Additionally, as Figure 1 shows, depending on composition, the samples can be molten at room temperature or they can be semicrystalline. Such wide range of thermal properties can lead to fine tuning mechanical properties and crystallinities to tailor applications.

3.1. Nucleation Kinetics Studied by PLOM

Counting the number of spherulites in PLOM experiments is the usual way of obtaining nucleation data by assuming that each spherulite grows from one heterogeneous nucleus. In this work, we studied the nucleation kinetics by determining the nucleation density as a function of time by PLOM, from which nucleation rates can be calculated.

Figure 2 shows four examples of plots of the nucleation density ρ_{nuclei} (nuclei/mm^3) as a function of time for neat PBS, neat PCL, and two sample copolymers. The rest of the data can be found in the Supplementary Information (Figure SI-1). The nucleation density increases almost linearly with time at short times, then it tends to saturate. The number of heterogeneous nuclei that are activated at longer times increases as nucleation temperature decreases, a typical behavior of polymer nucleation [20]. As expected, the nucleation density at any given time increases as T_c decreases, because the thermodynamic driving force for primary nucleation increases with supercooling [21].

Figure 3 shows plots of nucleation density versus temperature taken at a constant nucleation time of 100 s for neat PBS and PBS-rich copolymers (Figure 3a) and 10 min in the case of PCL and BS$_{11}$CL$_{89}$ copolymer (Figure 3b).

PBS exhibits the lowest nucleation density of all samples, therefore, the largest spherulites (see Figure 5 below). As the amount of CL comonomer increases in the PBS-rich copolymers (Figure 3a), the nucleation density increases, as well as the supercooling needed for nucleation. In the case of PBS-rich copolymers, Figure 3a (and Figure SI-1 in the SI) shows nucleation data for seven different copolymers, with compositions ranging from 91% to 45% PBS.

The dependence of nucleation density on supercooling, can be observed in Figure SI-2. The data presented in Figure 3a can be reduced to a supercooling range between 32 and 40 °C, i.e., only 8 °C. This means that a large part of the horizontal shift in the curves of Figure 3a (spanning nearly 60 °C in crystallization temperature) is due to changes in supercooling. These changes are caused by the variations in equilibrium melting temperatures with composition (see Figure 1).

PCL has a higher nucleation density than PBS when compared at equal supercoolings (see Figure SI-2b). When a small amount of BS comonomer is incorporated, as in random copolymer BS$_{11}$CL$_{89}$, the nucleation density increases significantly (Figure 3b). Due to the very high nucleation density of the other PCL-rich composition copolymers (with higher amounts of PBS), it was impossible to determine their nucleation kinetics. Examples of the microspherulitic morphologies obtained for such PCL-rich copolymers can be observed in Figure 5 below.

It is interesting to note than in both sides of the pseudo-eutectic point (i.e., the PBS-rich side represented in Figure 3a and the PCL-rich side represented in Figure 3b, see also Figure 1), the copolymers exhibit higher nucleation density than their corresponding homopolymers. This behavior could be somewhat analogous to what has been observed in long-chain branched polylactides (PLLAs) [22] or long-chain branched polypropylenes (PPs) with respect to linear analogs [23]. The interruption of crystallizable linear sequences with defects has been reported to increase nucleation density although the reasons are not clear. In the present case, the linear crystallizable sequence of PBS, for instance, is being changed by the introduction of randomly placed PCL repeat units. Even though the random copolyesters can form a single phase in the melt, there may be at the segmental level, some preference for PBS-PBS local chain segmental contacts in comparison to less favorable PBS-PCL contacts. We speculate that this may drive the enhancement of nucleation, but more in-depth studies would be needed to ascertain the exact reason for this behavior.

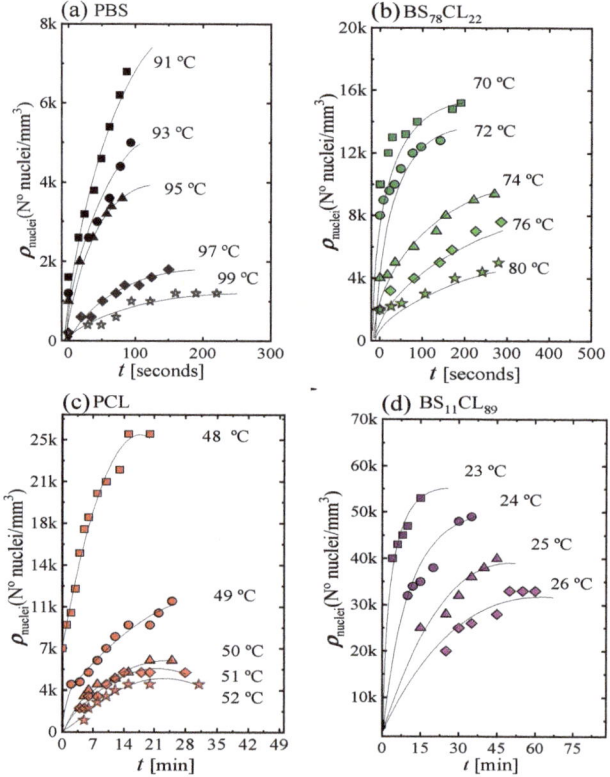

Figure 2. Nucleation kinetics data obtained by polarized light optical microscopy (PLOM): (**a**) PBS; (**b**) $BS_{78}CL_{22}$; (**c**) PCL; (**d**) $BS_{11}CL_{89}$. Nuclei density as a function of time at different crystallization temperatures for the indicated samples.

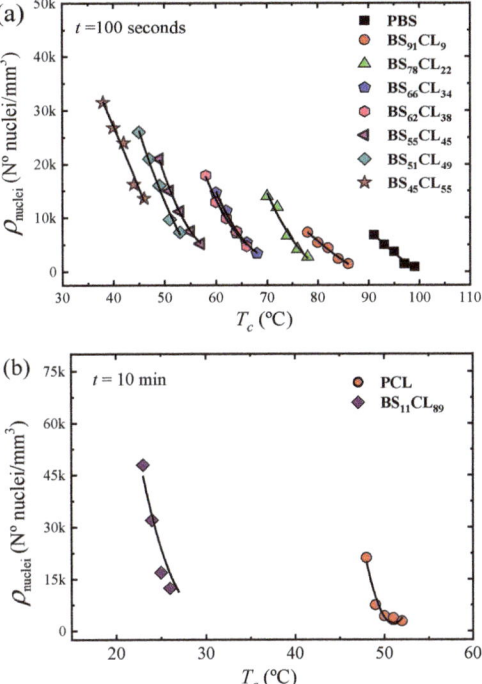

Figure 3. (a) Nuclei density during isothermal crystallization as a function of T_c at a constant time of 100 s for PBS-rich compositions and (b) at 10 min for PCL-rich compositions.

The Fisher–Turnbull nucleation theory [24] can be used to quantify the activation free energy of primary nucleation. This theory gives the steady-state rate of primary nucleation per unit volume and time, $I = dN/dt$, for a heterogeneous nucleation process on a preexisting flat surface (or heterogeneous nucleus) as:

$$logI = logI_0 - \frac{\Delta F^*}{2.3kT} - \frac{16\sigma\sigma_e(\Delta\sigma)T_m^{\circ 2}}{2.3kT(\Delta T)^2(\Delta H_v)^2}, \quad (1)$$

where I_0 is related to the diffusion of polymeric segments from the melt to the nucleation site, ΔF^* is a parameter proportional to the primary nucleation free energy, and σ and σ_e are the lateral and fold surface free energies, respectively. ΔT is the supercooling defined as $\Delta T = T_m^0 - T_c$ and T_m^0 is the equilibrium melting point. $\Delta\sigma$ is the interfacial free energy difference, given by:

$$\Delta\sigma = \sigma + \sigma_{s/c} - \sigma_{s/m}, \quad (2)$$

in which $\sigma_{s/c}$ is the crystal-substrate interfacial energy and $\sigma_{s/m}$ is the melt-substrate interfacial energy. Therefore, $\Delta\sigma$ can be considered proportional to the surface tension properties of the substrate, polymer crystal and polymer melt. The interfacial free energy difference is a convenient way to express the nucleating ability of the substrate towards the polymer melt.

In this work, the values of T_m^0 (listed in Table SI-1 and plotted in Figure 1) were obtained by isothermal crystallization DSC experiments followed by Hoffman–Weeks extrapolations (see Figure SI-3). ΔH_v is the volumetric melting enthalpy (J/m^3) and it was estimated by $\Delta H_V = \Delta H_m^0 \times \rho$, so that ρ = 1.26 g/cm^3 and ΔH_m^0 = 213 J/g for neat PBS [25] and ρ = 1.14 g/cm^3 and ΔH_m^0 = 139.5 J/g for neat PCL [26].

In this work, we employed the value of $\Delta H_m^0 = 213$ J/g for neat PBS and PBS-rich phase composition that has been recently obtained by some of us [25]. This value was determined employing a combined DSC and X-ray diffraction method using isothermal crystallization data. This experimentally extrapolated value is higher than that of $\Delta H_m^0 = 110$ J/g, estimated empirically by the group contribution method [27], but very close to the value of 210 J/g reported by Papageorgiou et al. [28].

The values of the nucleation rate I were calculated from the initial slope (i.e., at short measurement times, where linear trends were obtained) of the plots shown in Figure 2 and Figure SI-1. Figure 4a shows log I as a function of CL-unit molar fraction for a constant supercooling of $\Delta T = 40$ °C. The nucleation rate strongly depends on copolymer composition. Adding a comonomer randomly along the chain to either PBS or PCL largely increases the nucleation rate. In the PBS-rich composition side (to the left of the pseudo-eutectic point signaled by a vertical line in Figure 4a) the nucleation rate increases up to 7.5 times with respect to neat PBS, as the amount of PCL units in the random copolymer increases. Neat PCL nucleates faster than neat PBS. In the PCL-rich composition side, only one copolymer was measured (whose nucleation rate increased two-fold with respect to neat PCL), as increasing PBS content towards the pseudo-eutectic point increased nucleation rate so much that measurements were no longer possible.

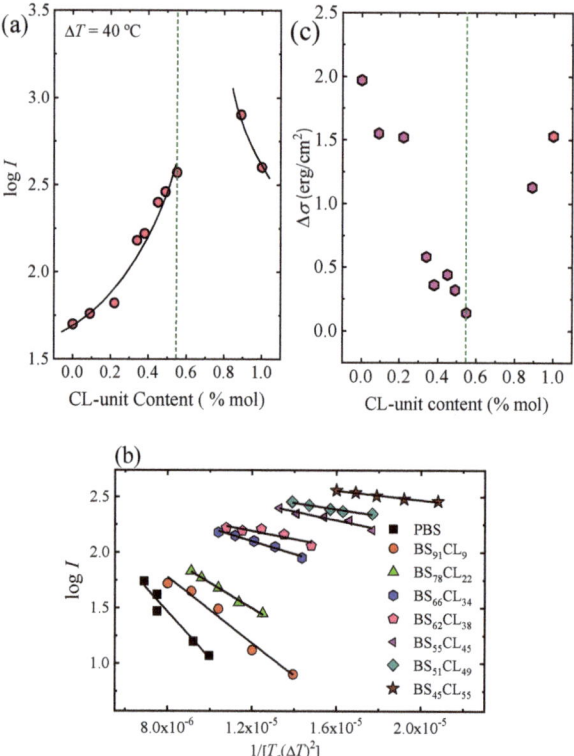

Figure 4. Nucleation rate (*I*) data: (**a**) log I as a function of copolymer composition, expressed as mol % of ε-caprolactone (CL) units, taken at a constant supercooling of $\Delta T = 40$ °C. The segmented vertical line is drawn to indicate the pseudo-eutectic composition. (**b**) Plot of *log I* versus $1/T_c(\Delta T)^2$ for PBS-rich compositions. The black lines represent fittings to the Turnbull–Fisher equation (Equation (1)). (**c**) Interfacial free energy difference ($\Delta \sigma$) as a function of composition. The segmented vertical line is drawn to indicate the pseudo-eutectic composition.

Figure 4b shows the Turnbull–Fisher plots for PBS and PBS-rich compositions based on Equation (1). Turnbull–Fisher plots for PCL and $BS_{11}CL_{89}$ copolymer are presented in the supplementary information (Figure SI-4). The nucleation data can be successfully fitted with the linearized version of Equation (1). From the slope, a value of the interfacial free energy difference ($\Delta\sigma$) can be obtained.

Small values of $\Delta\sigma$ are indicative of good nucleation efficiency since a lower amount of interfacial energy is required to form the crystal–substrate interface. Table 2 reports a value of $\Delta\sigma$ for PBS equal to 1.97 erg/cm^2. As seen in Figure 4c (and Table 2), this interfacial free energy difference progressively decreases in the copolymers as the amount of CL comonomer increases, indicating that the primary nucleation process is facilitated by copolymerization with PCL until the pseudo-eutectic point is reached. On the right-hand side of the pseudo-eutetic point in Figure 4c, PCL has a $\Delta\sigma$ value of 1.53 erg/cm^2, which is, as expected, smaller than that of PBS, as PCL has a larger nucleation density at equivalent supercoolings than PBS. The copolymer $B_{11}CL_{89}$ shows an even smaller value of $\Delta\sigma$, as the incorporation of PBS in the copolymer increases its nucleation capacity.

Table 2. Primary nucleation and growth isothermal kinetics data parameters according to Equations (1) and (3) derived from experimental results obtained by PLOM.

Copolyester	Nucleation, Equation (1)			Growth, Equation (3)			
	$\Delta\sigma$ (erg/cm^2)	$^a R^2$	K_g^G (K^2)	σ (erg/cm^2)	σ_e (erg/cm^2)	q (erg)	$^b R^2$
PBS	1.97	0.938	8.66 × 10^4	12.4	79.5	3.37 × 10^{-13}	0.983
$BS_{91}CL_9$	1.55	0.963	5.80 × 10^4	12.4	55.0	2.33 × 10^{-13}	0.994
$BS_{78}CL_{22}$	1.52	0.993	5.12 × 10^4	12.4	49.6	2.10 × 10^{-13}	0.973
$BS_{66}CL_{34}$	0.58	0.977	8.94 × 10^4	12.4	88.9	3.78 × 10^{-13}	0.982
$BS_{62}CL_{38}$	0.36	0.913	8.92 × 10^4	12.4	89.4	3.87 × 10^{-13}	0.982
$BS_{55}CL_{45}$	0.44	0.920	14.4 × 10^4	12.4	148.8	6.30 × 10^{-13}	0.999
$BS_{51}CL_{49}$	0.32	0.944	14.5 × 10^4	12.4	148.9	6.32 × 10^{-13}	0.999
$BS_{45}CL_{55}$	0.14	0.973	15.7 × 10^4	12.4	165.5	7.02 × 10^{-13}	0.974
$BS_{11}CL_{89}$	1.13	0.867	14.7 × 10^4	6.8	169.6	6.32 × 10^{-13}	0.999
PCL	1.53	0.999	10.4 × 10^4	6.8	112.0	4.17 × 10^{-13}	0.996

^{a} R^2 is the correlation coefficient for the fitting of the nucleation kinetics with the Turnbull–Fisher model (Equation (1)), $\log I$ vs. $1/T(\Delta T)^2$. ^{b} R^2 is the correlation coefficient for the fitting of the Lauritzen–Hoffman model (Equation (3)), $\ln G + U^*/R(T_c - T_0)$ vs. $1/f.Tc.\Delta T$.

3.2. Kinetics of Superstructural Growth (Secondary Nucleation) by PLOM

PBS, PCL, and all the random copolymers prepared in this work exhibited spherulitic superstructural morphologies. Examples of the spherulites obtained at a constant supercooling value of 40 °C can be observed in Figure 5. Both PBS and PCL exhibited well-developed spherulites without banding. PBS-rich copolymers that contain more than 34% PCL exhibit clear banding. This is consistent with previous works indicating that the addition of diluents (for PBS-rich compositions, crystallization occurs while PCL chains are in the liquid state) to several polyesters induces banding [29,30].

Isothermal crystallization experiments were performed to follow the growth of spherulites as a function of time using PLOM. The growth rate was calculated from the slope of spherulite radius versus time plots, which were always observed to be highly linear [18,31].

The experimental growth rates are plotted as a function of the isothermal crystallization temperatures employed in Figure 6a with a linear scale and in Figure 6b with a log scale, so that differences in G values for PBS-rich samples with PCL contents larger than 22% are observed. The incorporation of PCL repeat units in the random copolymers have a dramatic influence on the growth rate of the PBS-rich phase spherulites, as G decreases up to 3.5 orders of magnitude (Figure 6b). The decrease in G values with comonomer incorporation for the PBS rich copolymers is due to two reasons. Firstly, as in any isodimorphic copolymer, there is a competition between inclusion and exclusion of repeat units within the PBS crystal lattice, where exclusion typically predominates. Secondly, incorporation of PCL repeat units in the copolymer chains reduces T_g values (as shown in Figure 1), thereby causing a plasticization effect on the PBS-rich phase. In the case of the PCL-rich compositions, the spherulitic growth rate was determined for only one copolymer (i.e., $BS_{11}CL_{89}$), as in the other cases, as pointed out above, the nucleation rate and nucleation density were so high, that it was impossible to measure the extremely fast growth of very small spherulites. For this copolymer, the growth rate decreased in relative terms (see Figure 6c) by a factor of approximately 2.5 at a supercooling of 40 °C.

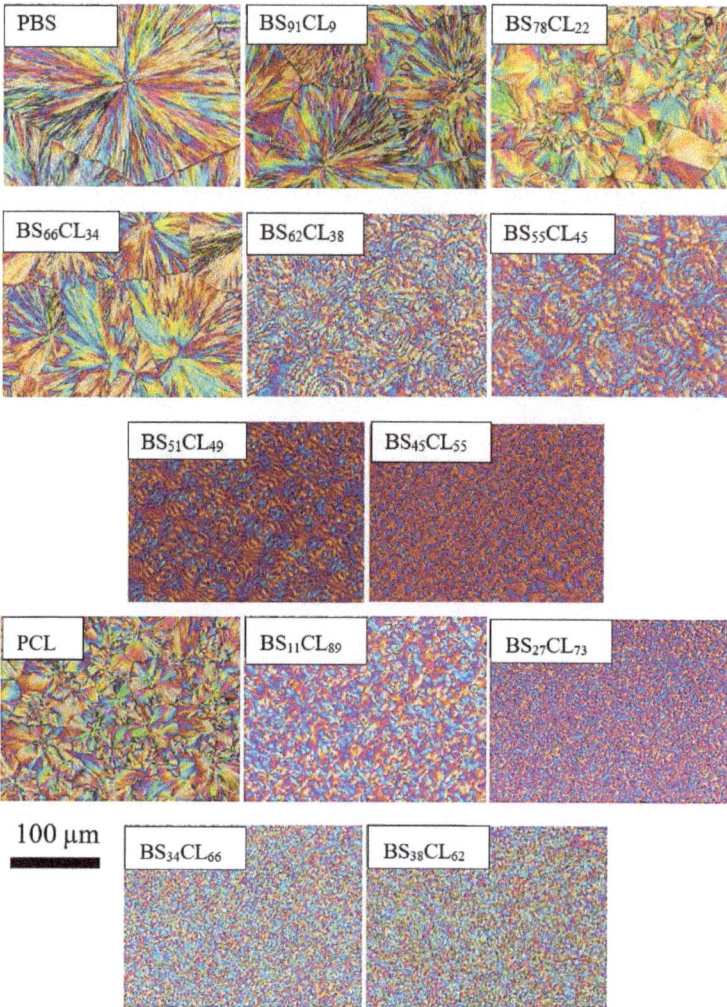

Figure 5. PLOM micrographs after isothermal crystallization at $\Delta T = 40$ °C for the indicated samples.

The data presented in Figure 6a are plotted as a function of supercooling in the Supplementary Information (Figure SI-5). The PBS-rich growth rate data is shifted horizontally and but there is no overlap in the y axis values. If we were dealing with a simple solvent effect, the growth rate curves at different compositions should completely overlap in a master curve when plotted as a function of supercooling. The lack of superposition is due to the fact that PCL repeat units are randomly incorporated and covalently bonded with the PBS repeat units. The interruption of crystallizable PBS repeat units (by the majority of PCL repeat units that are excluded from the crystals) makes more difficult the secondary nucleation process.

Figure 6c shows how the growth rate depends on composition at a constant supercooling of 40 °C. The trend is the opposite as that obtained for primary nucleation (compare Figure 6c with Figure 4a). In order to quantify the restrictions imposed by the comonomer on the crystallization of the major component, we employed the Lauritzen and Hoffman theory, as it allows the calculation of energetic terms related to the secondary nucleation process (i.e., growth process).

The Lauritzen and Hoffman (LH) nucleation and growth theory [32] was used to fit the spherulitic growth rate data as a function of isothermal crystallization temperature, according to the following equation:

$$G = G_0 \exp\left[\frac{-U^*}{R(T_c - T_0)}\right]\left[\frac{-K_g^G}{fT(T_m^0 - T_c)}\right], \quad (3)$$

where G_0 is the growth rate constant that includes all the terms that are temperature-insensitive, U^* is the transport activation energy which characterizes molecular diffusion across the interfacial boundary between melt and crystals (in this work, we employ a constant value of 1500 cal/mol). T_c is the crystallization temperature and T_0 is a hypothetical temperature at which all chain movements freeze (taken as $T_0 = T_g - 30$ °C); T_m^0 is the equilibrium melting temperature and f is a temperature correction factor given by the following expression: $f = 2T_c/(T_c + T_m^0)$.

The equilibrium melting temperatures T_m^0 were estimated by the Hoffman–Weeks linear extrapolation (Figure SI-3 and Table SI-1). The parameter K_g^G is proportional to the energy barrier for secondary nucleation or spherulitic growth and is given by:

$$K_g^G = \frac{jb_0\sigma\sigma_e T_m^0}{k\Delta h_f}, \quad (4)$$

where j is assumed to be equal to 2 for crystallization in the so-called Regime II, a regime where both secondary nucleation at the growth front and the rate of spread along the growing crystal face are comparable [26]. The other terms in the equation are the width of the chain b_0, the lateral surface free energy σ, the fold surface free energy σ_e, the Boltzman constant k, and the equilibrium latent heat of fusion, ΔH_m^0.

Plotting $lnG + \frac{-U}{R(Tc-T0)}$ versus $1/T_c(\Delta T)f$ (i.e., the Lauritzen and Hoffman plots) gives a straight line and its slope and intercept are equal to K_g^G and G_0 respectively. Examples of LH plots can be found in the Supplementary Information, Figure SI-6. Having the value of K_g^G, the magnitude of $\sigma\sigma_e$ can be calculated from Equation (5). In order to calculate separately the values of σ and σ_e, the following expression can be used [33]:

$$\sigma = 0.1\Delta h_f \sqrt{a_0 b_0}, \quad (5)$$

where $a_0 b_0$ is the cross sectional area of the chain. To obtain the parameters of the LH theory, the following values were used for neat PBS and BS-rich compositions [34,35]: $a_0 = 5.25$ Å and $b_0 = 4.04$ Å, and for neat PCL and CL-rich compositions [36]: $a_0 = 4.52$ Å and $b_0 = 4.12$ Å.

Finally, q, the work done by the macromolecule to form a fold is given by [33]:

$$q = 2a_0 b_0 \sigma_e. \quad (6)$$

The solid lines in Figure 6a,b correspond to fittings to Equation (3). Table 2 shows that K_g^G values (which are proportional to the energy barrier for spherulitic growth) for the PBS-rich crystal phase tend to increase as PCL repeat units are incorporated in the random copolymers until a maximum value is reached at the pseudo-eutectic point. Similar trends are observed for the fold surface free energy and for the work done to form folds.

A plot of fold surface free energy versus composition can be found in the Supplementary Information (Figure SI-7). These results quantitatively measure how comonomer incorporation makes difficult the spherulitic growth of the PBS-rich phase. A similar interpretation can be done to the mirror values presented in Table 2 for PCL and the $BS_{11}CL_{89}$ copolymer with respect to the PCL phase.

The results presented in the two sections above can be summarized by comparing Figure 4 with Figure 6. The incorporation of comonomers at each side of the eutectic causes an increase in the nucleation density and nucleation rate but at the same time a decrease in spherulitic growth rate. These two processes, primary nucleation and growth are combined when a semicrystalline polymer is crystallized from the melt. Their simultaneous effect can be ascertained by determining overall crystallization kinetics by DSC.

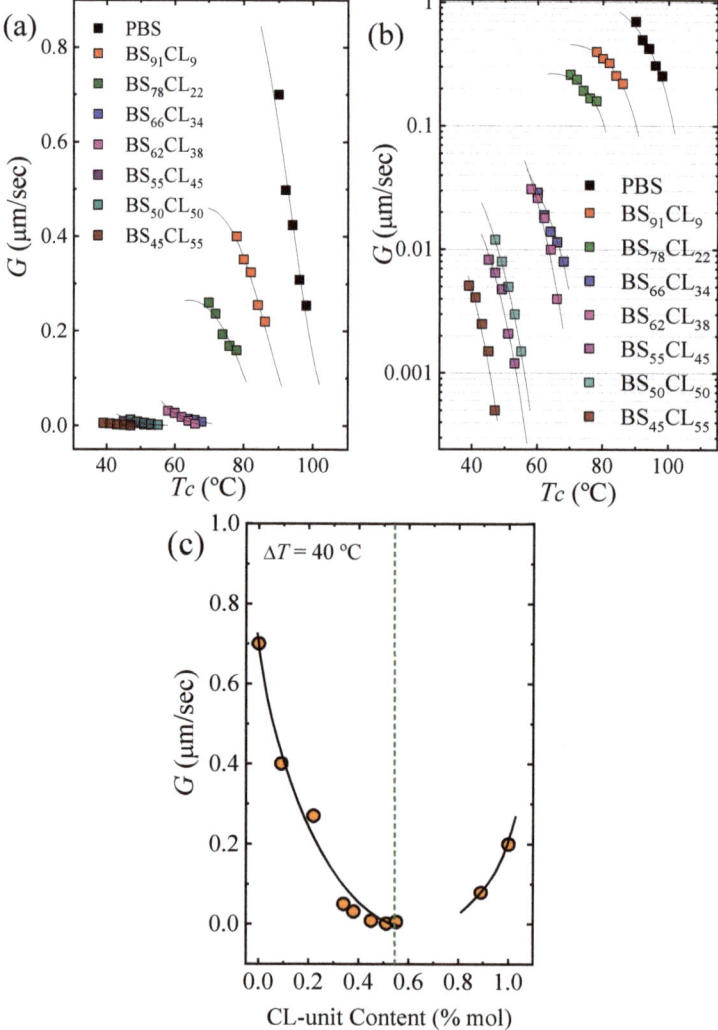

Figure 6. Spherulitic growth rates determined by PLOM. (**a**) Growth rate, G, as a function of T_c. (**b**) Same data as in (**a**) but G is plotted on a logarithmic scale. The black solid lines are fits to the experimental data performed with the Lauritzen and Hoffman theory (L-H). (**c**) G versus CL-unit content at ΔT = 40 °C. The black line is an arbitrary polynomial fit drawn to guide the eye.

3.3. Overall Crystallization Kinetics Studied by DSC

The overall isothermal crystallization kinetics considers both nucleation and growth, and can be conveniently determined by isothermal DSC experiments. Figure 7a,b shows the experimental overall crystallization rate expressed as the inverse of the crystallization half time ($\tau_{50\%}$). By DSC, we were able to determine the overall isothermal crystallization kinetics for both homopolymers and all copolymers (five copolymers where only the PBS-rich phase crystallized and four copolymers where only the PCL-rich phase can crystallize).

For the special composition at the eutectic point (i.e., $BS_{45}CL_{55}$) that shows two crystalline phases, PBS-rich phase and PCL-rich phase, we performed isothermal crystallization using different protocols. For the PBS-rich phase crystallization, isothermal DSC experiments were performed at temperatures where the PCL-rich phase is in the melt and cannot crystallize, while in the PCL-rich phase, a special protocol was adopted to previously crystallize the PBS-rich phase to saturation (see experimental part).

Figure 7a,b shows the strong dependence of the overall crystallization rate and the temperature range where measurements were possible on copolymer composition. In the case of the PBS and all PBS-rich compositions, the overall crystallization proceeds from a single-phase melt. Upon increasing PCL content, the amount of the crystallizable PBS-rich phase decreases and there will be more molten PCL component causing a plasticization ("solvent effect"). In addition, the effect of PCL exclusion in the PBS-rich crystal lattice may cause some further reduction in crystallization rate. Figure 7a shows that the temperature needed for crystallization decreases as PCL content in the copolymer increases, while the overall crystallization rate measured at the minimum T_c value possible tends to decrease.

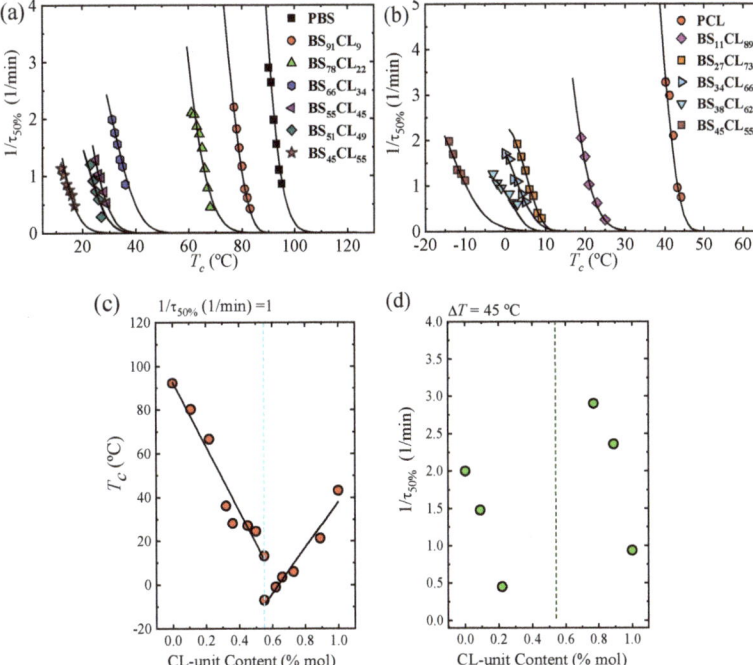

Figure 7. Overall crystallization versus isothermal crystallization temperature for neat PBS and PBS-rich compositions (**a**) and for neat PCL and PCL-rich compositions (**b**) versus T_c. Continuous lines correspond to the fitting of the Lauritzen–Hoffman theory with the parameters in Table 3. Changes of T_c versus CL-unit content in a constant rate ($1/\tau_{50\%} = 1$ min^{-1}) (**c**). Changes of inverse of half-crystallization time, $1/\tau_{50\%}$ versus CL-unit content in a constant supercooling degree, $\Delta T = 45\,°C$ (**d**).

Figure 7c plots the crystallization temperature needed to obtain the same overall crystallization rate of 1 min^{-1}. These T_c values monotonically decrease with PCL content until the pseudo-eutectic region is reached. On the PCL-rich side, Figure 7b,c shows similar results, as the crystallization temperatures needed to crystallize the PCL phase decrease as PBS repeat units are added to the copolymer.

To check if the supercooling is playing a major role upon changing composition, we plot the data contained in Figure 7a,b as a function of ΔT in Figure 8a,b, respectively. Surprisingly, the trends are quite different depending on the phase under consideration, or the composition range.

Figure 8a shows that the curves of PBS-rich overall growth rate data that originally spanned a T_c range of approximately 90 °C (in Figure 7a) are now within 30 °C in supercooling, attesting for the thermodynamic compensation of the solvent effect, as the PCL-rich phase is in the melt. In fact, the curves of PBS and BS$_{91}$CL$_9$ completely overlap, while that of BS$_{78}$CL$_{22}$ is relatively close to that of neat PBS. However, beyond 22% CL incorporation in the copolymer, the samples require much larger supercooling to crystallize. It is clear that the dominant factor to the left of the eutectic point is the growth rate, as the results presented in Figures 6a and 7a imply an overall crystallization rate reduction with PCL incorporation in the copolymer, both in terms of crystallization temperature or supercooling. In spite of the increase in nucleation density and nucleation rate (see Figures 3 and 4) with PCL incorporation in the copolymer, it is the very large decrease in growth rate (of up to three orders of magnitude, see Figure 6b) that dominates, leading to a decrease in overall crystallization rate (Figures 7a and 8a).

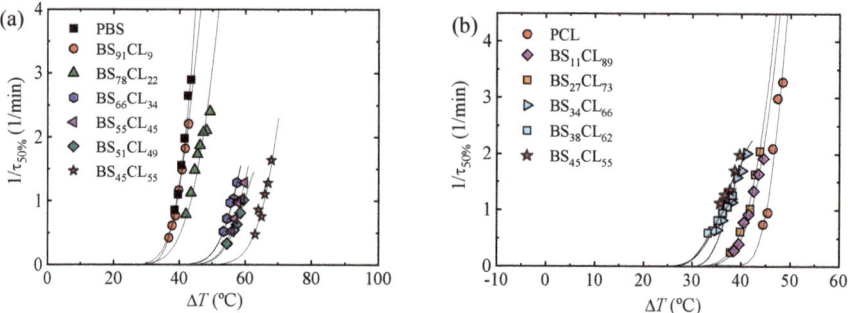

Figure 8. Overall crystallization for neat PBS and PBS-rich compositions (**a**) and for neat PCL and PCL-rich compositions (**b**) versus supercooling temperature.

Figure 7d represents the overall crystallization rate as a function of composition for a constant supercooling of 45 °C. In the PBS-rich side of the pseudo-eutectic region (left-hand side of Figure 7d), only three data points are plotted, as they are the only ones that could be measured at such constant supercooling value (check Figure 8b). The trend clearly shows a significant decrease in overall crystallization rate as CL unit content increases, as expected from Figures 7a and 8a and the discussion above. Even though the value of supercooling is not exactly the same (with only 5 °C difference), a comparison with Figures 4a and 6c clearly indicates that the PBS-rich phase overall crystallization is dominated by growth rate.

Figure 8b shows remarkable results for the PCL-rich phase overall crystallization. The $1/\tau_{50\%}$ curves versus temperature span a temperature range of 65 °C (Figure 6b). When they are plot as a function of supercooling, they only span 15 °C. However, they do not overlap, as would be expected for a simple solvent effect. In fact, the supercooling needed for crystallization of the PCL-rich phase remarkably decreases as PBS repeat units are included in the copolymer. The results indicate an acceleration of the overall crystallization rate (at constant supercooling) that can only be explained by the increase in both nucleation density and nucleation rate. We were only able to measure the increase in nucleation density and nucleation rate in Figures 3 and 4 for neat PCL and BS$_{11}$CL$_{89}$, as further incorporation of BS units increased the nucleation density so much that measurements by polarized optical microscopy of nucleation rate became impossible. Hence, we are convinced that primary nucleation enhancement upon PBS repeat unit incorporation in the copolymers is the reason behind the acceleration of the overall crystallization kinetics, when this is considered in terms of supercooling. In the right-hand side of the pseudo-eutectic point in Figure 7d, the increase in overall crystallization rate at a constant supercooling of 45 °C can be appreciated.

The Lauritzen and Hoffman theory can also be applied to fit the overall crystallization data presented above. Equation (3) has to be modified to employ, as a characteristic rate, the inverse of the half-crystallization time determined by DSC, as follows [37]:

$$1/\tau_{50\%} = 1/\tau_{50\%}^0 \exp\left[\frac{U}{R(T_c - T_0)}\right]\left[\frac{-K_g^\tau}{fT(T_m^0 - T_c)}\right], \tag{7}$$

where all the terms have been defined above, except for K_g^τ, which now represents a parameter proportional to the energy barrier for both primary nucleation and spherulitic growth. The superscript τ is used to indicate its origin (coming from DSC data, and hence from fitting $1/\tau_{50\%}$ versus crystallization temperature). In this way, it is different from K_g^G, defined in Equation (3), derived from growth rate data and therefore proportional just to the free energy barrier for secondary nucleation or growth. The solid lines in Figure 7a,b and Figure 8a,b represent the fits to the Lauritzen and Hoffman theory. Table 3 on the other hand reports all the relevant parameters.

There are no results in the literature regarding the isothermal crystallization of PBS-ran-PCL copolymers, therefore, we cannot compare the parameters reported in Tables 2 and 3 with literature values. In the case of the homopolymers, Wu et al. [38] and Papageorgiou et al. [39] reported K_g^τ values for neat PBS equal to 1.157×10^5 and 2.64×10^5 K^2, respectively, which are close to the value obtained in this work, i.e., 2.04×10^5 K^2. A value for K_g^G for neat PBS reported [39] equal to 1.88×10^5 has been reported, that is somewhat higher than that obtained in this work, i.e., 0.87×10^5 K^2. For neat PCL, the energetic parameters previously reported based on fits of the Lauritzen and Hoffman theories [40] are in the same order of magnitude as those reported here [33].

Figure 9 plots both K_g values, obtained by PLOM (K_g^G) and DSC (K_g^τ) as a function of CL-unit molar content. As expected, all K_g^τ values are larger than K_g^G values, as DSC measurements take into account both nucleation and growth, while PLOM measurements considered only growth (see more details in [37]).

In the case of K_g^G values, the trends observed are expected in view of the results obtained in Figure 6. The energy barrier for crystal growth increased with CL-unit molar content, since growth rate decreased as comonomer incorporation increased. On the other hand, when we analyzed the results obtained for K_g^τ in Figure 9, we noticed that there is a clear asymmetry depending on which side of the pseudo-eutectic region the material is. On the PBS-rich side (left-hand side of Figure 9), K_g^τ values rapidly increased upon CL units addition. This is expected from the results presented in Figure 7d, where a large decrease in overall crystallization rate for the PBS-rich side of the composition range can be observed.

In the case of the PCL-rich composition range, we would have expected a decrease in K_g^τ values with PCL content increases according to Figure 7d. Instead, we observe in Figure 9 that the energy barrier for both nucleation and growth does not significantly change with composition (see right-hand side of Figure 9). We have to remember that for the PCL-rich copolymers the situation is particularly complicated as the nucleation density and nucleation rate increase with CL-unit content but the growth rate decreases. Hence, even though according to Figure 7d the overall crystallization rate at constant supercooling seems to be dominated by primary nucleation, the values of K_g^τ are obtained from the slope of the Lauritzen and Hoffman plots that take into account the full range of supercoolings where the measurements were taken. Therefore, it seems that when the overall energy barrier is considered, there is a balance between nucleation and growth which keeps the K_g^τ values constant with composition.

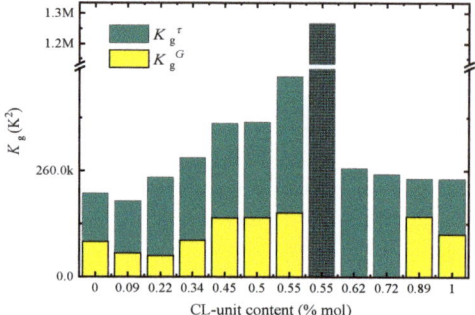

Figure 9. K_g versus CL-unit molar fraction that obtained for PLOM experiments (K_g^G) and DSC experiments (K_g^τ).

Table 3. Parameters obtained from fitting the DSC data presented in Figure 7a,b to the Lauritzen and Hoffman model (Equation (7)).

Copolyester	K_g^τ (K^2)	R^2	σ (erg/cm^2)	σ_e (erg/cm^2)	q (erg)
PBS	2.04×10^5	0.9697	12.4	186.9	7.93×10^{-13}
$BS_{91}CL_9$	1.85×10^5	0.9938	12.4	173.8	3.37×10^{-13}
$BS_{78}CL_{22}$	2.44×10^5	0.9514	12.4	236.3	10.0×10^{-13}
$BS_{66}CL_{34}$	2.92×10^5	0.9958	12.4	290.4	12.3×10^{-13}
$BS_{55}CL_{45}$	3.76×10^5	0.9775	12.4	387.3	16.4×10^{-13}
$BS_{51}CL_{49}$	3.79×10^5	0.9973	12.4	388.5	16.8×10^{-13}
$BS_{45}CL_{55}$ (BS − rich)	4.91×10^5	0.9502	12.4	518.7	22.0×10^{-13}
$BS_{45}CL_{55}$ (CL − rich)	12.5×10^5	0.9880	6.8	1377.0	51.3×10^{-13}
$BS_{38}CL_{62}$	2.66×10^5	0.9917	6.8	328.0	12.6×10^{-13}
$BS_{27}CL_{73}$	2.52×10^5	0.9845	6.8	305.9	11.4×10^{-13}
$BS_{11}CL_{89}$	2.41×10^5	0.9945	6.8	278.4	10.4×10^{-13}
PCL	2.40×10^5	0.9082	6.8	260.6	9.71×10^{-13}

R^2 is the correlation coefficient for the Lauritzen–Hoffman (Equation (7)) linear plots $\ln(1/\tau\, 50\%) + U^*/R(T_c - T_0)$ vs. $1/f \cdot T_c \cdot \Delta T$.

3.4. Double Crystallization at the Pseudo-Eutectic Point

For the copolymer whose composition is within the pseudo-eutectic point, i.e., BS$_{45}$CL$_{55}$, we performed isothermal crystallization in a wide range of crystallization temperatures T_c, to find the temperature region where only one of the phases, PBS or PCL, is able to crystallize. Figure 10 shows the heating DSC scan recorded at 10 °C/min for BS$_{45}$CL$_{55}$ sample after it was isothermally crystallized at the indicated T_c values.

At least five different endotherms can be found upon close examination of Figure 10 and we indicated with dashed lines how these endotherms approximately shift depending on the T_c values employed before heating the samples. The first melting peak T_{m1} is present in all melting curves and its location is almost at 7 °C higher than the crystallization temperature. This peak has been traditionally regarded as the melting of thin crystals formed during the secondary crystallization process [41]. The second peak, labeled T_{m2}, appeared at T_c values lower than −6 °C and corresponds to the melting of PCL-rich crystals. The third peak or T_{m3} labeled peak in Figure 10 highly depends on the isothermal crystallization temperature and corresponds to the melting of the PBS-rich crystals, which were formed during the isothermal crystallization.

In addition, a melting peak (T_{m5}) at around 50 °C and another one just below it (T_{m4}) were observed. The melting peak labeled T_{m5} corresponds to the melting of PBS crystals that have reorganized during the heating scan, and have a melting point which is almost constant at around 50 °C, regardless of the crystallization temperature [42,43]. The T_{m2} and T_{m3} peaks increase almost linearly with increasing

T_c. As shown in Figure 10, T_{m3} disappeared in the DSC heating curves where the crystallization temperature is less than −9 °C.

The morphologies obtained after isothermal crystallization at three selected temperatures can be observed in Figure 10b–d. As it will be shown below, WAXS experiments confirmed that at very low T_c values including −12 °C, only PCL-rich crystals can be formed. Figure 10d shows small spherulites that were formed at T_c = −4 °C with spherulites size around 10 μm. At T_c = −8 °C, where both PCL and PBS crystals can form, there are two crystals sizes, one with 4 μm radii (PBS crystals) and another with around 1.5 μm size (PCL crystals), see Figure 10d. Figure 10b shows only PCL crystals with small spherulites size (less than 1 μm) at T_c = −12 °C.

We performed in situ synchrotron WAXS experiments for the sample at the pseudo-eutectic point, to clarify the temperature range of crystallization of the PBS-rich and the PCL-rich phases and corroborate the assignment of the thermal transitions in Figure 10. These experiments were performed during isothermal crystallization (for 20 min) at three different T_c values chosen from three different crystallization regions in Figure 10.

Figure 11a–c shows selected real-time WAXS diffractograms for $BS_{45}CL_{55}$ (i.e., the sample at the pseudo-eutectic point) measured during isothermal crystallization at −12, −9, and −6 °C. If the sample shows characteristic reflections at q = 13.9 and 16.1 nm^{-1}, they correspond to the PBS (020) and (110) crystallographic planes. If the sample exhibits reflections at q = 15.3 and 17.4 nm^{-1}, they belong to the PCL (110) and (200) planes [10].

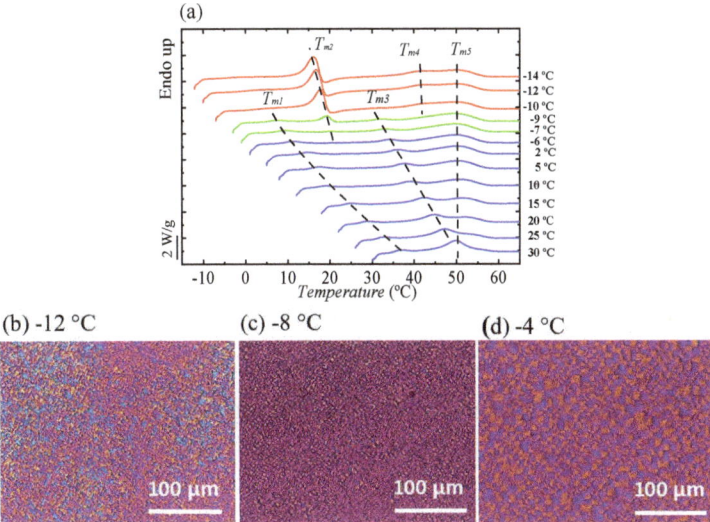

Figure 10. (a) DSC heating runs (at 10 °C/min) after isothermal crystallization at different temperatures. See text for the explanation of the color code employed. PLOM micrographs after isothermal crystallization at −12 °C (b), at −8 °C (c), and at −4 °C (d) for the $BS_{45}CL_{55}$ sample.

Changes in the crystallization temperature strongly affect the diffraction pattern at the pseudo-eutectic point. As can be seen in Figure 11, at −12 °C only the PCL-rich phase is able to crystalize (Figure 11a) while at −6 °C (Figure 11c) only the PBS-rich phase crystallizes. On the other hand, at the intermediate T_c value of −9 °C, both PBS-rich and PCL-rich phases can crystallize. If the DSC curves of Figure 10 are considered again, the WAXS assignments are consistent with the heating runs after crystallization for all samples crystallized at −9 °C and higher. In the case of low crystallization temperatures, i.e., below −9 °C, it should be noted that WAXS indicate that only the PCL-rich phase can crystallize. The DSC heating runs shown in Figure 10 also show melting transitions corresponding to the melting of PBS-rich phase. These PBS-rich phase crystals must be formed by cold-crystallization during the heating scan for the samples crystallized at −10, −12, and −14 °C in Figure 10. In fact, upon close examination of Figure 10, the end of a cold crystallization process can be observed just after the melting peak of the PCL-rich phase crystals.

Taking into account the WAXS and DSC results presented in Figures 10 and 11, the DSC curves in Figure 10 were plotted with a color code to indicate which phases can crystallize during isothermal crystallization depending on the T_c values employed. If the T_c values are −10 °C or lower, only the PCL-rich phase can crystallize, and the curves were arbitrarily plotted in red in Figure 10a. If the T_c values are between −9 and −7 °C (including these two temperatures), both the PCL-rich and the PBS-rich phases can crystallize (green curves in Figure 10). Finally, if the T_c temperatures are −6 °C and above, only the PBS-rich phase can crystallize (blue curves in Figure 10).

The pseudo-eutectic sample, $BS_{45}CL_{55}$, exhibits a very interesting phase behavior, as depending on the crystallization conditions, one or both phases can be formed. We have studied previously the nonisothermal crystallization of the same copolymers employed here [17]. It is interesting to note that under nonisothermal conditions, the cooling rate employed determines which phase can crystallize and also if one or two phases are formed. In this work, on the other hand, we show that one or two phases can be formed depending on the isothermal crystallization temperature chosen. Therefore, the properties of this isodimorphic copolyester with pseudo-eutectic composition can be tailored by varying both nonisothermal or isothermal crystallization conditions, a remarkable and novel behavior, as far as the authors are aware.

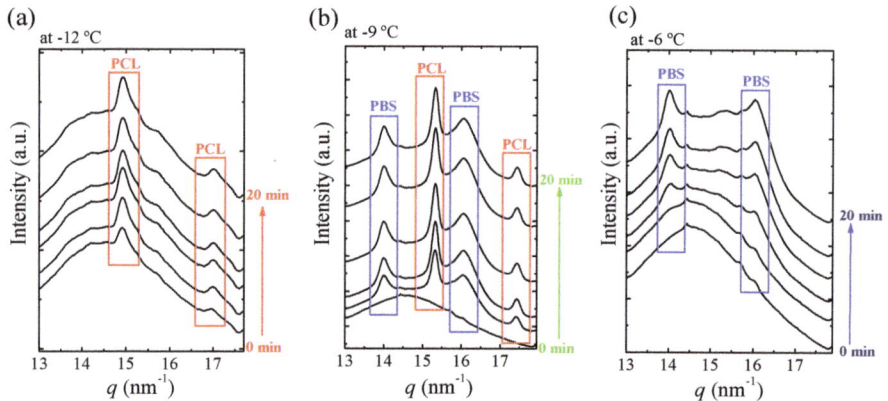

Figure 11. Wide angle X-ray diffraction (WAXS) diffraction patterns of $BS_{45}CL_{55}$ registered during isothermal crystallization at −12 °C (**a**), −9 °C (**b**), and −6 °C (**c**).

4. Conclusions

The complex isothermal nucleation, growth and overall crystallization of isodimorphic PBS-*ran*-PCL copolyesters were studied for the first time. The equilibrium melting temperatures show a very clear pseudo-eutectic point at a composition of $BS_{45}CL_{55}$. To the left of the pseudo-eutectic point (in a plot of T_m^0 versus CL-unit molar content) only the PBS-rich phase is able to crystallize, at the pseudo-eutectic point both PBS-rich and PCL-rich phases can crystallize and to the right of the pseudo-eutectic point only PCL-rich crystals are formed. With respect to the parent homopolymers, any comonomer incorporation on either side of the pseudo-eutectic point causes as increase in nucleation density and nucleation rate, as well as a decrease in spherulitic growth rate.

As a result, the overall crystallization rate determined by DSC was a strong function of composition and supercooling. For PBS-rich copolymers, the PBS-rich phase overall crystallization rate-determining-step was the spherulitic growth rate. On the other hand, for PCL-rich copolymers, the nucleation rate (which was always larger for mirror compositions) gained more importance in the control of the overall crystallization rate.

The crystallization of the isodimorphic copolyester with pseudo-eutectic composition can be tailored by varying the isothermal crystallization temperature, depending on which, either one or both phases are able to crystallize. Such remarkable property control allows the possibility of having a single copolyester with only PCL crystals, only PBS crystals, or both types of crystals, thus exhibiting very different thermal properties.

Supplementary Materials: The following materials are available online at http://www.mdpi.com/2073-4360/12/1/17/s1, Table SI-1. Equilibrium melting temperatures for CoP(BS$_x$CL$_y$) compositions and their corresponding homopolymers, Figure SI-1. Nucleation kinetics studies by PLOM. Nuclei density as a function of time at different crystallization temperature for PBS-rich phase samples: (a) BS$_{91}$CL$_9$, (b) BS$_{66}$CL$_{34}$, (c) BS$_{62}$CL$_{38}$, (d) BS$_{55}$CL$_{45}$, (e) BS$_{51}$CL$_{49}$, and (f) BS$_{45}$CL$_{55}$. T_c employed are chosen so that ΔT = 40, 38, 36, 34, 32 °C for all samples, Figure SI-2. Nuclei density during isothermal crystallization as a functional ΔT for PBS-rich (a) and for PCL-rich (b) copolyesters, Figure SI-3. Hoffman–Weeks plots for PBS-*ran*-PCL compositions. The black solid line represents the thermodynamic equilibrium line $T_m = T_c$, Figure SI-4. Plot of *log I* versus $1/T(\Delta T)^2$ and fitting to Turnbull–Fisher equation (Equation (1)) for PBS-rich (a) and PCL-rich (b) compositions, Figure SI-5. Spherulitic growth rates G determined by PLOM for neat PBS and PBS-rich (a) and for neat PCL and PCL-rich (b) copolymers as a function of supercooling, Figure SI-6. The fits to the Lauritzen–Hoffman equation using the free Origin plug-in developed by Lorenzo et al. and the experimental data for the (a-a′) PBS, (b-b′) BS$_{78}$CL$_{22}$, and (c-c′) BS$_{45}$CL$_{55}$, Figure SI-7. The σ_e value versus CL-unit molar fraction that obtained for PLOM experiments (σ_g^G) and DSC experiments (σ_g^τ).

Author Contributions: The general conceptualization of the work described here was performed by A.J.M. S.M.-G. and A.M.d.I. designed the synthetic route of the copolymers and their molecular characterization. The experiments were performed by M.S. The general supervision of experimental measurements at the UPV/EHU labs was done by A.M. and at UPC by A.M.d.I. M.Z. and A.J.M. supervised calculations and fittings of experimental data. The paper was written by M.S. and A.J.M. and revised by all co-authors. All authors have read and agreed to the published version of the manuscript.

Funding: This research was funded by (a) MINECO through project MAT2017-83014-C2-1-P, (b) ALBA synchrotron facility through granted proposal 2018082953 (c) the Basque Government through grant IT1309-19 and (d) European Union´s Horizon 2020 research and innovation programme under the Marie Sklodowska-Curie grant agreement No 778092

Acknowledgments: MS gratefully acknowledges the award of a PhD fellowship by POLYMAT Basque Center for Macromolecular Design and Engineering.

Conflicts of Interest: The authors declare no conflict of interest.

References

1. Lendlein, A.; Sisson, A. *Handbook of Biodegradable Polymers: Isolation, Synthesis, Characterization and Applications*; Wiley-VCH: Weinheim, Germany, 2011.
2. Tanaka, M.; Sato, K.; Kitakami, E.; Kobayashi, S.; Hoshiba, T.; Fukushima, K. Design of biocompatible and biodegradable polymers based on intermediate water concept. *Polym. J.* **2015**, *47*, 114–121. [CrossRef]
3. Sisti, L.; Totaro, G.; Marchese, P. *PBS Makes Its Entrance into the Family of Biobased Plastics*; John Wiley & Sons: Hobocan, NJ, USA, 2016.
4. Pan, P.; Inoue, Y. Polymorphism and isomorphism in biodegradable polyesters. *Prog. Polym. Sci.* **2009**, *34*, 605–640. [CrossRef]
5. Ikada, Y.; Tsuji, H. Biodegradable polyesters for medical and ecological applications. *Macromol. Rapid Commun.* **2000**, *21*, 117–132. [CrossRef]
6. Williams, C.K. Synthesis of functionalized biodegradable polyesters. *Chem. Soc. Rev.* **2007**, *36*, 1573–1580. [CrossRef] [PubMed]
7. Fakirov, S. *Biodegradable Polyesters*; John Wiley & Sons: Auckland, New Zealand, 2015.
8. Manavitehrani, I.; Fathi, A.; Badr, H.; Daly, S.; Negahi Shirazi, A.; Dehghani, F. Biomedical applications of biodegradable polyesters. *Polymers* **2016**, *8*, 20. [CrossRef] [PubMed]
9. Arandia, I.; Mugica, A.; Zubitur, M.; Mincheva, R.; Dubois, P.; Müller, A.J.; Alegría, A. The complex amorphous phase in poly (butylene succinate-ran-butylene azelate) isodimorphic copolyesters. *Macromolecules* **2017**, *50*, 1569–1578. [CrossRef]

10. Ciulik, C.; Safari, M.; Martínez de Ilarduya, A.; Morales-Huerta, J.C.; Iturrospe, A.; Arbe, A.; Müller, A.J.; Muñoz-Guerra, S. Poly(butylene succinate-ran-ε-caprolactone) copolyesters: Enzymatic synthesis and crystalline isodimorphic character. *Eur. Polym. J.* **2017**, *95*, 795–808. [CrossRef]
11. Yu, Y.; Wei, Z.; Liu, Y.; Hua, Z.; Leng, X.; Li, Y. Effect of chain length of comonomeric diols on competition and miscibility of isodimorphism: A comparative study of poly(butylene glutarate-co-butylene azelate) and poly(octylene glutarate-co-octylene azelate). *Eur. Polym. J.* **2018**, *105*, 274–285. [CrossRef]
12. Pérez-Camargo, R.A.; Arandia, I.; Safari, M.; Cavallo, D.; Lotti, N.; Soccio, M.; Müller, A.J. Crystallization of isodimorphic aliphatic random copolyesters: Pseudo-eutectic behavior and double-crystalline materials. *Eur. Polym. J.* **2018**, *101*, 233–247. [CrossRef]
13. Mandelkern, L. *Crystallization of Polymers. Volume 1, Equilibrium Concepts*; Cambridge University Press: Cambridge, UK, 2002; Volume 1.
14. Allegra, G.; Bassi, I. Isomorphism in synthetic macromolecular systems. In *Fortschritte der Hochpolymeren-Forschung*; Springer: Heidelberg/Berlin, Germany, 1969; pp. 549–574.
15. Yu, Y.; Sang, L.; Wei, Z.; Leng, X.; Li, Y. Unique isodimorphism and isomorphism behaviors of even-odd poly(hexamethylene dicarboxylate) aliphatic copolyesters. *Polymer* **2017**, *115*, 106–117. [CrossRef]
16. Ye, H.-M.; Wang, R.-D.; Liu, J.; Xu, J.; Guo, B.-H. Isomorphism in poly (butylene succinate-co-butylene fumarate) and its application as polymeric nucleating agent for poly(butylene succinate). *Macromolecules* **2012**, *45*, 5667–5675. [CrossRef]
17. Safari, M.; Martínez de Ilarduya, A.; Mugica, A.; Zubitur, M.; Muñoz-Guerra, S.N.; Müller, A.J. Tuning the Thermal Properties and Morphology of Isodimorphic Poly[(butylene succinate)-ran-(ε-caprolactone)] Copolyesters by Changing Composition, Molecular Weight, and Thermal History. *Macromolecules* **2018**, *51*, 9589–9601. [CrossRef]
18. Müller, A.J.; Michell, R.M.; Lorenzo, A.T. *Isothermal Crystallization Kinetics of Polymers*; John Wiley & Sons: Hoboken, NJ, USA, 2016.
19. Lorenzo, A.T.; Arnal, M.L.; Albuerne, J.; Müller, A.J. DSC isothermal polymer crystallization kinetics measurements and the use of the Avrami equation to fit the data: Guidelines to avoid common problems. *Polym. Test.* **2007**, *26*, 222–231. [CrossRef]
20. Sharples, A. *Overall Kinetics of Crystallization*; Edward Arnold: London, UK, 1966.
21. Mandelkern, L. *Crystallization of Polymers: Volume 2, Kinetics and Mechanisms*; Cambridge University Press: Cambridge, UK, 2004.
22. Brüster, B.; Montesinos, A.; Reumaux, P.; Pérez-Camargo, R.A.; Mugica, A.; Zubitur, M.; Müller, A.J.; Dubois, P.; Addiego, F. Crystallization kinetics of polylactide: Reactive plasticization and reprocessing effects. *Polym. Degrad. Stab.* **2018**, *148*, 56–66. [CrossRef]
23. Tian, J.; Yu, W.; Zhou, C. Crystallization kinetics of linear and long-chain branched polypropylene. *J. Macromol. Sci. B* **2006**, *45*, 969–985. [CrossRef]
24. Wunderlich, B. *Macromolecular Physics, Crystal Nucleation, Growth, Annealing*; Academic Press: New York, NY, USA, 1976; Volume 2.
25. Arandia, I.; Zaldua, N.; Maiz, J.; Pérez-Camargo, R.A.; Mugica, A.; Zubitur, M.; Mincheva, R.; Dubois, P.; Müller, A.J. Tailoring the isothermal crystallization kinetics of isodimorphic poly(butylene succinate-ran-butylene azelate) random copolymers by changing composition. *Polymer* **2019**, *183*, 121863. [CrossRef]
26. Pitt, C.; Chasalow, F.; Hibionada, Y.; Klimas, D.; Schindler, A. Aliphatic polyesters. I. The degradation of poly(ε-caprolactone) in vivo. *J. Appl. Polym. Sci.* **1981**, *26*, 3779–3787. [CrossRef]
27. Van Krevelen, D.W.; Te Nijenhuis, K. *Properties of Polymers: Their Correlation with Chemical Structure; Their Numerical Estimation and Prediction from Additive Group Contributions*; Elsevier: Amsterdam, The Netherlands, 2009.
28. Papageorgiou, G.Z.; Bikiaris, D.N. Crystallization and melting behavior of three biodegradable poly(alkylene succinates). A comparative study. *Polymer* **2005**, *46*, 12081–12092. [CrossRef]
29. Crist, B.; Schultz, J.M. Polymer spherulites: A critical review. *Prog. Polym. Sci.* **2016**, *56*, 1–63. [CrossRef]
30. Woo, E.M.; Lugito, G. Origins of periodic bands in polymer spherulites. *Eur. Polym. J.* **2015**, *71*, 27–60. [CrossRef]
31. Magill, J. Review spherulites: A personal perspective. *J. Mater. Sci.* **2001**, *36*, 3143–3164. [CrossRef]

32. Hoffman, J.D.; Davis, G.T.; Lauritzen, J.I. The rate of crystallization of linear polymers with chain folding. In *Treatise on Solid State Chemistry*; Springer: Washington, DC, USA, 1976; pp. 497–614.
33. Mark, J.E. *Physical Properties of Polymers Handbook*, 2nd ed.; Springer: Cincinnati, OH, USA, 2007.
34. Ichikawa, Y.; Kondo, H.; Igarashi, Y.; Noguchi, K.; Okuyama, K.; Washiyama, J. Crystal structures of α and β forms of poly(tetramethylene succinate). *Polymer* **2000**, *41*, 4719–4727. [CrossRef]
35. Ren, M.; Song, J.; Song, C.; Zhang, H.; Sun, X.; Chen, Q.; Zhang, H.; Mo, Z. Crystallization kinetics and morphology of poly(butylene succinate-co-adipate). *J. Polym. Sci. Polym. Phys.* **2005**, *43*, 3231–3241. [CrossRef]
36. Phillips, P.J.; Rensch, G.J.; Taylor, K.D. Crystallization studies of poly(ε-caprolactone). I. Morphology and kinetics. *J. Polym. Sci. Polym. Phys.* **1987**, *25*, 1725–1740. [CrossRef]
37. Lorenzo, A.T.; Müller, A.J. Estimation of the nucleation and crystal growth contributions to the overall crystallization energy barrier. *J. Polym. Sci. Polym. Phys.* **2008**, *46*, 1478–1487. [CrossRef]
38. Chen, Y.-A.; Tsai, G.-S.; Chen, E.-C.; Wu, T.-M. Crystallization behaviors and microstructures of poly(butylene succinate-co-adipate)/modified layered double hydroxide nanocomposites. *J. Mater. Sci.* **2016**, *51*, 4021–4030. [CrossRef]
39. Papageorgiou, G.Z.; Achilias, D.S.; Bikiaris, D.N. Crystallization kinetics of biodegradable poly(butylene succinate) under isothermal and non-isothermal conditions. *Maromol. Chem. Phys.* **2007**, *208*, 1250–1264. [CrossRef]
40. Müller, A.J.; Albuerne, J.; Marquez, L.; Raquez, J.-M.; Degee, P.; Dubois, P.; Hobbs, J.; Hamley, I.W. Self-nucleation and crystallization kinetics of double crystalline poly(p-dioxanone)-b-poly (e-caprolactone) diblock copolymers. *Faraday Discuss.* **2005**, 231–252. [CrossRef]
41. Papageorgiou, G.Z.; Papageorgiou, D.G.; Chrissafis, K.; Bikiaris, D.; Will, J.; Hoppe, A.; Roether, J.A.; Boccaccini, A.R. Crystallization and melting behavior of poly(butylene succinate) nanocomposites containing silica-nanotubes and strontium hydroxyapatite nanorods. *Ind. Eng. Chem. Res.* **2013**, *53*, 678–692. [CrossRef]
42. Wang, Y.; Chen, S.; Zhang, S.; Ma, L.; Shi, G.; Yang, L. Crystallization and melting behavior of poly(butylene succinate)/silicon nitride composites: The influence of filler's phase structure. *Thermochim. Acta* **2016**, *627*, 68–76. [CrossRef]
43. Yasuniwa, M.; Satou, T. Multiple melting behavior of poly(butylene succinate). I. Thermal analysis of melt-crystallized samples. *J. Polym. Sci. Polym. Phys.* **2002**, *40*, 2411–2420. [CrossRef]

© 2019 by the authors. Licensee MDPI, Basel, Switzerland. This article is an open access article distributed under the terms and conditions of the Creative Commons Attribution (CC BY) license (http://creativecommons.org/licenses/by/4.0/).

Article

Assessment of the Mechanical and Thermal Properties of Injection-Molded Poly(3-hydroxybutyrate-*co*-3-hydroxyhexanoate)/Hydroxyapatite Nanoparticles Parts for Use in Bone Tissue Engineering

Juan Ivorra-Martinez [1], Luis Quiles-Carrillo [1], Teodomiro Boronat [1], Sergio Torres-Giner [2,*] and José A. Covas [3,*]

1. Technological Institute of Materials (ITM), Universitat Politècnica de València (UPV), Plaza Ferrándiz y Carbonell 1, 03801 Alcoy, Spain; juaivmar@doctor.upv.es (J.I.-M.); luiquic1@epsa.upv.es (L.Q.-C.); tboronat@dimm.upv.es (T.B.)
2. Novel Materials and Nanotechnology Group, Institute of Agrochemistry and Food Technology (IATA), Spanish National Research Council (CSIC), Calle Catedrático Agustín Escardino Benlloch 7, 46980 Paterna, Spain
3. Institute for Polymers and Composites, University of Minho, 4804-533 Guimarães, Portugal
* Correspondence: storresginer@iata.csic.es (S.T.-G.); jcovas@dep.uminho.pt (J.A.C.); Tel.: +34-963-900-022 (S.T.-G.); +35-1253-510-320 (J.A.C.)

Received: 18 May 2020; Accepted: 17 June 2020; Published: 21 June 2020

Abstract: In the present study, poly(3-hydroxybutyrate-*co*-3-hydroxyhexanoate) [P(3HB-*co*-3HHx)] was reinforced with hydroxyapatite nanoparticles (nHA) to produce novel nanocomposites for potential uses in bone reconstruction. Contents of nHA in the 2.5–20 wt % range were incorporated into P(3HB-*co*-3HHx) by melt compounding and the resulting pellets were shaped into parts by injection molding. The addition of nHA improved the mechanical strength and the thermomechanical resistance of the microbial copolyester parts. In particular, the addition of 20 wt % of nHA increased the tensile (E_t) and flexural (E_f) moduli by approximately 64% and 61%, respectively. At the highest contents, however, the nanoparticles tended to agglomerate, and the ductility, toughness, and thermal stability of the parts also declined. The P(3HB-*co*-3HHx) parts filled with nHA contents of up to 10 wt % matched more closely the mechanical properties of the native bone in terms of strength and ductility when compared with metal alloys and other biopolymers used in bone tissue engineering. This fact, in combination with their biocompatibility, enables the development of nanocomposite parts to be applied as low-stress implantable devices that can promote bone reconstruction and be reabsorbed into the human body.

Keywords: P(3HB-*co*-3HHx); nHA; nanocomposites; mechanical properties; bone reconstruction

1. Introduction

Bone fracture is one of the most common injuries. Bone regeneration encompasses three stages, namely inflammation, bone production, and bone remodeling [1]. During the latter, it is extremely important to expose the bone to the natural load-bearing conditions associated to its function [2]. Currently, titanium alloys such as Ti-6Al-4V are the most used for the manufacture of orthopedic fixing devices and bone implants due to their excellent biocompatibility and high mechanical resistance [3]. However, they prevent the bone from being subjected to the required mechanical loadings [4]. Indeed, while natural bone has a modulus ranging between 8 to 25 GPa, metals have a modulus of 110–210 GPa, which results in the load being imparted onto the device rather than the bone which then causes a localized decrease in bone mineral density [5]. Meanwhile, metal ion leaching increases inflammation and irritation around the implant [6]. As a result, there is often a need for a second surgery to remove the fixation device, leading to higher medical costs and greatly increased patient discomfort. A current alternative is the use of fixation devices that metabolize in the human body after fulfilling their function [7]. In particular, the use of biopolymers with biocompatibility and reabsorption capacities is very promising [8]. Biocompatibility involves the capability of a given substance to perform with a suitable host response in a particular use. Furthermore, no substance or material can be "biocompatible" if it releases cytotoxic substances. The degradation process of a given biopolymer within the human body consists of two phases. First, the biopolymer chains break, either as a consequence of hydrolysis or due to the action of a body enzyme. Thereafter, the human body assimilates the fragments. For this purpose, either a phagocytosic or metabolic process develops [9]. Surface porosity, shape, and tissue environment, including chemical build-up of the materials, play a significant role in biocompatibility [10,11].

For the past few decades, polymers of the polyhydroxyalkanoates (PHAs) family have been paving the way for the development of new biomedical products. These microbial biopolyesters degrade when exposed to marine sediment, soil or compost. A vast number of microorganisms secrete extracellular PHA-hydrolyzing enzymes, so-called PHA depolymerases, to degrade PHA into their oligomers and monomers, which subsequently act as nutrients inside the cells [12]. Their potential as alternatives for the manufacture of a wide range of medical devices, such as absorbable sutures, surgical pins or staples, is well recognized on account of their biodegradable nature as well as disintegration by surface erosion [13]. Broadly, the biocompatibility of PHA materials can be differentiated into two categories, immunocompatibility and nonallergic response. The former involves the extent of antigenic resemblance between the tissues of various individuals that determines the acceptance or rejection of allografts. PHAs are essentially immunocompatible for use in medical applications, that is, their materials should not elicit harsh immune responses upon introduction into the soft tissues or blood of a host organism [14]. Indeed, 3-hydroxybutyrate (3HB), the main monomeric constituent of most PHAs, is a result of cellular metabolism that is formed by oxidation of fatty acid within the liver cells and it is a usual component of human blood [15]. Other previous studies have also revealed that PHA did not elicit an allergic response or any hypersensitive immune reaction [10,16].

Depending on the number of carbon atoms in the monomers, PHAs can be classified as short-chain-length PHAs (*scl*-PHAs; 3–5 C-atoms) and medium-chain-length PHAs (*mcl*-PHAs; 6–14 C-atoms). Generally, *scl*-PHAs are rigid and brittle, while *mcl*-PHAs have higher flexibility and toughness [17]. Poly(3-hydroxybutyrate) (PHB) is the simplest and most common member of the PHA family. However, the high brittleness of PHB and other *scl*-PHAs, such as poly(3-hydroxybutyrate-*co*-3-hydroxyvalerate) (PHBV) with less than 15 mol % fraction of 3-hydroxyvalerate (3HV), restricts their application in bone fixing devices [18]. In this regard, poly(3-hydroxybutyrate-*co*-3hydroxyhexanoate) [P(3HB-*co*-3HHx)], also referred as PHBH, represents a recent addition to the group of PHAs for biomedical applications. The introduction of the *mcl* 3-hydroxyhexanoate (3HHx) co-monomer into the polymer backbone of PHB significantly increases the flexibility and reduces stiffness [19]. Therefore, the macroscopic properties of P(3HB-*co*-3HHx) vary with the proportion of each monomer in the copolyester [20], in which the higher the 3HHx content, the higher the ductility [21]. Apart from the changes in the mechanical properties, the most remarkable transformation that P(3HB-*co*-3HHx) brings along is its ability to undergo enzymatic degradation by lipase [22], which is not seen in either PHB or PHBV. Prior experiments have shown that materials based on P(3HB-*co*-3HHx) and other *mcl*-PHAs have good biocompatibility for chondrocytes [23], nerve cells [24] as well as osteoblast and fibroblast cells [25,26]. This property should make P(3HB-*co*-3HHx) a suitable choice for several tissue engineering applications since it adds a further variable that can be used to tailor its degradation [27].

While PHAs are biocompatible substrates for cell propagation and are potentially an effective template for the repair of osseous and chondral defects, there is still a need to improve the mechanical strength, thermal resistance, and biological response of these biomaterials in order to make them more suitable for bone tissue engineering. Osteoconductive fillers can be introduced into polymer matrices with the aim of improving the mechanical properties and also accelerating the bone repair process by favoring the growth of bone cells inside the pores [28,29]. For example, calcium orthophosphates (CaPO$_4$) have bioactive properties that increase bone cell proliferation, the so-called osteoinduction [30]. As a rule, both the mechanical resistance and bioactivity of composites prepared with collagen, chitin and/or gelatin, increase with increasing CaPO$_4$ content [31]. Hydroxyapatite, Ca$_5$(PO$_4$)$_3$OHCa$_5$(PO$_4$)$_3$OH, which is the principal crystalline constituent of bone, shows a high degree of biocompatibility and good osteoconductive and osteoinductive properties. Therefore, hydroxyapatite nanoparticle or nanohydroxyapatite (nHA) is the most widely used "bioceramic" for the manufacture of medical devices and dental implants [32]. This fact is exemplified by the production of prostheses for cranial reconstruction using poly(methyl methacrylate) (PMMA)/nHA composites [33]. Indeed, nHA exists in the human bone in the form of nanometer-sized threads, thus ensuring biocompatibility. At present, it is mostly used to produce surface coatings, as its biomimetic mineralization enables the production of biomaterials with biomimetic compositions and hierarchical micro/nanostructures that closely mimic the extracellular matrix of native bone tissue [34,35].

Due to the well-known high bioactive properties in terms of bone regeneration of PHA- and nHA-based composites, this study aims to determine the physical properties of injection-molded parts made of P(3HB-co-3HHx)/nHA composites, for potential use as bone resorbable devices. To this end, different contents of nHA were incorporated into P(3HB-co-3HHx) and the mechanical, thermal, and thermomechanical properties were analyzed and compared to some metal alloy-based solutions currently available in the biomedical field. As a first, the parts showed sufficient dimensional and thermal stability for bone tissue engineering and their elasticity was nearer to that of the natural bone when compared to the metal alloys used for bone implants.

2. Materials and Methods

2.1. Materials

P(3HB-co-3HHx) copolymer was supplied by Ercros S.A. (Barcelona, Spain) as ErcrosBio PH110. The ratio of 3HHx in the copolyester is ~10 mol % and its number average molecular weight (M$_n$) is 1.22×10^5 g/mol. It shows a melt flow index (MFI) of 1 g/10 min (2.16kg/160 °C) according to the ISO 1133-2 standard and a true density of 1.20 g/cm^3 following the UNE EN ISO 1183-1 standard. Hydroxyapatite synthetic nanopowder was procured from Sigma-Aldrich S.A. (Madrid, Spain) with commercial reference 677418. According to the manufacturer, it presents the following properties: particle size < 200 nm, surface area > 9.4 m^2/g by Brunauer-Emmett-Teller (BET) analysis, purity ≥ 97%, and molecular weight (M$_W$) of 502.31 g/mol.

2.2. Preparation and Processing of P(3HB-co-3HHx)/nHA Parts

Both P(3HB-co-3HHx) pellets and nHA powder were dried separately for at least 6 h at 80 °C in a dehumidifying oven from Industrial Marsé S.A. (Barcelona, Spain). The materials were then pre-mixed manually in closed zip-bags at the ratios presented in Table 1.

Table 1. Code and composition of the samples prepared according to the weight content (wt %) of poly(3-hydroxybutyrate-co-3-hydroxyhexanoate) [P(3HB-co-3HHx)] and hydroxyapatite nanoparticles (nHA).

Sample	P(3HB-co-3HHx) (wt %)	nHA (wt %)
P(3HB-co-3HHx)	100	0
P(3HB-co-3HHx) + 2.5 nHA	97.5	2.5
P(3HB-co-3HHx) + 5 nHA	95	5
P(3HB-co-3HHx) + 10 nHA	90	10
P(3HB-co-3HHx) + 20 nHA	80	20

The different P(3HB-co-3HHx) and nHA mixtures weighing 800 g were melt-compounded using a co-rotating twin-screw extruder from Dupra S.L. (Castalla, Spain). It features two screws with a diameter (D) of 25 mm and a length-to-diameter ratio (L/D) of 24, while the modular barrel is equipped with 4 individual heating zones coupled to a strand die. Further details of the extruder can be found elsewhere [36]. Extrusion was performed with a screw speed of 20–25 rpm to prevent material degradation due to shear-induced viscous dissipation, a feed of 1.2 kg/h, and a barrel set temperature profile of 110–120–130–140 °C from hopper to die. The extruded filaments were cooled down in an air stream and pelletized using an air-knife unit.

Test parts for characterization were obtained by injection molding. The equipment (Meteor 270/75, Mateu & Solé, Barcelona, Spain) was operated with a barrel set temperature profile of 115–120–125–130 °C from hopper to nozzle, with the mold kept at 60 °C. An injection time of 1 s was used to avoid material degradation by shear-induced viscous dissipation. The clamping force was 75 tons and the cooling time was set at 60 s. Parts with a thickness of approximately 4 mm were obtained for characterization. Since P(3HB-co-3HHx) develops secondary crystallization with time, the parts were allowed to age for 14 days at room temperature prior to characterization.

2.3. Mechanical Tests

Uniaxial tensile tests were performed according to the ISO 527-2: 2012 standard using a universal testing machine ELIB-50 (Ibertest S.A., Madrid, Spain) fitted with a load cell of 5 kN and using a 3542-050M-050-ST extensometer from Epsilon Technology Corporation (Jackson, WY, USA). Flexural properties were determined following the ISO 178: 2011 standard using the same equipment. Both tests were carried out at 5 mm/min using 150 mm × 10 mm × 4 mm parts. Charpy impact tests were performed following the ISO 179-1: 2010 standard. Samples with a V-shaped notch with a radius of 0.25 mm and dimensions 80 mm × 10 mm × 4 mm were subjected to the impact of a 1-J pendulum impact tester from Metrotec S.A. (San Sebastián, Spain). Shore hardness was measured with a 673-D durometer (J. Bot Instruments, Barcelona, Spain), following the ISO 868: 2003 standard. At least six parts were tested for each mechanical test.

2.4. Thermal Tests

Samples weighing 5–10 mg were analyzed by differential scanning calorimetry (DSC) in a Q200 from TA Instruments (New Castle, DE, USA) to study the thermal transitions. The samples were subjected to a three-stage thermal cycle in which the samples were first heated from −50 to 200 °C and cooled down to −50 °C in order to eliminate the thermal history and then reheated to 200 °C. All the heating and cooling scans were performed at 10 °C/min. Testing was performed under inert atmosphere using a nitrogen flow of 50 mL/min. The degree of crystallinity (X_{C_max}) was calculated using Equation (1) [37]:

$$Xc_{max} = \left[\frac{\Delta H_m}{\Delta H_m \cdot (1-w)} \right] \cdot 100\% \qquad (1)$$

where ΔH_m (J/g) corresponds to the melting enthalpy of P(3HB-co-3HHx), ΔH_m^0 (J/g) is the theoretical value of a fully crystalline of P(3HB-co-3HHx), taken as 146 J/g [38], an $1-w$ indicates the weight fraction of P(3HB-co-3HHx) in the sample.

Thermogravimetric analysis (TGA) was performed to determine the thermal stability of the injection-molded parts. Samples weighing 10–20 mg were heated from 30 to 700 °C at a heating rate of 20 °C/min in a TGA 100 from Linseis Messgeräte GmbH (Selb, Germany) under nitrogen atmosphere with a flow rate of 25 mL/min. All thermal tests were carried out in triplicate.

2.5. Thermomechanical Tests

Injection-molded parts sizing 10 mm × 5 mm × 4 mm were subjected to a temperature sweep from −70 to 100 °C at a heating rate of 2 °C/min using a DMA-1 from Mettler-Toledo S.A. (Barcelona, Spain).

Dynamic thermomechanical analysis (DMTA) was carried out in bending mode with a maximum bending strain of 10 μm at a frequency of 1 Hz and a force of 0.02 N.

The dimensional stability of the parts was studied by thermomechanical analysis (TMA) in a Q400 thermomechanical analyzer from TA Instruments (New Castle, DE, USA). The applied force was set to 0.02 N and the temperature program was scheduled from −70 to 70 °C in air atmosphere (50 mL/min) at a constant heating rate of 2 °C/min. All thermomechanical tests were performed in triplicate.

2.6. Microscopy

The fracture surfaces of the injection-molded parts after the Charpy impact tests were analyzed by field-emission scanning electron microscopy (FESEM) (Oxford Instruments, Abingdon, UK) with an electron acceleration voltage of 2 kV. A gold-palladium coating was applied through sputtering (SC7620, from Quorum Technologies Ltd, East Sussex, UK). Additionally, to visualize the dispersion of nHA in the P(3HB-co-3HHx) matrix, the fracture surfaces were attacked with 6M hydrochloric acid (HCl) (37% purity, Panreac AppliChem, Barcelona, Spain) for 12 h to selectively remove nHA prior to observation [39].

2.7. Statistical Analysis

Statistical evaluation of the mechanical, thermal, and thermomechanical properties of P(3HB-co-3HHx)/nHA parts was carried out with the open source R software (http://www.r-project.org) with a Shapiro–Wilk test regarding a normal distribution for n < 1000. Tukey tests were performed to determine significant differences between the data on normally distributed data. In order to establish the non-parametric relationship between mechanical properties and nHA content in the parts, the Spearman's correlation test was followed. The number of tested samples for each test is included in Table 2 and the level of significance was established as $p < 0.05$ in all cases.

Table 2. Number of tested samples (n) for each injection-molded poly(3-hydroxybutyrate-co-3-hydroxyhexanoate) [P(3HB-co-3HHx)]/hydroxyapatite nanoparticles (nHA) parts and the type of statistical test performed for each testing method with level of significance (p).

Testing Method	n	Normality Test	p	Significance Test	p
Tensile	6	Shapiro–Wilk	0.05	Tukey	0.05
Flexural	6	Shapiro–Wilk	0.05	Tukey	0.05
Hardness	7	Shapiro–Wilk	0.05	Tukey	0.05
Impact strength	8	Shapiro–Wilk	0.05	Tukey	0.05
DSC	3	-	-	Kruskal-Wallis	0.05
TGA	3	-	-	Kruskal-Wallis	0.05
DMTA	3	-	-	Kruskal-Wallis	0.05
TMA	3	-	-	Kruskal-Wallis	0.05

DSC = differential scanning calorimetry; TGA = thermogravimetric analysis; DMTA = dynamic thermomechanical analysis; TMA = thermomechanical analysis.

3. Results and Discussion

3.1. Mechanical Characterization of the P(3HB-co-3HHx)/nHA Parts

The data collected for the mechanical properties from the tensile, flexural, hardness, and impact Charpy tests of the neat P(3HB-co-3HHx) and P(3HB-co-3HHx)/nHA composite parts produced with the different compositions is summarized in Table 3. Figures 1 and 2 display the effect of nHA incorporation on the tensile and flexural properties, respectively, whereas Table 4 shows the correlation coefficient (r_s) and p for each mechanical property according to the Spearman's test.

Table 3. Mechanical properties of the injection-molded parts of poly(3-hydroxybutyrate-co-3-hydroxyhexanoate) [P(3HB-co-3HHx)]/hydroxyapatite nanoparticles (nHA) in terms of maximum tensile stress (σ_{max}), tensile modulus (E_t), elongation at break (ε_b), maximum flexural stress (σ_f), flexural modulus (E_f), Shore D hardness, and impact strength.

Part	σ_{max} (MPa)	E_t (MPa)	ε_b (%)	σ_f (MPa)	E_f (MPa)	Shore D Hardness	Impact Strength (kJ/m^2)
P(3HB-co-3HHx)	17.7 ± 1.1	1022.3 ± 59.2	19.4 ± 0.8	24.1 ± 1.9	735.3 ± 43.5	64.2 ± 0.8	5.1 ± 0.3
P(3HB-co-3HHx) + 2.5 nHA	16.1 ± 0.5 *	1097.0 ± 52.3 *	12.9 ± 0.3 *	25.6 ± 1.0 *	744.1 ± 26.3	64.0 ± 0.9	3.5 ± 0.2 *
P(3HB-co-3HHx) + 5 nHA	15.8 ± 0.6	1113.1 ± 28.3	12.5 ± 0.8	26.4 ± 1.8	813.2 ± 24.1 *	65.2 ± 0.8 *	2.6 ± 0.2 *
P(3HB-co-3HHx) + 10 nHA	15.5 ± 0.5	1398.2 ± 68.3 *	10.4 ± 0.6 *	26.7 ± 0.5	919.8 ± 38.6 *	65.8 ± 1.1	2.2 ± 0.1 *
P(3HB-co-3HHx) + 20 nHA	14.2 ± 0.2 *	1681.4 ± 56.3 *	6.5 ± 0.7 *	26.9 ± 1.9	1182.5 ± 45.2 *	69.4 ± 0.5 *	1.7 ± 0.2 *

* Indicates a significant difference compared with the previous sample ($p < 0.05$). Level of significance (p) values are included in Table S1.

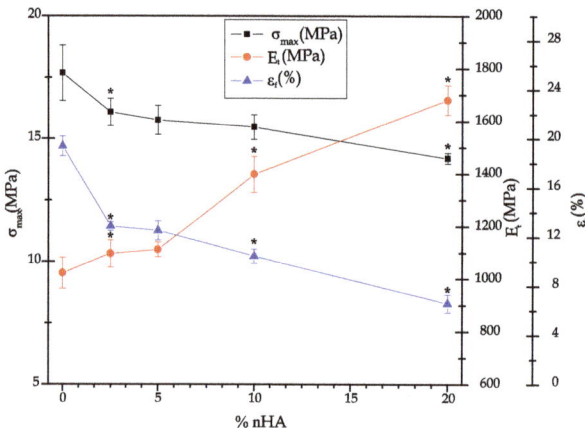

Figure 1. Evolution of the maximum tensile stress (σ_{max}), tensile modulus (E_t), elongation at break (ε_b) in the injection-molded parts of poly(3-hydroxybutyrate-co-3-hydroxyhexanoate) [P(3HB-co-3HHx)] with the content of hydroxyapatite nanoparticles (nHAs). * Indicates a significant difference compared with the previous sample ($p < 0.05$)

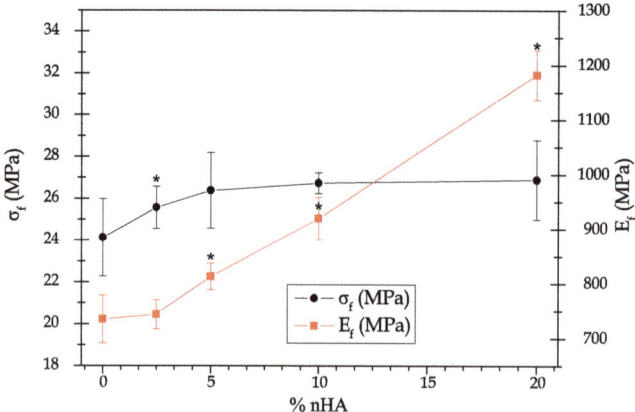

Figure 2. Evolution of the maximum flexural stress (σ_f) and flexural modulus (E_f) in the injection-molded parts of poly(3-hydroxybutyrate-co-3-hydroxyhexanoate) [P(3HB-co-3HHx)] with the content of hydroxyapatite nanoparticles (nHAs). * Indicates a significant difference compared with the previous sample ($p < 0.05$).

Table 4. Spearman's test correlation coefficient (r_s) and level of significance (p) for each mechanical property.

Mechanical Properties	r_s	p
σ_{max} (MPa)	−0.917	0.028
E_t (MPa)	+0.988	0.002
ε_b (%)	−0.903	0.035
σ_f (MPa)	+0.782	0.118
E_f (MPa)	+0.993	0.001
Shore D hardness	+0.977	0.004
Impact strength (kJ/m^2)	−0.839	0.032

The tensile properties of the injection-molded P(3HB-co-3HHx) parts were relatively similar to those reported by Giubilini et al. [40], although the here-prepared materials were slightly less mechanically resistant and more ductile. These differences could be related to the 3HHx monomer content in the copolyester as well as to differences in processing. One can observe in both Table 3 and Figure 1 that the values of σ_{max} and ε_b decreased, while those of E_t increased with increasing nHA concentration in the nanocomposite parts. The Spearman's test confirmed the existence of a trend between the tensile properties of the nanocomposites and the nHA content, showing a negative r_s trend (inversely proportional correlation) for σ_{max} and ε_b and a positive trend (directly proportional correlation) for E_t, while in all cases $p < 0.05$. In particular, the addition of 20 wt % of nHA produced a slight decrease of σ_{max} from 17.7 to 14.4 MPa, but an increase of nearly 64% in E_t (from approximately 1 to 1.7 GPa) accompanied with a significant loss of ductility (ε_b was reduced from 19.4 to 6.5%). The reduction in stress was probably caused by the poor interface adhesion between biopolymer and nanofiller. Higher interfacial adhesion can probably be promoted through the pretreatment of nHA with silanes [41], but it could negatively affect the biocompatibility of the parts. The increase in E_t was anticipated, since nHA forms highly rigid structures. Furthermore, as it will be discussed during the thermal characterization, the addition of nHA could promote higher degrees of crystallinity and, hence, higher stiffness. Although similar results have been reported earlier [42,43], the here-prepared parts showed higher ductility due to the use of a more flexible PHA. The decrease observed in stiffness with increasing nHA content can be attributed to insufficient wetting and impregnation of the nanoparticles by the polymer matrix, mainly due to particle agglomeration during manufacture or processing of the materials [44]. However, melt-mixing methodologies using co-rotating twin-screw extruders, as adopted here, can generally yield well-dispersed nanocomposites [45]. Ductility loss was expected since the presence of nHA can prompt polymer crystallinity, hindering chain mobility due to adsorption of biopolymer chains on the surface of the nanoparticles [46,47].

In Figure 2, it can be seen that the addition of nHA to P(3HB-co-3HHx) increased both σ_f and E_f, particularly the latter. The former increased up to a content of 5 wt % of nHA and then became insensitive to higher nanoparticle contents, since the values showed no significant differences. Indeed, the Spearman's test showed a positive correlation ($r_s > 0$) for both E_f and σ_f, however, for the latter, the statistical hypothesis should be rejected as p was higher than 0.05. Contrarily, the addition of 20 wt % of nHA caused an increase of approximately 60% of E_f, as similarly observed above for E_t. The resultant increase in mechanical strength can be related to the intrinsic high values of compressive strength and modulus of nHA, which are in the ranges of 500–1000 MPa and 80–110 GPa, respectively [48,49].

In comparison with the mechanical values of other degradable and non-degradable materials, the P(3HB-co-3HHx)/nHA parts produced in this study showed intermediate values to most biodegradable polymers and metal alloys. For instance, the E_t values of poly(ε-caprolactone) (PCL) and PLA materials range between 400–600 MPa [50] and 2–3 GPa [51,52], respectively, while other biodegradable copolyesters such as poly(butylene adipate-co-terephthalate) (PBAT) show significantly lower values [53]. However, PLA is a brittle polymer, which can limit its application in bone fixation devices, or any other biomedical device that would be subjected to local flexural stress or impacts. The values attained are relatively similar to those of poly(lactic-co-glycolic acid) (PLGA), that is,

1.4–2.8 GPa [54]. Indeed, PLGA is widely used in biomedical and pharmaceutical applications, but it shows longer degradation times, which can extend up to 12 months [55]. Regarding metal alloys, the E_t values of the most widely used stainless steels for implant fixing devices and screws, that is, SUS316L stainless steel and cobalt-chrome (Co-Cr) alloys, are around 180 GPa and 210 GPa, respectively [56]. Lower values have been reported for titanium (Ti) and its light alloys, such as Ti-6Al-4V ELI, which are also widely used for making implant devices, having a value of around 110 GPa [57]. As shown above, in comparison to metal alloys, the elasticity of the P(3HB-co-3HHx)/nHA composites prepared in this study is nearer to that of the natural bone, which is in the 8–25 GPa range [5]. Thus, from a mechanical point of view, their use in bone scaffolds and resorbable plates or screws looks promising.

As expected, hardness increased with the presence of nHA that, due to its ceramic nature, is highly rigid. The increase was significant at nHA contents higher than 2.5 wt % and this effect was statistically corroborated by Spearman's test, showing a positive trend with an r_s value of ~0.98. In addition, molecular mobility could be reduced due to the presence of the nanoparticles [58]. In particular, the incorporation of 20 wt % of nHA yielded an increase of 8% in hardness. A similar increase in Shore D hardness was reported by Ferri et al. [39] for PLA after the incorporation of nHA. In particular, it increased from 73.9, for neat PLA, up to 78.4, for the PLA composite containing 30 wt % of nHA. As also anticipated, the impact strength of the nanocomposites diminished significantly with increasing nHA content with significant differences between the samples, which was confirmed by the negative correlation obtained by the Spearman's test ($r_s \simeq -0.84$). For instance, the nanocomposite parts containing 20 wt % of nHA revealed an impact strength approximately three times lower than that of the neat P(3HB-co-3HHx) part, that is, it reduced from 5.1 to 1.7 kJ/m^2. Lower values of impact strength were reported for V-notched injection-molded pieces of PLA, that is, 2.1 kJ/m^2 [51]. In addition, significantly higher values have been described for Ti-6Al-4V, with a Rockwell hardness C (HRC) of 38 and approximately 112 kJ/m^2 impact strength [59]. In the case of natural bone, toughness varies widely with age and type. For instance, the impact strength of the femora ranges from 4 to 70 kJ/m^2 [60]. Therefore, the various mechanicals tests revealed a clear tendency towards a decrease in ductility and an increase in stiffness of the injection-molded parts with increasing nHA content, which are closer to those of the natural bone.

In summary, the here-developed P(3HB-co-3HHx)/nHA parts showed an improvement of the stiffness determined in terms of E_t and E_f, in which a positive trend was observed in both cases ($r_s > 0$). The ductile properties, that is, ε_b and impact strength, showed negative trends ($r_s < 0$), which was ascribed to a chain mobility reduction that also contributed to a hardness increase of the nanocomposite, showing a positive trend in the Spearman's test.

3.2. Thermal Characterization of the P(3HB-co-3HHx)/nHA Parts

Figure 3 displays the DSC curves for the neat P(3HB-co-3HHx) part and the P(3HB-co-3HHx)/nHA composite parts with different nanoparticle contents. Table 5 presents the thermal properties obtained from the second heating scan, after erasing the thermal history of the sample. At approximately 0 °C, one could observe a step change in the base lines, which corresponded to the glass transition temperature (T_g) of P(3HB-co-3HHx). This second-order thermal transition was located at −0.3 °C for the neat biopolymer and it was significantly unaffected by the presence of nHA. The exothermic peaks located between 40 and 70 °C corresponded to the cold crystallization temperature (T_{cc}) of P(3HB-co-3HHx). In the case of the neat biopolymer part, this peak was located at 49.8 °C. It could be observed that the values of T_{cc} increased with increasing nHA content until 10 wt %, and then slightly decreased at the highest content tested, that is, 20 wt %. These results suggested that low nHA contents impaired the movement of P(3HB-co-3HHx) chains and, hence, hindered the crystallization process. A similar thermal behavior during the analysis of the second heating curves was recently observed by Senatov et al. [61], who associated the presence of nHA to a decrease in the molecular chain mobility of the biopolymer that impeded the crystallization process. Finally, the crystalline P(3HB-co-3HHx) domains melted in the thermal range from 100 to 150 °C in two peaks. Furthermore, the occurrence of a broad melting region suggested the presence of heterogeneous crystallites with different degrees of perfection, commonly produced in PHAs with relatively high comonomer contents [62]. The thermogram of neat P(3HB-co-3HHx) revealed two melting temperatures (T_{m1} and T_{m2}) at approximately 113 and 140 °C. Similar thermal properties were reported by Zhou et al. [63] for P(3HB-co-3HHx) with 11 mol % content of 3HHx, who also observed a double-melting peak phenomenon in the DSC heating curves of this copolyester. The presence of two melting peaks have been previously ascribed to the melting–recrystallization–melting process of P(3HB-co-3HHx) [64]. During this process, imperfect crystals melt at lower temperatures and the amorphous regions order into packed spherulites with thicker lamellar thicknesses that, thereafter, melt at higher temperatures. Alternatively, the melting peaks attained at low temperatures, that is, 110–115 °C, could also relate to the crystalline phase of the 3HHx-rich fractions. Lastly, one could observe that the melting profile of P(3HB-co-3HHx) was nearly unaffected by the nHA presence, indicating that the nanoparticles did not significantly influence the crystallization process.

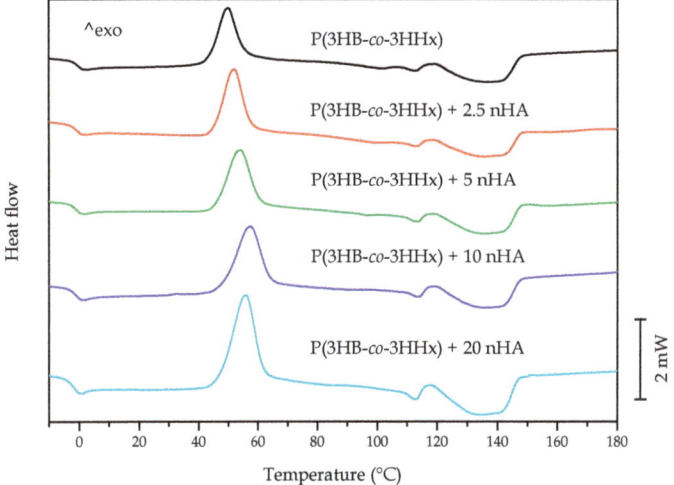

Figure 3. Differential scanning calorimetry (DSC) thermograms taken during second heating of the injection-molded poly(3-hydroxybutyrate-co-3-hydroxyhexanoate) [P(3HB-co-3HHx)]/hydroxyapatite nanoparticle (nHA) parts.

Table 5. Thermal properties of the injection-molded poly(3-hydroxybutyrate-co-3-hydroxyhexanoate) [P(3HB-co-3HHx)]/hydroxyapatite nanoparticle (nHA) parts in terms of glass transition temperature (T_g), cold crystallization temperature (T_{cc}), melting temperatures (T_{m1} and T_{m2}), cold crystallization enthalpy (ΔH_{cc}), melting enthalpy (ΔH_m), and maximum degree of crystallinity ($\chi_{c\,max}$).

Part	T_g (°C)	T_{cc} (°C)	T_{m1} (°C)	T_{m2} (°C)	ΔH_{cc} (J/g)	ΔH_m (J/g)	$X_{c\,max}$ (%)
P(3HB-co-3HHx)	−0.3 ± 0.1	49.8 ± 0.5	112.9 ± 0.5	139.7 ± 0.3	20.7 ± 0.5	31.2 ± 0.4	21.4 ± 1.8
P(3HB-co-3HHx) + 2.5 nHA	−0.2 ± 0.1	52.1 ± 0.4	113.6 ± 0.3	140.9 ± 0.4 *	29.8 ± 0.4 *	35.0 ± 0.5 *	24.3 ± 2.4 *
P(3HB-co-3HHx) + 5 nHA	−0.4 ± 0.2	54.1 ± 0.2	113.8 ± 0.4	139.4 ± 0.2	31.8 ± 0.6 *	40.1 ± 0.3 *	28.9 ± 2.2 *
P(3HB-co-3HHx) + 10 nHA	−0.4 ± 0.2	57.4 ± 0.3 *	114.3 ± 0.2	139.0 ± 0.3	29.0 ± 0.5 *	32.5 ± 0.4 *	24.7 ± 1.5 *
P(3HB-co-3HHx) + 20 nHA	−0.3 ± 0.1	55.9 ± 0.4 *	113.4 ± 0.5	139.1 ± 0.4	25.9 ± 0.4 *	29.0 ± 0.1 *	24.8 ± 1.4

* Indicates a significant difference compared with the previous sample ($p < 0.05$). Level of significance (p) values are included in Table S2.

In addition to the characteristic values of T_g, T_{cc}, and T_m, the enthalpies corresponding to the cold crystallization (ΔH_{cc}) and melting (ΔH_m) enthalpies were collected from the DSC curves. The latter parameter was used to determine the maximum degree of crystallinity, that is, X_{C_max}, which gives more information about the effect of the additives on the biopolymer, since it does not consider the crystals formed during cold crystallization. It can be seen that P(3HB-co-3HHx) showed a maximum degree of crystallinity of 21.4%. One can also observe that crystallinity varied significantly with nHA content. In particular, as nHA was gradually incorporated in higher percentages, the crystallinity increased steadily up to a maximum of nearly 29% at 5 wt % of nHA and then it slightly decreased to values close to 25% for nHA contents of 10 and 20 wt %. This result, in combination with the slightly higher T_{cc} and T_m values, suggests that the nanoparticles hindered the formation of crystals at low temperatures, but the crystals formed were slightly more perfect and more mass crystallized. This is in agreement with previous studies that concluded that the introduction of nHA into biopolyesters has an effect on the ordering of their molecular chains by acting as a nucleating agent [61,65].

Figure 4 presents the thermogravimetric data for all the materials, while Table 6 gathers the main thermal stability parameters obtained from the TGA curves. Thermal degradation of P(3HB-co-3HHx) was observed to occur through a one-step process, which is in agreement with the values reported by Li et al. [20], who showed that the thermal stability of the microbial copolyester was as high as 225 °C with almost no mass loss. The temperature at 5% mass loss ($T_{5\%}$) showed no significant differences with nHA contents of up to 5 wt %, but a significant decrease was observed for higher loadings. The temperature at which the maximum mass loss rate occurred (T_{deg}) increased from 296.7 °C, for the neat P(3HB-co-3HHx) part, to 300.9 °C, for the part of P(3HB-co-3HHx) filled with 2.5 %wt of nHA. This increase in thermal stability has been previously ascribed to the formation of strong hydrogen interactions and Van der Walls forces between the inorganic nanoparticles and the biopolymer chains during the melt-mixing process [66]. The values of T_{deg} remained nearly constant, showing no significant differences for nHA contents from 2.5 to 10 wt %, but it significantly decreased to 295.6 °C in the part filled with 20 wt % of nHA. The onset of degradation was also reduced for the most filled sample, showing a $T_{5\%}$ value of 254.8 °C, which represents a reduction of approximately 18 °C in comparison to the unfilled P(3HB-co-3HHx) sample and its nanocomposites at low contents. These results further indicate that the nanoparticles formed aggregates at high contents, which created volumetric gradients of concentration [66]. In this regard, Bikiaris et al. [65] suggested that when high amounts of nanosized filler aggregates are formed, the structure shifts from nanocomposite to microcomposite and, thus, the shielding effect of the nanosized particles is lessened. In addition, Chen et al. [67] reported that high loadings of nHA in PHBV lower the onset degradation temperature since they can catalyze thermal decomposition. In any case, low nHA loadings (<10 wt %) slightly improved the thermal stability of P(3HB-co-3HHx) parts and their thermal stability is considered to be high enough for bone tissue engineering and biomedical applications, which can require thermal sterilization methods such as dry heat sterilization (160 °C for 2 h) and steam sterilization (121 °C for 20–60 min) [68]. However, the relatively low T_m of P(3HB-co-3HHx) would limit the use of these techniques for sterilization and the resultant implantable biomedical devices should be sterilized at low temperatures using ethylene oxide (EO) gas, gamma radiation or ozone. Finally, it can be observed that the residual mass at 700 °C increased gradually with the nHA content due to the high thermal stability of the mineral nanoparticles.

Table 6. Main thermal degradation parameters of the injection-molded poly(3-hydroxybutyrate-co-3-hydroxyhexanoate) [P(3HB-co-3HHx)]/hydroxyapatite nanoparticle (nHA) parts in terms of onset temperature of degradation ($T_{5\%}$), degradation temperature (T_{deg}), and residual mass at 700 °C.

Part	$T_{5\%}$ (°C)	T_{deg} (°C)	Residual Mass (%)
P(3HB-co-3HHx)	272.5 ± 2.3	296.7 ± 1.4	2.6 ± 0.3
P(3HB-co-3HHx) + 2.5 nHA	272.2 ± 1.7	300.9 ± 2.2	5.1 ± 0.5 *
P(3HB-co-3HHx) + 5 nHA	272.4 ± 1.3	299.9 ± 1.7	7.6 ± 0.4 *
P(3HB-co-3HHx) + 10 nHA	262.3 ± 1.8 *	299.6 ± 1.8	12.2 ± 0.7 *
P(3HB-co-3HHx) + 20 nHA	254.8 ± 1.3 *	295.6 ± 1.6 *	22.1 ± 0.8 *

* Indicates a significant difference compared with the previous sample ($p < 0.05$). Level of significance (p) values are included in Table S3.

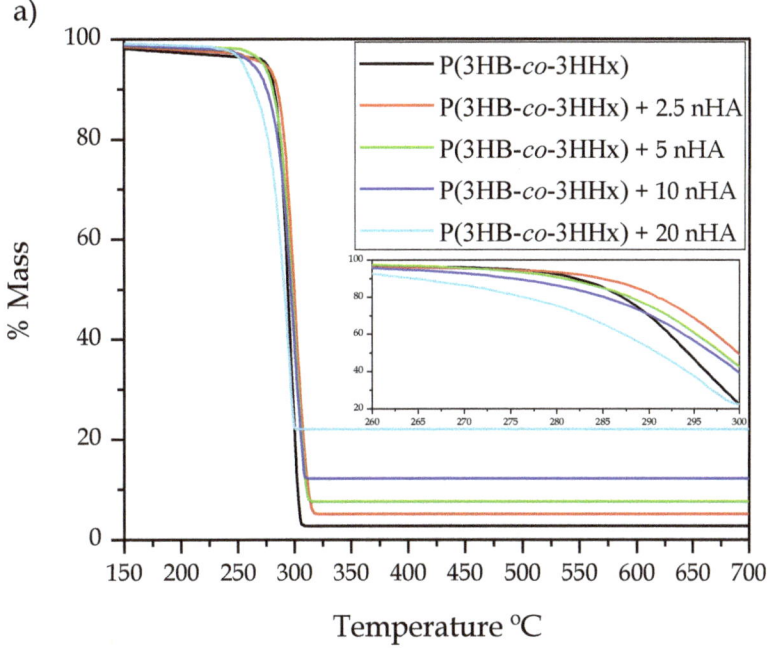
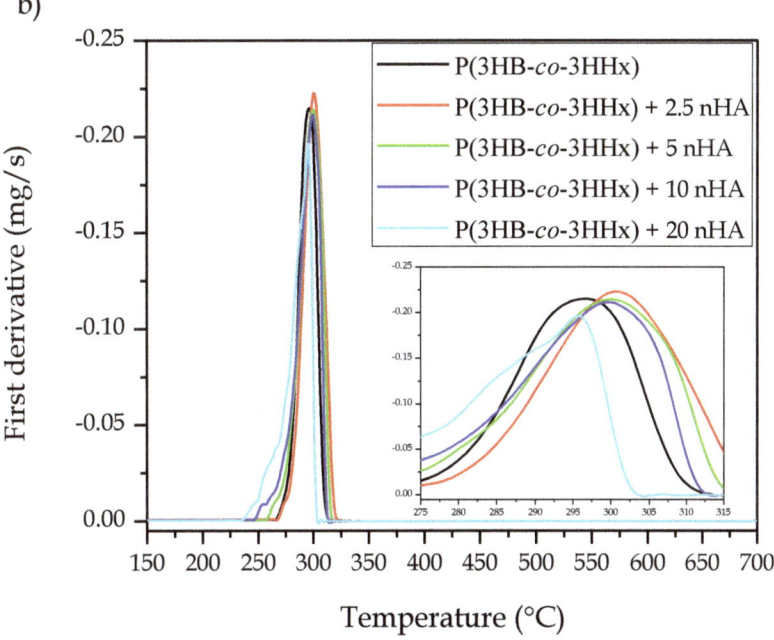

Figure 4. (**a**) Thermogravimetric analysis (TGA) and (**b**) first derivate thermogravimetric (DTG) curves of the injection-molded poly(3-hydroxybutyrate-*co*-3-hydroxyhexanoate) [P(3HB-*co*-3HHx)]/ hydroxyapatite nanoparticle (nHA) parts.

3.3. Thermomechanical Characterization of the P(3HB-co-3HHx)/nHA Parts

DMTA was carried out on the injection-molded composite parts in order to understand the role played by nHA on the viscoelastic behavior of P(3HB-co-3HHx)/nHA. Figure 5 illustrates the DMTA curves of the neat P(3HB-co-3HHx) part and the P(3HB-co-3HHx)/nHA composite parts with different nanoparticle contents. Figure 5a gathers the evolution of the storage moduli (E′) in the temperature sweep from −40 to 80 °C at a frequency of 1 Hz. The T_g values and the corresponding values of E′ at −40, 37, and 70 °C are presented in Table 7, since the first and last temperatures are representative of the stored elastic energy of the amorphous phase of P(3HB-co-3HHx) in its glassy and rubber states, respectively, whereas the middle one corresponds to the actual temperature of the human body. It can be observed that all the P(3HB-co-3HHx)-based parts presented a similar thermomechanical profile. In particular, the samples showed high E′ values, that is, high stiffness, at temperatures below 0 °C and then E′ sharply decreased. This thermomechanical change was produced because the temperature exceeded the alpha (α)-relaxation of the biopolymer, which is related to its T_g. One can also observe that the rate of decrease of E′ reduced somewhat when the temperature reached approximately 40 °C due to the occurrence of cold crystallization. The values of E′ at −40, 37, and 70 °C of the neat P(3HB-co-3HHx) part were 1909.9, 519.2, and 210.5 MPa, respectively. The E′ value attained at 37 °C was in accordance with the mechanical data presented in Section 3.1, which indicated that only the P(3HB-co-3HHx) parts filled with the highest nHA contents, that is, 15 and 20 wt %, showed significantly higher values. However, the results also indicated that the parts crystallized during the ageing process since the thermomechanical changes during and after cold crystallization were relatively low. As expected, the E′ values progressively increased with increasing the nHA content, given the high stiffness of the nanoparticles. It is worth noting that the reinforcing effect was more noticeable at higher temperatures since the amorphous phase of P(3HB-co-3HHx) was in the rubber state. Indeed, at higher temperatures, the thermomechanical response of all the P(3HB-co-3HHx) composite parts was significantly different, dependent upon the nHA content. For instance, at −40 °C the E′ value increased from 1935.2 MPa for the nanocomposite part containing 2.5 wt % of nHA, to 2100.4 MPa for the part filled with 20 wt % of nHA, whereas these values increased from 212.3 MPa to 333.1 MPa at 70 °C.

The loss tangent or dynamic damping factor (tan δ) curves are shown in Figure 5b. Since the position of the tan δ peak gives an indication of the biopolymer's T_g, these values were also included in Table 7. In the case of the neat P(3HB-co-3HHx) part, the tan δ peak was located at 10.7 °C, which is similar to that reported by Valentini et al. [69]. It is worth mentioning that, in all cases, the tan δ peaks were approximately 10 °C higher than the T_g values. Since tan δ represents the ratio of the viscous to the elastic response of a viscoelastic material, this indicates that part of the applied load was dissipated by energy dissipation mechanisms such as segmental motions, which are related to T_g, but part of the energy was also stored and released upon removal of the load at higher temperatures. One can observe that the incorporation of nHA shifted slightly the position of the tan δ peaks and also reduced their intensity for the highest nHA loadings, that is, 10 and 20 wt %. Decreasing tan δ peaks intensity indicated that the nanocomposite parts showed a more elastic response and, hence, presented more potential to store the applied load rather than dissipating it [70]. This reduction is directly related to the higher E′ values attained due to nanoparticle reinforcement and it confirmed that nHA imposed restrictions on the molecular motion of the P(3HB-co-3HHx) chains, resulting in a material with more elastic behavior [71]. It also correlated well with the DSC results shown above, indicating that P(3HB-co-3HHx) developed more crystallinity in the nanocomposite parts due to the nucleating effect of nHA and, thus, the less amorphous phase underwent glass transition.

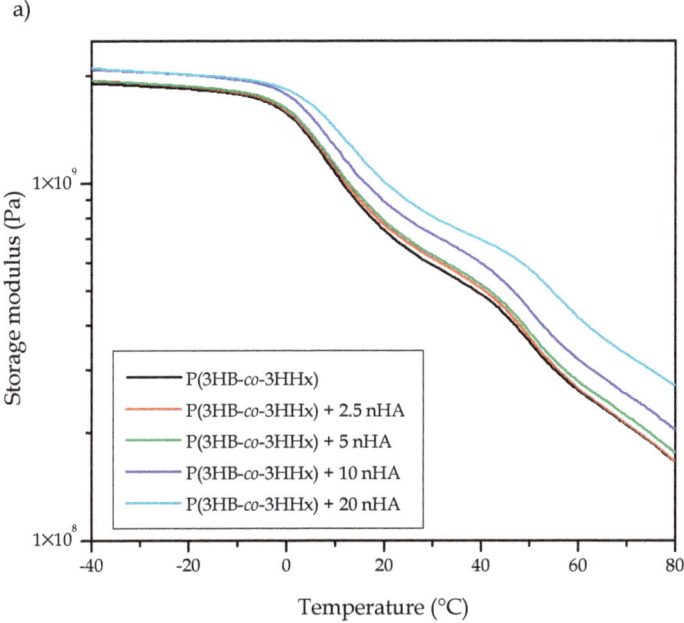
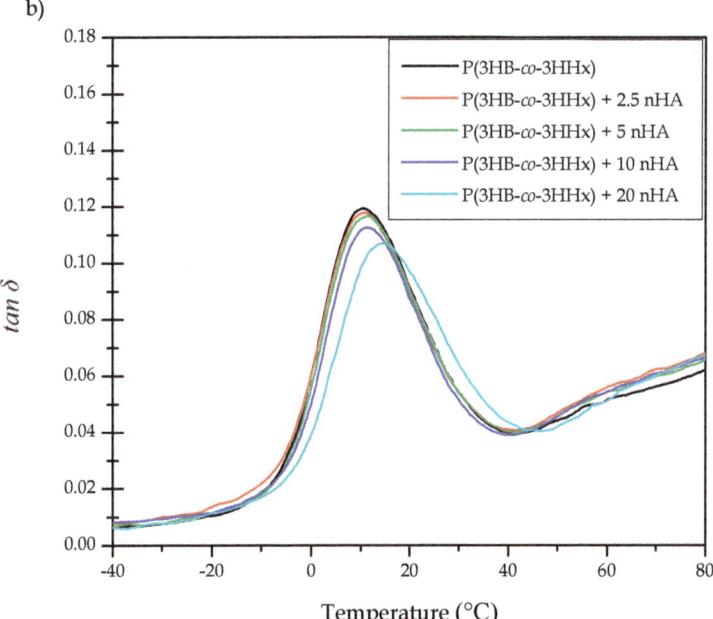

Figure 5. Evolution as a function of temperature of the (**a**) storage modulus and (**b**) dynamic damping factor (*tan δ*) of the injection-molded hydroxybutyrate-*co*-3-hydroxyhexanoate) [P(3HB-*co*-3HHx)]/ hydroxyapatite nanoparticles (nHA) parts.

Table 7. Thermomechanical properties of the injection-molded poly(3-hydroxybutyrate-*co*-3-hydroxyhexanoate) [P(3HB-*co*-3HHx)]/hydroxyapatite nanoparticles (nHA) parts in terms of dynamic damping factor (*tan δ*) peak, glass transition temperature (T_g), storage modulus (E') measured at −40, 37, and 70 °C, and coefficient of linear thermal expansion (CLTE) below and above T_g.

Part	DMTA					TMA	
	tan δ Peak (°C)	E' at −40 °C (MPa)	E' at 37 °C (MPa)	E' at 70 °C (MPa)	T_g (°C)	CLTE (μ/m·°C)	
						Below T_g	Above T_g
P(3HB-*co*-3HHx)	10.7 ± 0.4	1909.9 ± 50.2	519.2 ± 14.2	210.5 ± 2.5	−0.6 ± 0.2	64.3 ± 1.1	177.2 ± 4.6
P(3HB-*co*-3HHx) + 2.5 nHA	10.9 ± 0.2	1935.2 ± 41.7 *	544.1 ± 26.3 *	212.3 ± 3.1	−0.3 ± 0.2	61.3 ± 0.4 *	176.1 ± 7.2 *
P(3HB-*co*-3HHx) + 5 nHA	11.2 ± 0.3	1940.1 ± 82.5 *	557.8 ± 17.1 *	222.5 ± 4.6 *	−0.1 ± 0.1	59.3 ± 0.5	175.0 ± 0.8
P(3HB-*co*-3HHx) + 10 nHA	11.4 ± 0.5	2090.3 ± 74.6 *	639.0 ± 18.1 *	256.8 ± 5.1 *	−0.4 ± 0.2	58.2 ± 0.4	170.2 ± 3.8 *
P(3HB-*co*-3HHx) + 20 nHA	14.5 ± 0.4 *	2100.4 ± 65.1 *	728.8 ± 26.6 *	333.1 ± 3.4 *	−0.3 ± 0.2	56.7 ± 0.7	159.1 ± 5.7 *

* Indicates a significant difference compared with the previous sample ($p < 0.05$). Level of significance (p) values are included in Table S4.

The effect of temperature on the dimensional stability of the P(3HB-*co*-3HHx)/nHA parts was also determined by TMA. The coefficient of linear thermal expansion (CLTE), both below and above T_g, was obtained from the change in dimensions versus temperature and it is also included in Table 7 along with the T_g values. In all cases, lower CLTE values were attained in the parts below T_g, due to the lower mobility of the P(3HB-*co*-3HHx) chains of the amorphous regions in the glassy state. As anticipated, both below and above T_g, the CLTE values decreased significantly with increasing nHA content due to the increasing replacement of the soft biopolymer matrix by a ceramic material with a considerably lower CLTE value, that is, 13.6 μm/m·°C [72]. As a result, the CLTE value below T_g was reduced from 64.3 μm/m·°C for the neat P(3HB-*co*-3HHx) part, to 56.7 μm/m·°C for the nanocomposite part filled with 20 wt % of nHA. Similarly, above T_g, it decreased from 177.2 to 159.1 μm/m·°C, respectively. This thermomechanical response was slightly better than that of the PLA/nHA composites, in which the CLTE values below T_g decreased from 73 to 71 μm/m·°C after the incorporation of 20 wt % of nHA into PLA μm/m·°C [73]. These results point out that the nanocomposite parts prepared herein show excellent dimensional stability against temperature exposition. However, it is also worth mentioning that, as expected, the CLTE of Ti-based materials was significantly lower, having a mean value of 8.7 μm/m· °C [72].

3.4. Morphological Characterization of the P(3HB-co-3HHx)/nHA Parts

Figure 6 shows the samples before and after the various processing steps. Combining melt compounding and injection molding represents a cost-competitive melt-processing methodology to produce a large number of parts using nanocomposites. According to this route, the P(3HB-*co*-3HHx) pellets and the nHA powder were pre-mixed and fed together to the co-rotating twin-screw extruder. In this way, pellets of nanocomposites containing different contents of dispersed nHA particles were obtained. They were subsequently injection molded into dumbbell bars. All parts were defect-free and had a bright surface, the nHA content influencing their color; neat P(3HB-*co*-3HHx) parts were yellow pale, typical of microbial PHA, while the presence of nanoparticles induced a whiter color.

Figure 6. Processing steps carried out to prepare the poly(3-hydroxybutyrate-*co*-3-hydroxyhexanoate) [P(3HB-*co*-3HHx)]/hydroxyapatite nanoparticle (nHA) parts; from left to right: as-received P(3HB-*co*-3HHx) pellets and nHA powder, compounded pellets of the nanocomposite, injection-molded parts.

Figure 7 shows the FESEM image, taken at 10,000×, of the nHA powder. The nanoparticles show a flake-like morphology based on plates with sizes 60–120 nm and mean cross-sections of approximately 30 nm. This particular morphology of nHA has been reported to occur at pH values below 9, due to the solution environment changes by the OH⁻ ions during synthesis using polyethylene glycol (PEG) as a template [74].

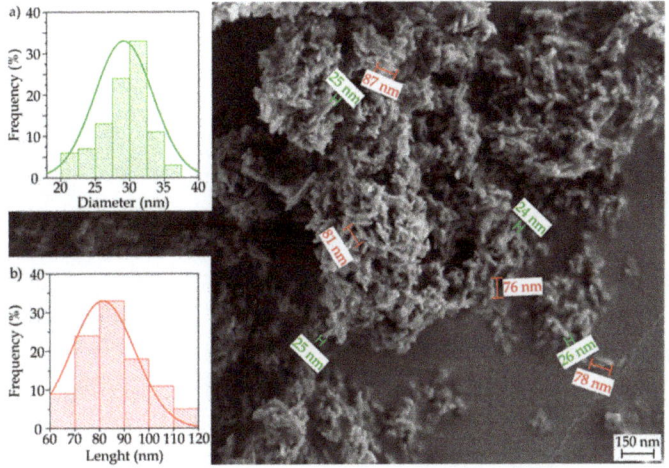

Figure 7. Field-emission scanning electron microscopy (FESEM) images of the hydroxyapatite nanoparticles (nHA) powder. Image was taken at 10,000× with scale marker 150 nm.

Micrographs obtained by FESEM of the fracture surfaces of the injection-molded parts of P(3HB-*co*-3HHx) and the various P(3HB-*co*-3HHx)/nHA composites after the Charpy impact tests are gathered in Figure 8. The fracture surface of the neat P(3HB-*co*-3HHx) part, shown in Figure 8a, indicated that the material presented a relatively high toughness, since it yielded a rough surface with the presence of multiple microcracks and some holes. Some microparticles could be seen in the inset FESEM micrograph taken at higher magnification, which could be related to the presence of nucleating agents and/or fillers added by the manufacturer, such as boron nitride (BN). In this regard, Türkez, et al. [75] have recently demonstrated that BN nanoparticles show slight cytotoxicity potential. In particular, contents below 100 mg/L did not lead to lethal response on human primary alveolar epithelial cells (HPAEpiC), suggesting their safe and effective use in both pharmacological and medical applications. Figure 8b–e gather the fracture surfaces of the P(3HB-*co*-3HHx)/nHA composite parts. The morphological characteristics of the fracture surfaces for the nanocomposites filled with low nHA contents, that is, 2.5 and 5 wt %, remained very similar to that of neat P(3HB-*co*-3HHx)/nHA. In all cases, the nanoparticles were relatively well dispersed and distributed within the biopolymer matrix. However, at higher contents, the nanoparticles tended to form some microaggreagates and the resultant fracture surfaces were smoother, indicating that the nanocomposites were more brittle.

Due to the low nHA particle size and the presence of BN and/or additives in the P(3HB-*co*-3HHx) matrix, selective separation was carried out on the fracture surfaces of the nanocomposite parts, in order to better evaluate the dispersion of the nanoparticles. Figure 9 presents the FESEM images of the fracture surfaces subjected to treatment with 6 M HCl for 12 h. The voids and holes formed in the surfaces were related to removed/dissolved nHA and the overall void size and distribution gave an indication of the original particle dispersion. The micrographs revealed that some microholes were produced after the selective attack on the P(3HB-*co*-3HHx) parts filled with 10 and 20 wt % of nHA, which should correspond to nHA aggregates, whereas the nanocomposites containing low nanoparticle loadings showed nano-sized holes well distributed along the biopolymer matrix, which suggested an efficient dispersion. Agglomeration was particularly noticeable for the nanocomposite part containing 20 wt % of nHA, thus indicating that the presence of aggregates could induce particle debonding during fracture, as a result of the dissimilar mechanical strength and rigidity of the ceramic nanoparticles and biopolymer matrix. Therefore, the present results correlate well with the mechanical and thermal properties described above, in which nHA loadings of up to 10 wt % increased the mechanical and thermal performance of the P(3HB-co-3HHx) parts, whereas the highest nHA content impaired the overall properties due to nanoparticle aggregation.

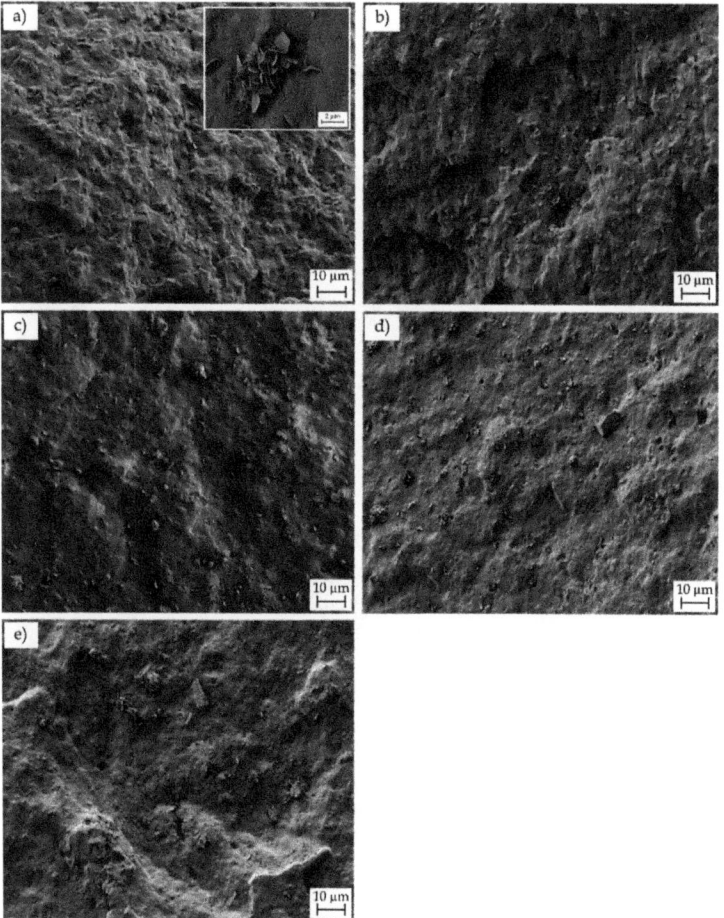

Figure 8. Field-emission scanning electron microscopy (FESEM) images of the fracture surfaces of the injection-molded poly(3-hydroxybutyrate-co-3-hydroxyhexanoate) [P(3HB-co-3HHx)]/hydroxyapatite nanoparticle (nHA) parts of: (**a**) neat P(3HB-co-3HHx); (**b**) P(3HB-co-3HHx) + 2.5 nHA; (**c**) P(3HB-co-3HHx) + 5 nHA; (**d**) P(3HB-co-3HHx) + 10 nHA; (**e**) P(3HB-co-3HHx) + 20 nHA. Images were taken at 500× and with scale markers of 10 μm. Inset image showing the detail of the microparticles was taken at 2500× with scale marker of 2 μm.

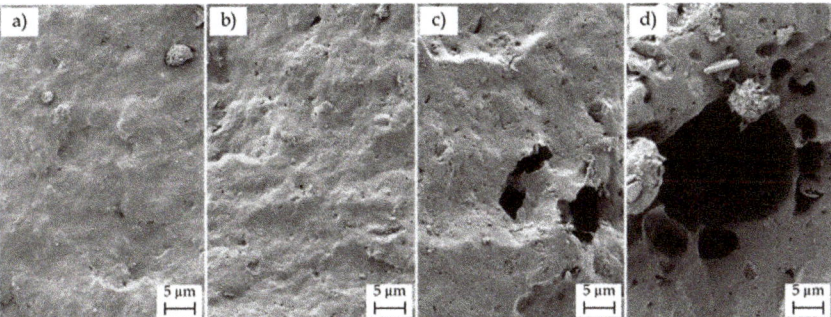

Figure 9. Field-emission scanning electron microscopy (FESEM) images of the fracture surfaces of the injection-molded poly(3-hydroxybutyrate-*co*-3-hydroxyhexanoate) [P(3HB-*co*-3HHx)]/hydroxyapatite nanoparticles (nHA) parts after selective attack with 6M hydrochloric acid (HCl) for 12 h: (**a**) P(3HB-*co*-3HHx) + 2.5 nHA; (**b**) P(3HB-*co*-3HHx) + 5 nHA; (**c**) P(3HB-*co*-3HHx) + 10 nHA; (**d**) P(3HB-*co*-3HHx) + 20 nHA. Images were taken at 1000× with scale marker of 5 μm.

4. Conclusions

One of the most exciting areas of new material development in the biomedical device community is resorbable polymers. As bone scaffolds, biodegradable and biocompatible polymers will maintain their strength until the liquid in contact begins the dissolution process, eventually leading to their complete elimination from the body, thus avoiding a second surgery for their removal. The herein-prepared injection-molded composite parts of P(3HB-*co*-3HHx)/nHA showed a better matching of mechanical and thermomechanical performance than metal alloys to replace natural bone. While natural bone has a modulus ranging from about 8–25 GPa, the herein-prepared injection-molded parts showed E_t values from approximately 1 up to 1.7 GPa and ε_b values ranging from 6.5 to 19.4%. The incorporation of up to 10 wt % of nHA also improved slightly the thermal stability of the P(3HB-*co*-3HHx) parts and their thermal stability was considered to be high enough for bone tissue engineering, taking into account that nonthermal sterilization methods would be required. These balanced properties in terms of strength and ductility offer the biomedical industry a material that can accomplish different applications in bone reconstruction, for which high-stress materials are not needed, such as bone screws and small orthopedic plates or rods. Future works will explore the potential use of the P(3HB-*co*-3HHx)/nHA composites as drug delivery systems.

Supplementary Materials: The following tables are available online at http://www.mdpi.com/2073-4360/12/6/1389/s1, Table S1: Level of significance (*p*) values for Table 1; Table S2: Level of significance (*p*) values for Table 2; Table S3: Level of significance (*p*) values for Table 3; Table S4: Level of significance (*p*) values for Table 4.

Author Contributions: Conceptualization, S.T.-G. and L.Q.-C.; methodology, J.A.C. and J.I.-M.; validation, J.A.C., J.I.-M. and L.Q.-C.; formal analysis, L.Q.-C. and J.I.-M.; investigation, J.A.C., T.B., L.Q.-C. and J.I.-M.; data curation, L.Q.-C.; writing—original draft preparation, L.Q.-C. and T.B.; writing—review and editing, S.T.-G., J.I.-M., J.A.C.; supervision, S.T.-G., J.A.C. and T.B.; project administration, T.B. All authors have read and agreed to the published version of the manuscript.

Funding: This research work was funded by the Spanish Ministry of Science and Innovation (MICI) project numbers RTI2018-097249-B-C21 and MAT2017-84909-C2-2-R and the POLISABIO program with grant number 2019-A02.

Acknowledgments: L.Q.-C. wants to thank GVA for his FPI grant (ACIF/2016/182) and the Spanish Ministry of Education, Culture, and Sports (MECD) for his FPU grant (FPU15/03812). S.T.-G. acknowledges MICI for his Juan de la Cierva–Incorporación contract (IJCI-2016-29675). J.I.-M. wants to thank Universitat Politècnica de València for his FPI grant (PAID-2019- SP20190011). Microscopy services of the Universitat Politècnica de València (UPV) are acknowledged for their help in collecting and analyzing the FESEM images. Authors also thank Ercros S.A. for kindly supplying ErcrosBio® PH110.

Conflicts of Interest: The authors declare no conflict of interest.

References

1. Ghiasi, M.S.; Chen, J.; Vaziri, A.; Rodriguez, E.K.; Nazarian, A. Bone fracture healing in mechanobiological modeling: A review of principles and methods. *Bone Rep.* **2017**, *6*, 87–100. [CrossRef] [PubMed]
2. Pankaj, P.; Xie, S. The risk of loosening of extramedullary fracture fixation devices. *Injury* **2019**, *50*, S66–S72. [CrossRef] [PubMed]
3. Mugnai, R.; Tarallo, L.; Capra, F.; Catani, F. Biomechanical comparison between stainless steel, titanium and carbon-fiber reinforced polyetheretherketone volar locking plates for distal radius fractures. *Orthop. Traumatol. Surg. Res.* **2018**, *104*, 877–882. [CrossRef]
4. Shayesteh Moghaddam, N.; Jahadakbar, A.; Amerinatanzi, A.; Skoracki, R.; Miller, M.; Dean, D.; Elahinia, M. Fixation release and the bone bandaid: A new bone fixation device paradigm. *Bioengineering* **2017**, *4*, 5. [CrossRef]
5. Heimbach, B.; Tonyali, B.; Zhang, D.; Wei, M. High performance resorbable composites for load-bearing bone fixation devices. *J. Mech. Behav. Biomed. Mater.* **2018**, *81*, 1–9. [CrossRef]
6. Sun, B.; Tian, H.-Y.; Zhang, C.-X.; An, G. Preparation of biomimetic-bone materials and their application to the removal of heavy metals. *AIChE J.* **2013**, *59*, 229–240. [CrossRef]
7. Gotman, I.; Swain, S.K.; Sharipova, A.; Gutmanas, E.Y. Bioresorbable Ca-phosphate-polymer/metal and Fe-Ag nanocomposites for macro-porous scaffolds with tunable degradation and drug release. *AIP Conf. Proc.* **2016**, *1783*, 020062.
8. Perale, G.; Hilborn, J. *Bioresorbable Polymers for Biomedical Applications: From Fundamentals to Translational Medicine*; Woodhead Publishing: Sawston, UK, 2017.
9. Pertici, G. 1—Introduction to bioresorbable polymers for biomedical applications. In *Bioresorbable Polymers for Biomedical Applications*; Perale, G., Hilborn, J., Eds.; Woodhead Publishing: Sawston, UK, 2017; pp. 3–29.
10. Saito, T.; Tomita, K.; Juni, K.; Ooba, K. In vivo and in vitro degradation of poly(3-hydroxybutyrate) in pat. *Biomaterials* **1991**, *12*, 309–312. [CrossRef]
11. Williams, S.F.; Martin, D.P.; Horowitz, D.M.; Peoples, O.P. PHA applications: Addressing the price performance issue I. Tissue engineering. *Int. J. Biol. Macromol.* **1999**, *25*, 111–121. [CrossRef]
12. Singh, A.K.; Mallick, N. Biological system as reactor for the production of biodegradable thermoplastics, polyhydroxyalkanoates. In *Industrial Biotechnology: Sustainable Production and Bioresource Utilization*; CRC Press: Boca Raton, FL, USA, 2016; pp. 281–323.
13. Rodriguez-Contreras, A. Recent advances in the use of polyhydroyalkanoates in biomedicine. *Bioengineering* **2019**, *6*, 82. [CrossRef] [PubMed]
14. Shrivastav, A.; Kim, H.Y.; Kim, Y.R. Advances in the applications of polyhydroxyalkanoate nanoparticles for novel drug delivery system. *BioMed Res. Int.* **2013**, *2013*, 581684. [CrossRef]
15. Zinn, M.; Witholt, B.; Egli, T. Occurrence, synthesis and medical application of bacterial polyhydroxyalkanoate. *Adv. Drug Deliv. Rev.* **2001**, *53*, 5–21. [CrossRef]
16. Shishatskaya, E.I.; Volova, T.G.; Gitelson, I.I. In vivo toxicological evaluation of polyhydroxyalkanoates. *Dokl. Biol. Sci.* **2002**, *383*, 109–111. [CrossRef] [PubMed]
17. Liu, M.-H.; Chen, Y., Jr.; Lee, C.-Y. Characterization of medium-chain-length polyhydroxyalkanoate biosynthesis by Pseudomonas mosselii TO7 using crude glycerol. *Biosci. Biotechnol. Biochem.* **2018**, *82*, 532–539. [CrossRef] [PubMed]
18. Yoshie, N.; Saito, M.; Inoue, Y. Effect of chemical compositional distribution on solid-state structures and properties of poly(3-hydroxybutyrate-co-3-hydroxyvalerate). *Polymer* **2004**, *45*, 1903–1911. [CrossRef]
19. Ozdil, D.; Wimpenny, I.; Aydin, H.M.; Yang, Y. 13—Biocompatibility of biodegradable medical polymers. In *Science and Principles of Biodegradable and Bioresorbable Medical Polymers*; Zhang, X., Ed.; Woodhead Publishing: Sawston, UK, 2017; pp. 379–414.
20. Li, J.; Yin, F.; Li, D.; Ma, X.; Zhou, J. Mechanical, thermal, and barrier properties of PHBH/cellulose biocomposite films prepared by the solution casting method. *Bioresources* **2019**, *14*, 1219–1228.
21. Baidurah, S.; Murugan, P.; Sen, K.Y.; Furuyama, Y.; Nonome, M.; Sudesh, K.; Ishida, Y. Evaluation of soil burial biodegradation behavior of poly(3-hydroxybutyrate-co-3-hydroxyhexanoate) on the basis of change in copolymer composition monitored by thermally assisted hydrolysis and methylation-gas chromatography. *J. Anal. Appl. Pyrolysis* **2019**, *137*, 146–150. [CrossRef]

22. Misra, S.K.; Valappil, S.P.; Roy, I.; Boccaccini, A.R. Polyhydroxyalkanoate (PHA)/Inorganic Phase Composites for Tissue Engineering Applications. *Biomacromolecules* **2006**, *7*, 2249–2258. [CrossRef] [PubMed]
23. Nagao, Y.; Takasu, A.; Boccaccini, A.R. Anode-Selective Electrophoretic Deposition of a Bioactive Glass/Sulfone-Containing Click Polyester Composite. *Macromolecules* **2012**, *45*, 3326–3334. [CrossRef]
24. Luckarift, H.R.; Sizemore, S.R.; Farrington, K.E.; Roy, J.; Lau, C.; Atanassov, P.B.; Johnson, G.R. Facile Fabrication of Scalable, Hierarchically Structured Polymer/Carbon Architectures for Bioelectrodes. *ACS Appl. Mater. Interfaces* **2012**, *4*, 2082–2087. [CrossRef]
25. Leite, Á.J.; Mano, J.F. Biomedical applications of natural-based polymers combined with bioactive glass nanoparticles. *J. Mater. Chem. B* **2017**, *5*, 4555–4568. [CrossRef] [PubMed]
26. Kawarazaki, I.; Takasu, A. Synthesis of Unsaturated Nonionic Poly(ester-sulfones) via Acyclic Diene Metathesis (ADMET) Polymerization and Anode-Selective Electrophoretic Deposition. *Macromol. Chem. Phys.* **2016**, *217*, 2595–2600. [CrossRef]
27. Kehail, A.A.; Boominathan, V.; Fodor, K.; Chalivendra, V.; Ferreira, T.; Brigham, C.J. In Vivo and In Vitro Degradation Studies for Poly(3-hydroxybutyrate-co-3-hydroxyhexanoate) Biopolymer. *J. Polym. Environ.* **2017**, *25*, 296–307. [CrossRef]
28. GrØNdahl, L.; Jack, K.S. 5—Composite materials for bone repair. In *Biomedical Composites*; Ambrosio, L., Ed.; Woodhead Publishing: Sawston, UK, 2010; pp. 101–126.
29. Wypych, G. *Functional Fillers: Chemical Composition, Morphology, Performance, Applications*; ChemTec Publishing: Scarborough, ON, Canada, 2018.
30. Dorozhkin, S.V. Functionalized calcium orthophosphates (CaPO4) and their biomedical applications. *J. Mater. Chem. B* **2019**, *7*, 7471–7489. [CrossRef]
31. Dorozhkin, S.V. Biocomposites and hybrid biomaterials based on calcium orthophosphates. *Biomatter* **2011**, *1*, 3–56. [CrossRef] [PubMed]
32. Ma, B.; Han, J.; Zhang, S.; Liu, F.; Wang, S.; Duan, J.; Sang, Y.; Jiang, H.; Li, D.; Ge, S.; et al. Hydroxyapatite nanobelt/polylactic acid Janus membrane with osteoinduction/barrier dual functions for precise bone defect repair. *Acta Biomater.* **2018**, *71*, 108–117. [CrossRef]
33. Iratwar, D.S.W.; Pisulkar, D.S.; Patil, D.A. Cranioplasty for prosthetic cranial reconstruction of skull defects using pre-frabricated polymethyl methacrylate grafts reinforced by hydroxyapatite particles intraoperatively in rural population of Central India region: A pilot study. *Int. J. Med. Biomed. Stud.* **2019**, *3*. [CrossRef]
34. Tsiapalis, D.; De Pieri, A.; Biggs, M.; Pandit, A.; Zeugolis, D.I. Biomimetic Bioactive Biomaterials: The Next Generation of Implantable Devices. *ACS Biomater. Sci. Eng.* **2017**, *3*, 1172–1174. [CrossRef]
35. Galindo, T.G.P.; Chai, Y.; Tagaya, M. Hydroxyapatite Nanoparticle Coating on Polymer for Constructing Effective Biointeractive Interfaces. *J. Nanomater.* **2019**, *2019*, 6495239. [CrossRef]
36. Agüero, Á.; Garcia-Sanoguera, D.; Lascano, D.; Rojas-Lema, S.; Ivorra-Martinez, J.; Fenollar, O.; Torres-Giner, S. Evaluation of Different Compatibilization Strategies to Improve the Performance of Injection-Molded Green Composite Pieces Made of Polylactide Reinforced with Short Flaxseed Fibers. *Polymers* **2020**, *12*, 821. [CrossRef]
37. Rojas-Lema, S.; Quiles-Carrillo, L.; Garcia-Garcia, D.; Melendez-Rodriguez, B.; Balart, R.; Torres-Giner, S. Tailoring the Properties of Thermo-Compressed Polylactide Films for Food Packaging Applications by Individual and Combined Additions of Lactic Acid Oligomer and Halloysite Nanotubes. *Molecules* **2020**, *25*, 1976. [CrossRef] [PubMed]
38. Mahmood, H.; Pegoretti, A.; Brusa, R.S.; Ceccato, R.; Penasa, L.; Tarter, S.; Checchetto, R. Molecular transport through 3-hydroxybutyrate co-3-hydroxyhexanoate biopolymer films with dispersed graphene oxide nanoparticles: Gas barrier, structural and mechanical properties. *Polym. Test.* **2020**, *81*, 106181. [CrossRef]
39. Ferri, J.; Jordá, J.; Montanes, N.; Fenollar, O.; Balart, R. Manufacturing and characterization of poly (lactic acid) composites with hydroxyapatite. *J. Thermoplast. Compos. Mater.* **2018**, *31*, 865–881. [CrossRef]
40. Giubilini, A.; Sciancalepore, C.; Messori, M.; Bondioli, F. New biocomposite obtained using poly(3-hydroxybutyrate-co-3-hydroxyhexanoate) (PHBH) and microfibrillated cellulose. *J. Appl. Polym. Sci.* **2020**, *137*, 48953. [CrossRef]
41. Rehman, S.; Khan, K.; Mujahid, M.; Nosheen, S. Synthesis of nano-hydroxyapatite and its rapid mediated surface functionalization by silane coupling agent. *Mater. Sci. Eng. C* **2016**, *58*, 675–681. [CrossRef]

42. Medeiros, G.S.; Muñoz, P.A.; de Oliveira, C.F.; da Silva, L.C.; Malhotra, R.; Gonçalves, M.C.; Rosa, V.; Fechine, G.J. Polymer Nanocomposites Based on Poly (ε-caprolactone), Hydroxyapatite and Graphene Oxide. *J. Polym. Environ.* **2020**, *28*, 331–342. [CrossRef]
43. Pielichowska, K.; Król, K.; Majka, T.M. Polyoxymethylene-copolymer based composites with PEG-grafted hydroxyapatite with improved thermal stability. *Thermochim. Acta* **2016**, *633*, 98–107. [CrossRef]
44. Thorvaldsen, T.; Johnsen, B.B.; Olsen, T.; Hansen, F.K. Investigation of theoretical models for the elastic stiffness of nanoparticle-modified polymer composites. *J. Nanomater.* **2015**, *2015*, 281308. [CrossRef]
45. Joseph, R.; McGregor, W.; Martyn, M.; Tanner, K.; Coates, P. Effect of hydroxyapatite morphology/surface area on the rheology and processability of hydroxyapatite filled polyethylene composites. *Biomaterials* **2002**, *23*, 4295–4302. [CrossRef]
46. Kalfus, J.; Jancar, J. Viscoelastic response of nanocomposite poly (vinyl acetate)-hydroxyapatite with varying particle shape—Dynamic strain softening and modulus recovery. *Polym. Compos.* **2007**, *28*, 743–747. [CrossRef]
47. Lin, E.K.; Kolb, R.; Satija, S.K.; Wu, W.-l. Reduced polymer mobility near the polymer/solid interface as measured by neutron reflectivity. *Macromolecules* **1999**, *32*, 3753–3757. [CrossRef]
48. Heise, U.; Osborn, J.; Duwe, F. Hydroxyapatite ceramic as a bone substitute. *Int. Orthop.* **1990**, *14*, 329–338. [CrossRef] [PubMed]
49. Oonishi, H. Orthopaedic applications of hydroxyapatite. *Biomaterials* **1991**, *12*, 171–178. [CrossRef]
50. Yu, H.; Matthew, H.W.; Wooley, P.H.; Yang, S.Y. Effect of porosity and pore size on microstructures and mechanical properties of poly-ε-caprolactone-hydroxyapatite composites. *J. Biomed. Mater. Res. B* **2008**, *86*, 541–547. [CrossRef]
51. Quiles-Carrillo, L.; Montanes, N.; Pineiro, F.; Jorda-Vilaplana, A.; Torres-Giner, S. Ductility and toughness improvement of injection-molded compostable pieces of polylactide by melt blending with poly(ε-caprolactone) and thermoplastic starch. *Materials* **2018**, *11*, 2138. [CrossRef]
52. Lascano, D.; Moraga, G.; Ivorra-Martinez, J.; Rojas-Lema, S.; Torres-Giner, S.; Balart, R.; Boronat, T.; Quiles-Carrillo, L. Development of Injection-Molded Polylactide Pieces with High Toughness by the Addition of Lactic Acid Oligomer and Characterization of Their Shape Memory Behavior. *Polymers* **2019**, *11*, 2099. [CrossRef]
53. Quiles-Carrillo, L.; Montanes, N.; Lagaron, J.; Balart, R.; Torres-Giner, S. In situ compatibilization of biopolymer ternary blends by reactive extrusion with low-functionality epoxy-based styrene–acrylic oligomer. *J. Polym. Environ.* **2019**, *27*, 84–96. [CrossRef]
54. Sheikh, F.A.; Ju, H.W.; Moon, B.M.; Lee, O.J.; Kim, J.H.; Park, H.J.; Kim, D.W.; Kim, D.K.; Jang, J.E.; Khang, G. Hybrid scaffolds based on PLGA and silk for bone tissue engineering. *J. Tissue Eng. Regen. Med.* **2016**, *10*, 209–221. [CrossRef]
55. Ji, Y.; Xu, G.P.; Zhang, Z.P.; Xia, J.J.; Yan, J.L.; Pan, S.H. BMP-2/PLGA delayed-release microspheres composite graft, selection of bone particulate diameters, and prevention of aseptic inflammation for bone tissue engineering. *Ann. Biomed. Eng.* **2010**, *38*, 632–639. [CrossRef]
56. Bender, S.; Chalivendra, V.; Rahbar, N.; El Wakil, S. Mechanical characterization and modeling of graded porous stainless steel specimens for possible bone implant applications. *Int. J. Eng. Sci.* **2012**, *53*, 67–73. [CrossRef]
57. Niinomi, M.; Nakai, M. Titanium-based biomaterials for preventing stress shielding between implant devices and bone. *Int. J. Biomater.* **2011**, *2011*, 836587. [CrossRef] [PubMed]
58. Fouad, H.; Elleithy, R.; Alothman, O.Y. Thermo-mechanical, wear and fracture behavior of high-density polyethylene/hydroxyapatite nano composite for biomedical applications: Effect of accelerated ageing. *J. Mater. Sci. Technol.* **2013**, *29*, 573–581. [CrossRef]
59. Popovich, A.; Sufiiarov, V.; Borisov, E.; Polozov, I.A. Microstructure and mechanical properties of Ti-6Al-4V manufactured by SLM. *Key Eng. Mater.* **2015**, *651–653*, 677–682. [CrossRef]
60. Li, X.; Guo, C.; Liu, X.; Liu, L.; Bai, J.; Xue, F.; Lin, P.; Chu, C. Impact behaviors of poly-lactic acid based biocomposite reinforced with unidirectional high-strength magnesium alloy wires. *Prog. Nat. Sci. Mater. Int.* **2014**, *24*, 472–478. [CrossRef]
61. Senatov, F.S.; Niaza, K.V.; Zadorozhnyy, M.Y.; Maksimkin, A.V.; Kaloshkin, S.D.; Estrin, Y.Z. Mechanical properties and shape memory effect of 3D-printed PLA-based porous scaffolds. *J. Mech. Behav. Biomed. Mater.* **2016**, *57*, 139–148. [CrossRef]

62. Melendez-Rodriguez, B.; Castro-Mayorga, J.L.; Reis, M.A.M.; Sammon, C.; Cabedo, L.; Torres-Giner, S.; Lagaron, J.M. Preparation and Characterization of Electrospun Food Biopackaging Films of Poly(3-hydroxybutyrate-co-3-hydroxyvalerate) Derived from Fruit Pulp Biowaste. *Front. Sustain. Food Syst.* **2018**, *2*. [CrossRef]
63. Zhou, J.; Ma, X.; Li, J.; Zhu, L. Preparation and characterization of a bionanocomposite from poly(3-hydroxybutyrate-co-3-hydroxyhexanoate) and cellulose nanocrystals. *Cellulose* **2019**, *26*, 979–990. [CrossRef]
64. Yu, H.-Y.; Qin, Z.-Y.; Wang, L.-F.; Zhou, Z. Crystallization behavior and hydrophobic properties of biodegradable ethyl cellulose-g-poly(3-hydroxybutyrate-co-3-hydroxyvalerate): The influence of the side-chain length and grafting density. *Carbohydr. Polym.* **2012**, *87*, 2447–2454. [CrossRef]
65. Bikiaris, D. Can nanoparticles really enhance thermal stability of polymers? Part II: An overview on thermal decomposition of polycondensation polymers. *Thermochim. Acta* **2011**, *523*, 25–45. [CrossRef]
66. Pandele, A.M.; Constantinescu, A.; Radu, I.C.; Miculescu, F.; Ioan Voicu, S.; Ciocan, L.T. Synthesis and Characterization of PLA-Micro-structured Hydroxyapatite Composite Films. *Materials (Basel)* **2020**, *13*, 274. [CrossRef]
67. Chen, D.Z.; Tang, C.Y.; Chan, K.C.; Tsui, C.P.; Yu, P.H.F.; Leung, M.C.P.; Uskokovic, P.S. Dynamic mechanical properties and in vitro bioactivity of PHBHV/HA nanocomposite. *Compos. Sci. Technol.* **2007**, *67*, 1617–1626. [CrossRef]
68. Tipnis, N.P.; Burgess, D.J. Sterilization of implantable polymer-based medical devices: A review. *Int. J. Pharm.* **2018**, *544*, 455–460. [CrossRef] [PubMed]
69. Valentini, F.; Dorigato, A.; Rigotti, D.; Pegoretti, A. Polyhydroxyalkanoates/Fibrillated Nanocellulose Composites for Additive Manufacturing. *J. Polym. Environ.* **2019**, *27*, 1333–1341. [CrossRef]
70. Torres-Giner, S.; Montanes, N.; Fenollar, O.; García-Sanoguera, D.; Balart, R. Development and optimization of renewable vinyl plastisol/wood flour composites exposed to ultraviolet radiation. *Mater. Des.* **2016**, *108*, 648–658. [CrossRef]
71. Quiles-Carrillo, L.; Boronat, T.; Montanes, N.; Balart, R.; Torres-Giner, S. Injection-molded parts of fully bio-based polyamide 1010 strengthened with waste derived slate fibers pretreated with glycidyl- and amino-silane coupling agents. *Polym. Test.* **2019**, *77*. [CrossRef]
72. Mohan, L.; Durgalakshmi, D.; Geetha, M.; Narayanan, T.S.; Asokamani, R. Electrophoretic deposition of nanocomposite (HAp+ TiO$_2$) on titanium alloy for biomedical applicationsi. *Ceram. Int.* **2012**, *38*, 3435–3443. [CrossRef]
73. Ferri, J.; Motoc, D.L.; Bou, S.F.; Balart, R. Thermal expansivity and degradation properties of PLA/HA and PLA/βTCP in vitro conditioned composites. *J. Therm. Anal. Calorim.* **2019**, *138*, 2691–2702. [CrossRef]
74. Zuo, G.; Wei, X.; Sun, H.; Liu, S.; Zong, P.; Zeng, X.; Shen, Y. Morphology controlled synthesis of nano-hydroxyapatite using polyethylene glycol as a template. *J. Alloys Compd.* **2017**, *692*, 693–697. [CrossRef]
75. Türkez, H.; Arslan, M.E.; Sönmez, E.; Açikyildiz, M.; Tatar, A.; Geyikoğlu, F. Synthesis, characterization and cytotoxicity of boron nitride nanoparticles: Emphasis on toxicogenomics. *Cytotechnology* **2019**, *71*, 351–361. [CrossRef]

© 2020 by the authors. Licensee MDPI, Basel, Switzerland. This article is an open access article distributed under the terms and conditions of the Creative Commons Attribution (CC BY) license (http://creativecommons.org/licenses/by/4.0/).

Article

Development and Characterization of Polyester and Acrylate-Based Composites with Hydroxyapatite and Halloysite Nanotubes for Medical Applications

Elena Torres [1], Ivan Dominguez-Candela [2], Sergio Castello-Palacios [3], Anna Vallés-Lluch [3] and Vicent Fombuena [2],*

- [1] Textile Industry Research Association (AITEX), Plaza Emilio Sala 1, 03801 Alcoy, Spain; etorres@aitex.es
- [2] Technological Institute of Materials (ITM), Universitat Politècnica de València (UPV), Plaza Ferrándiz y Carbonell 1, 03801 Alcoy, Spain; ivdocan@epsa.upv.es
- [3] Centre for Biomaterials and Tissue Engineering, Universitat Politècnica de València (UPV), Camí de Vera s/n, 46022 Valencia, Spain; sercaspa@iteam.upv.es (S.C.-P.); avalles@ter.upv.es (A.V.-L.)
- * Correspondence: vifombor@upv.es

Received: 14 July 2020; Accepted: 28 July 2020; Published: 29 July 2020

Abstract: We aimed to study the distribution of hydroxyapatite (HA) and halloysite nanotubes (HNTs) as fillers and their influence on the hydrophobic character of conventional polymers used in the biomedical field. The hydrophobic polyester poly (ε-caprolactone) (PCL) was blended with its more hydrophilic counterpart poly (lactic acid) (PLA) and the hydrophilic acrylate poly (2-hydroxyethyl methacrylate) (PHEMA) was analogously compared to poly (ethyl methacrylate) (PEMA) and its copolymer. The addition of HA and HNTs clearly improve surface wettability in neat samples (PCL and PHEMA), but not that of the corresponding binary blends. Energy-dispersive X-ray spectroscopy mapping analyses show a homogenous distribution of HA with appropriate Ca/P ratios between 1.3 and 2, even on samples that were incubated for seven days in simulated body fluid, with the exception of PHEMA, which is excessively hydrophilic to promote the deposition of salts on its surface. HNTs promote large aggregates on more hydrophilic polymers. The degradation process of the biodegradable polyester PCL blended with PLA, and the addition of HA and HNTs, provide hydrophilic units and decrease the overall crystallinity of PCL. Consequently, after 12 weeks of incubation in phosphate buffered saline the mass loss increases up to 48% and mechanical properties decrease above 60% compared with the PCL/PLA blend.

Keywords: biomedical polymers; hydroxyapatite; halloysite; mechanical properties

1. Introduction

Tissue engineering has been exploring new methods to replace missing human tissues through biomaterials-based scaffolds, usually engineered to drive cell growth and provide shape to the creation of the new tissue. However, we should consider alternatives when selecting materials for bone fracture remodeling since conventional materials used for these applications include metallic prosthesis, bone grafts, or polymers. Currently, biopolymers are being intensively studied to replace both metal prostheses and autologous bone grafts because metal prostheses induce poor bone regeneration with formation of fragile porous bone [1] and, although autologous bone grafts induce the growth of strong bone, donor bone is needed, requiring additional chirurgical interventions, eventually causing infections [2]. Thus, polymers having easy processability to obtain desired geometries and special functionalities to accelerate bone growth will be the best option to treat bone fracture remodeling. Accordingly, biopolymers used as scaffolds for tissue engineering applications need to overcome two significant challenges. First, the biodegradation process should be controlled with non-toxic degradation by-products eliminated through natural pathways. Secondly, the material should maintain its structural and mechanical properties to avoid malformation of the new regenerated bone while healing [3].

Among all the biopolymers used for tissue engineering, bioabsorbable aliphatic polyesters are the dominant scaffolding materials because of their biodegradability properties. Biodegradable aliphatic polyesters, containing the ester functional group in their main chain, undergo hydrolytic cleavage generating oligomers, which will be subsequently assimilated into the surrounding environment. Poly (ε-caprolactone) (PCL), poly (lactic acid) (PLA) and poly (hydroxybutyrate) (PHB), approved by the U.S. Food and Drug Administration, are the most studied polyesters due to their easy processability and their tunability regarding crystallinity, thermal transition and mechanical strength properties [4,5]. The degradation rate and mechanical properties of biopolymers will be affected by the hydrophobicity, crystallinity and acidity of the selected polymer.

The hydrophobic biopolymer PCL is extensively used in drug delivery devices showing excellent biological activity. Accordingly, studies focused on its use as a scaffold and internal fixation system, although its low mechanical properties cannot meet the structural requirements of the host tissue. Consequently, an appropriate addition of fillers or blends could provide an adequate mechanical stiffness to resist in vivo stresses, preventing new tissue deformation [6–8]. As examples, Lowry et al. [9] tested PCL composites as internal fixation devices, observing a higher strength when using a PCL/bone complex compared with bony humerus healed with a stainless-steel implant. These observations are in concordance with the studies developed in the last decade by Rudd and co-workers [10–14].

Introduction of specific bioceramics can also confer new functions, such as higher biological activity. Thereupon, hydroxyapatite (HA) is broadly use as an inexpensive filler in tissue engineering [15–17] because of its osteoconductive properties, low inflammatory response and low toxicity in humans [18,19], based on the mineral phase of the human bone being mainly composed (around 60 wt %) of HA [20]. Therefore, introduction of HA into a polymer induces the formation of an apatite layer with similar characteristics to those of the bone mineral phase [21], inasmuch as HA improves cell attachment [22,23], inducing the differentiation of mesenchymal cells into osteoblasts, which accelerates bone formation [8]. Different authors observed both mechanical and biological improvement of biopolymer matrices with the addition of HA [24,25].

The PCL low stiffness can also be attributed to using halloysite nanotubes (HNTs) [26,27] which are an inexpensive biocompatible clay extensively used in biomedicine for drug delivery due to their tubular shape. HNTs also support cell adhesion, ascribed to HNTs surface nano-roughness, which acts as an anchor frame [28], and the interaction between silanol groups present on the HNTs surface [29] with hydroxyl and amino groups present on proteins.

In a previous study [30], mechanical and thermal properties of PCL were studied by modifying the additive percentage of the bioactive fillers HA and HNTs. Accordingly, the additive threshold was stablished in 7.5 wt % of HNTs and 20 wt % of HA achieving a noticeable improvement in mechanical properties with the simultaneous addition of the two fillers. As a result, the flexural modulus improved up to 112.3% reaching values of 886.8 MPa (standard deviation = 42.1), and Young's modulus increased to 109.3% with its greatest value at 449.6 MPa (standard deviation = 17.12). Knowing that HA promotes the formation of a layer of new bone, and that HA and HNTs alter hydrophobicity behavior, in a second study, [23], biological properties such as cell viability, proliferation and morphology supplied by both fillers were studied and compared on different pairs of polymers with similar chemical nature but different hydrophobicity. Accordingly, the hydrophobic polyester PCL was modified when it blended with poly (lactic acid (PLA) and combined with HA nanoparticles and HNTs. However, the hydrophilic poly (2-hydroxyethyl methacrylate) (PHEMA) was copolymerized as monomer with ethyl methacrylate (EMA) and also combined with HA and HNTs. These polymers, although dissimilar to PCL and PLA in terms of chemical nature and biodegradability, were chosen for comparison purposes because they are used for hard tissue applications. Initially, the in vitro biological development of polymers with different hydrophobicity showed that cells preferably proliferate on moderately hydrophobic surfaces (PCL/PLA). However, over longer culture periods, cell proliferation increased on more hydrophilic materials (P(HEMA-co-EMA)). Inorganic nanoparticles (HA and HNTs) improve cell viability and proliferation compared to the raw materials. We assumed that reduced cell spreading on hydrophobic surfaces at long culture times might occur as a consequence of two effects: protein absorption competition and the steric hindrance effect (solvation).

We acknowledge that the contributions of the previous studies [23,30] need to be accomplished by monitoring the degradation rate of biodegradable polyesters modified with HA and HNT. Biomedical polymers, after implantation, undergo significant changes regarding mechanical properties influenced by their degradation process. Considering that 75% of the human body is composed of water, hydrolytic degradation of aliphatic polyesters is an interesting feature for tissue engineering materials. Bone remodeling implies time-limited applications, which requires the elimination or degradation of the biopolymer after use to restore the surrounding living medium.

For all the above-mentioned reasons, we studied the bioactivity of polyester (PCL, PCL/PLA and PLA) and acrylates (PHEMA, P(HEMA-co-EMA) sets and their HA- and HNT-based nanocomposites, as well as the degradability of the polyester-based set. To determine the correct distribution of the fillers, a study was conducted using SEM-EDS and an evaluation of their wettability by measuring the contact angle. Finally, to demonstrate if the loads introduced in the nanocomposites diffuse to the environment, we evaluated the mechanical properties of the nanocomposites using tensile and flexion tests.

2. Experimental

2.1. Materials

Poly (ε-caprolactone) (PCL), with trade name CAPA 6500, was provided by Solvay Interox (Solvay Interox, Warrington, UK). CAPA 6500 is a high-molecular-weight thermoplastic linear polyester derived from its own lactone monomer. PLA Ingeo™ biopolymer 6201D is a thermoplastic available in pellet form with a glass transition temperature of 55–60 °C and a melting point of 155–170 °C. NatureWorks LLC (Nature Works LLC, Minnetonka, MN, USA). Hydroxyapatite (HA) with chemical formula ($HCa_5O_{13}P_3$), halloysite nanotubes (HNTs) ($Al_2Si_2O_5$ $(OH)_4$ $2H_2O$), ethyl methacrylate (EMA) with 99% purity, and hydroxyl-2-ethyl methacrylate (HEMA), with a minimum of purity of 96%, were supplied by Sigma-Aldrich (Madrid, Spain). Benzoin and ethylene glycol dimethacrylate 98% (EGDMA) were used as ultraviolet initiator and crosslinking agent during the preparation of HEMA/EMA compounds. Both were also supplied by Sigma Aldrich.

2.2. Preparation of the Polymer-Based Hybrids

The first set of materials based on PCL and PLA were received in pellet form and dried prior to their preparation in an air oven at 50 and 60 °C, respectively, to remove humidity. In parallel, HA and HNTs were dried separately in a vacuum oven for 48 h at 200 and 80 °C. The proportions detailed in Table 1 were weighed and pre-mixed in a zipper bag. By a twin screw co-rotating extruder with different temperature profiles, the mixtures were mechanically homogenized. Specifically, for PCL-based compounds, the temperature profile was 65/75/85/90 °C, and for PLA based-compounds, we used temperatures of 170/173/17/180 °C. After the extrusion process, the samples were cooled to room temperature and pelletized. Again, prior to the injection process, the different compounds were dried under the same conditions as mentioned above. The injection was carried out in a Meteor 270/75 injection molding machine (Mateu and Solé, Barcelona, Spain) using as temperature profiles of the extruder: 80/80/85/85/90 °C for PCL compounds and 170/173/175/180 °C for PLA compounds. Next, 13 mm diameter samples were punched out.

Table 1. Composition and coding of PCL, PCL/PLA, PHEMA, and P(HEMA-co-EMA) composites.

Code	Composition (wt %)			
	PCL	PLA	HA	HNTs
PCL	100	-	-	-
PCL_20HA	80	-	20	-
PCL_20HA_7.5HNTs	72.5	-	20	7.5
PCL/PLA	50	50	-	-
PCL/PLA_20HA	40	40	20	-
PCL/PLA_20HA_7.5HNTs	36.25	36.25	20	7.5

Code	Composition (wt %)			
	HEMA	EMA	HA	HNTs
PHEMA	100	-	-	-
PHEMA_20HA	80	-	20	-
PHEMA_20HA_7.5HNTs	72.5	-	20	7.5
P(HEMA-co-EMA)	50	50	-	-
P(HEMA-co-EMA)_20HA	40	40	20	-
P(HEMA-co-EMA)_20HA_7.5HNTs	36.25	36.25	20	7.5

The second set of compounds, based on ethyl methacrylate (EMA) and hydroxyl-2-ethyl methacrylate (HEMA), was obtained by simultaneous polymerization of the monomers, as summarized in Table 1. Using a 1:1 monomer ratio between HEMA and EMA to obtain the copolymer, the mixtures were stirred with 1 wt % benzoin and 0.5 wt % EGDMA. The corresponding ratios of HA and HNTs were added and stirred for 15 min. Each mixture was injected into a glass template for polymerization in an ultraviolet oven for 24 h and subsequently a 24 h post-polymerization process in an oven at 90 °C was required. The samples were immersed in boiling ethanol and cut into 13 mm diameter samples.

2.3. Contact Angle Measurements

The water contact angles (WCAs) of the nanocomposites were measured on the surface of the dry samples in the sessile drop mode. An Easy Drop Standard goniometer model FM140 (110/220 V, 50/60 Hz) supplied by Krüss GmbH (Hamburg, Germany) was used for this purpose. To determine the water contact angle, we used the Drop Shape Analysis SW21 (DSA1) software. A minimum of five replicates of each sample were analyzed, yielding a standard deviation of less than 5%.

2.4. Mechanical Properties

Tensile properties of PCL/PLA blends loaded with HA and HNTs were obtained using a universal test machine (Ibertest ELIB 30, SAE Ibertest, Madrid, Spain) according to ISO 527. Assays were carried out with a 5 kN load cell and a crosshead speed of 10 mm·min^{-1}. Moreover, to determine the Young's modulus more accurately, an axial extensometer IB/MFQ-R2 from Ibertest (Madrid, Spain) coupled to the universal test machine was used. The Young's modulus was calculated in each case from the stress–strain initial slope and averaged from five replicates.

2.5. Hydroxyapatite Nucleation

Hydroxyapatite nucleation was followed on three replicates per sample and time point. First, a simulated body fluid (SBF) solution with an ion concentration close to that of human blood plasma was prepared by the method proposed by Kokubo and coworkers [31,32]. To obtain the SBF, we prepared two solutions. Solution 1 consisted of 1.599 g of NaCl (Scharlau, 99% pure), 0.045 g of KCl (Scharlau 99% pure, Barcelona, Spain), 0.110 g of $CaCl_2·6H_2O$ (Fluka 99% pure, Madrid, Spain) and 0.061 g of $MgCl_2·6H_2O$ (Fluka) in deionized ultrapure water (Scharlau) up to 100 mL. Solution 2 was prepared by dissolving 0.032 g of $Na_2SO_4·10H_2O$ (Fluka), 0.071 g of $NaHCO_3$ (Fluka) and 0.046 g of $K_2HPO_4·3H_2O$ (Aldrich, 99% pure) in water up to 100 mL. Both solutions were buffered at pH 7.4 by adding the necessary amounts of aqueous 1 Mtris-hydroxymethyl aminomethane, $(CH_2OH)_3CNH_2$ (Aldrich), and 1 M hydrochloric acid (HCl, Aldrich, 37% pure). Next, both solutions were mixed to obtain SBF with the following molar ion concentrations: 142 Na^+, 5.0 K^+, 1.5 Mg^{2+}, 2.5 Ca^{2+}, 148.8 Cl^-, 4.2 HCO_3^-, 1.0 HPO_4^{2-} and 0.5 SO_4^{2-} mM. Samples were immersed in individual vials containing 10 mL of SBF solution with hydrazine (NaH_2) to prevent bacterial proliferation. The vials were placed in an incubator at 37 °C and 5% CO_2. A set of samples were withdrawn after 7 and 14 days.

2.6. Cell Seeding

NIH 3T3 fibroblast cells were expanded in the presence of 4.5 g L^{-1} glucose supplemented with 10% fetal bovine serum (Thermo Fisher, Gibco, Waltham, MS, USA) and 1% penicillin/streptomycin (P/S; Thermo Fisher, Gibco) in Dulbecco's modified Eagle medium (DMEM; Thermo Fisher, Gibco) at 37 °C in a 5% CO_2 incubator until confluence. After reaching confluence (3 days), cells were withdrawn from the culture flask. To proceed, 5 mL of versene solution (0.48 mM) formulated in 0.2 g ethyldiaminotetraacetic acid (EDTA) per liter of phosphate buffered saline (PBS) supplied by ThermoFisher (Gibco), were added for 5 min at 37 °C, and then removed. After, to neutralize the versene solution, 10 mL of DMEM was added, and the suspensions were centrifugated at 1000 rpm for 5 min. Then, the cells were resuspended in 1 mL medium, counted, diluted and seeded on the samples at a density of 2×10^4 cells cm^{-2}.

2.7. Morphological Analysis

A ZEISS FESEM ULTRATM 55 scanning electron microscopy (SEM) device was used to analyze the morphology of the HA coatings and the NIH 3T3 fibroblast cells and their layout on the surfaces. The morphology of the HA coatings was studied by SEM and energy-dispersive X-ray spectroscopy (EDS) images obtained to validate the formation of a hydroxyapatite layer and the Ca/P ratio. To this end, the samples were sputter-coated with carbon under vacuum through a BALL-TEC/SCD 005 sputter coater. The mapping spectra were taken at 15 kV of acceleration voltage and 5 mm working distance; a secondary electron detector was used. Silicon was used as optimization standard. The mappings were taken at a magnification of 5000×.

In the study of NIH 3T3 fibroblast cells, arrangements were analyzed after 1 and 14 days of incubation. After each period, the culture medium was removed to rinse the samples in phosphate buffer (PB; Affymetrix, Santa Clara, CF, USA) and samples fixed with 4% paraformaldehyde solution during 30 h at 37 °C. A vacuum system was used to remove the water and to avoid any deformations on cell morphology. For this purpose, samples were rinsed in PBS twice and carefully frozen in liquid nitrogen and transferred to a freezer-dryer for drying.

2.8. Degradation of PCL and PCL/PLA Based Hybrids

Degradation of PCL and PCL/PLA loaded with HA and HNTs was followed in vitro at 37 °C using PBS (0.01 M (NaCl 0.138 M; KCl 0.0027 M) with a pH 7.4, at 25 °C was supplied by Sigma Aldrich). Due to the stability of thermostable compounds based on PHEMA, this study was only carried out on compounds based on PCL. With the aim of accelerating the process, samples were previously immersed in a 2M NaOH solution for 24 h. Three replicates of each composition were immersed in individual tubes with 10 mL of Dulbecco's Phosphate Buffered Saline (DPBS, Sigma Aldrich) (pH 7.4) with screw caps and maintained at 37 °C in an incubator. Each sample was removed after 4, 8 and 12 weeks, rinsed thoroughly with deionized water, and dried in an oven at 35 °C for 12 h.

The weight loss and the mechanical integrity of the materials were evaluated. An electronic balance with a resolution of 0.1 mg was used to determine the mass loss as follows:

$$\% \ Mass \ loss = \frac{Mi - Mf}{Mi} \times 100 \tag{1}$$

where Mi is the initial mass and Mf is the final mass of the dry sample.

3. Results and Discussion

The water contact angle formed in the range of 40°–70° on a polymeric surface is known to influence cell attachment, since chemical surface interactions are a key factor during the bio-adhesion process [33]. Polymers with contact angles in this range, with a different chemical nature, were thus selected for this analysis, and their hydrophilicity was slightly modified by blending or copolymerizing with others of the same family. Therefore, polymer chemical surfaces were modified by blending hydrophilic/hydrophobic polymers and/or filling the polymer matrix with HA or HNT to analyze their role in wettability.

From the results in Figure 1, we can discern that the most hydrophobic sample was PCL with a contact angle of 105°. Blending PCL with the more hydrophilic PLA, the surface wettability improved 25.3%, with a value of 83.7°. Conversely, PHEMA was at the hydrophilic end, with a contact angle of 59.7°, and the copolymerization with the more hydrophobic EMA decreased surface wettability up to 73.7°. The addition of HA and HNTs clearly improved surface wettability on neat samples (PCL and PHEMA), but this effect was scarcely observed in mixed samples (PCL/PLA and P(HEMA-co-EMA)), considering the standard deviation. This effect could be attributed to the intermediate wettability of these samples, with contact angles in the vicinities of 80°, together with their heterogeneous composition at the nanoscale.

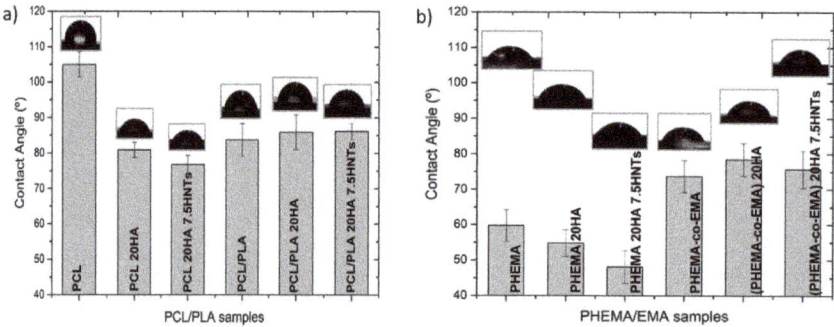

Figure 1. Variation of contact angle in compounds based on (**a**) PCL/PLA and (**b**) PHEMA/EMA.

After determining the wettability of the samples and the influence of the addition of HA and HNTs fillers, a SEM-EDS analysis was conducted to firstly determine if the loads were homogeneously dispersed and, secondly, to assess the formation of a hydroxyapatite layer resulting from the incubation in SBF at 37 °C. In addition, the EDS analysis (EDS spectra not shown) allowed the quantification of the Ca/P ratio and the comparison with that of the stoichiometric HA ($Ca_{10}(PO_4)_6(OH)_2$), Ca/P = 1.67 [34].

Figure 2 shows the images obtained after 7 and 14 days in SBF on the samples based on PCL/PLA. After seven days, the PCL-based hybrids did not efficiently induce apatite growth. Precipitation on PCL and PCL/PLA compounds without needle conformation corresponded to the salt dissolved in SBF medium, usually NaCl. The samples with HA filler provided nucleation sites, and the silanol groups (Si–OH) present in HNTs provide favorable locations for apatite nucleation. We speculated that the electrostatic interaction drives the formation of calcium silicate [35], since comparing pure polymers (especially PCL and PCL/PLA blends) with HA- and HNTs-modified materials, the nucleation efficiency increases with the filler. We already showed that both polar carboxyl groups and hydroxyl groups induce apatite nucleation [36].

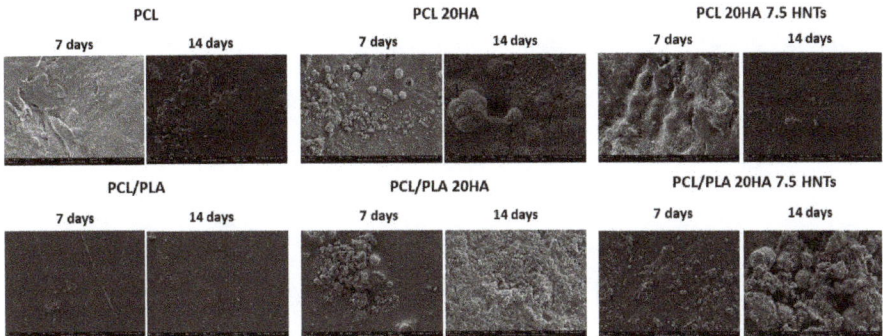

Figure 2. SEM images (5000×) taken for hydroxyapatite nucleation analysis on samples based on PCL/PLA.

However, Figure 3 summarizes SEM images of samples based on PHEMA/EMA. The polymethacrylate-based hybrids induced efficient apatite growth, with the exception of pure PHEMA, on which only scattered precipitated salts were observed on the surface. On the contrary, the P(HEMA-co-EMA) surface showed plenty of precipitates forming large and clear cauliflowers with intricate needle-shaped crystals [36]. As observed, once apatite nucleates on a location, it grows radially outward [37], creating cauliflower or hemispherical structures combined to form a continuous layer. Due to the weak hydrophilicity of P(HEMA-co-EMA), the biological activity is higher than PHEMA, where the number of polar groups available for nucleation per unit volume on the surface is greater. Therefore, the P(HEMA-co-EMA) surface adsorbs Ca^{2+} ions from the SBF solution more efficiently, thereby increasing the concentration of Ca^{2+} ions on the surface, and also forming Ca–P nucleation sites [38]. The first layer of apatite molecules generates the cauliflower aggregates from the secondary nucleation, observed especially in P(HEMA-co-EMA). This process induces a spherical growth perpendicular to the surface structure, which leads to the formation of clusters or grape-like structures [21].

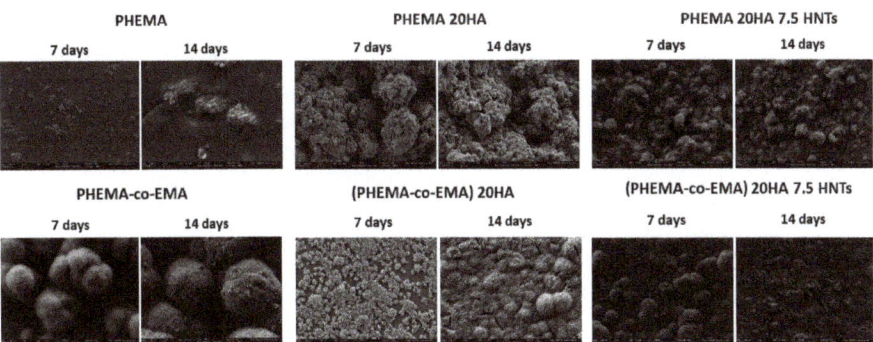

Figure 3. SEM images (5000×) taken for hydroxyapatite nucleation analysis on samples based on PHEMA/EMA.

Figure 4 reveals the assessment of the Ca/P ratio to verify the formation of hydroxyapatite layers. In most determinations, the Ca/P atomic ratio remained in acceptable values between 1.3 and 2, which pointed to the formation of calcium and phosphate deposits resembling physiological apatite structures [34]. Particularly, the highest Ca/P ratios were obtained in the PCL, PHEMA and PHEMA 20HA samples after an incubation period of seven days, all with values higher than 1.8. After 14 days of incubation, the Ca/P ratio tended to the physiological ratio of 1.67 in most samples. However,

PHEMA did not show calcium on the surface after 14 days of incubation. The hydrophilic character of this polymer probably hinders the deposition of salts on its surface and their evolution toward HA. This observation coincides with that of [36], where its closely-related poly (hydroxyethyl acrylate) (PHEA) did not induce an efficient apatite growth, whereas its copolymer with EMA did.

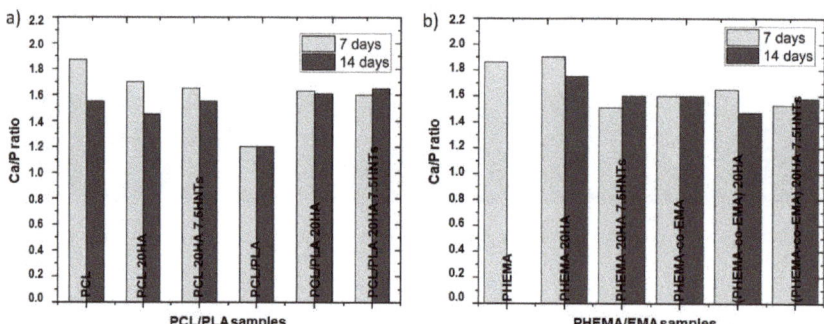

Figure 4. Ca/P ratio after 7 and 14 days of incubation in SBF obtained by EDS on nanocomposites based on: (**a**) PCL and PCL/PLA, (**b**) PHEMA and P(HEMA-co-EMA).

In the SEM morphological images (Figure 5), we see the in vitro biological development of polymers with different hydrophobicity. At initial stages (day 1), cells preferably proliferated and colonized moderately on the hydrophobic surface (PCL/PLA, with contact angle of 83.7°, as summarized in Figure 1). Thus, cells appear round, where interactions occur primarily between them or with the extracellular matrix, thus resulting in a monolayer of cells with few bonding sites with the polymer surface. When longer incubation periods were analyzed (14 days), cell proliferation was favored in more hydrophilic polymers, as is the case of P(HEMA-co-EMA), with a contact angle of 73.7°. In these samples, cells exhibited a flatter morphology, establishing contact with the polymer surface. With the addition of HA and HNTs inorganic fillers, in general terms, we observed an increase in proliferation compared with the raw materials. As different authors have concluded, this improvement can be attributed to the generation of new reactive sites with Ca^{2+} and PO_4^{3-} groups present in HA that bind with negative carboxylate and positive amino groups in proteins, respectively [39–42]. However, due to the presence of silanol groups (Si–OH) located at the surface of HNTs, the formation of hydrogen bonds between HNTs and proteins is allowed [29]. The results showed a greater proliferation at initial stages on moderately hydrophobic polymers (PCL/PLA), whereas over longer culture periods, more hydrophilic polymers P(HEMA-co-EMA) seem to improve cell proliferation, in concordance with the results by Zhou et al. [42]. Using polyvinyl alcohol (PVA), a highly hydrophilic polymer, the authors concluded that highly polar OH groups on neat PVA films might account for the delayed attachment of bone cells. Thus, we can possibly assume that reduced cell spreading on hydrophobic surfaces at long culture times might occur as a consequence of the protein absorption competition and the steric hindrance effect (solvation).

Regarding the mechanical properties, PLA is a hard polymer with good mechanical properties, its brittleness being its main disadvantage. However, PCL is a very soft polymer with a slow degradation rate. After mixing PLA and PCL, the synergy results in retaining the advantages of each polymer. Compared to pure PLA, the mixture has greater flexibility, hydrophobicity and crystallinity, which translate into a slower degradation rate (several months to years) [43]. To study if the PCL crosslinking factor (crystallinity) limits the mobility of the chains and, therefore, the degradation rate, an analysis of the mass loss and mechanical properties after 4, 8 and 12-weeks of incubation in PBS was conducted. The PCL crystallinity tends to reduce the chain mobility and thus its hydrolysis by means of the hindered access of enzymes to the polymer matrix [23].

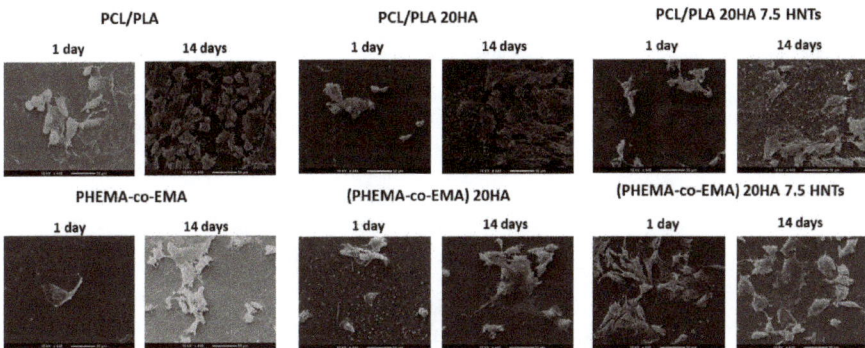

Figure 5. SEM images (400×) taken for cell proliferation at 1 and 14 days on samples based on PCL/PLA and PHEMA/EMA.

From the results shown in Figure 6, the PCL-based material did not show significant mass loss after degradation in PBS at 37 °C for 12 weeks. Although the mass loss of all PCL-based samples gradually increased, compared with samples containing inorganic fillers, the mass loss of pure PCL increased rapidly during the 12-week degradation process. The lower mass loss was observed in samples with HA because HA contains hydroxyl groups, which can neutralize the medium by reacting with degraded acid by-products, thereby reducing the effect of acid catalysis on PCL hydrolysis [44]. Conversely, PCL/PLA blends showed a maximum mass loss rate of the samples when using the two fillers, 41% after 12 weeks. Blending PCL with PLA provides hydrophilic units and reduces the overall crystallinity of PCL, thereby improving the accessibility of water molecules and ester bonds, and increasing the hydrolysis rate [45]. The addition of fillers provides hydrophobicity, and the nano-roughness of the surface promotes interaction with water molecules and thus causes hydrolytic cracking. Therefore, the evolution of mass loss is related to the mechanical properties, which were followed through the analysis of their Young's moduli, gathered in Figure 7.

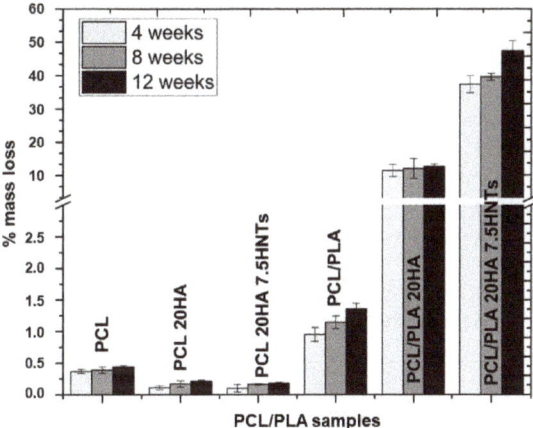

Figure 6. Results of degradation: mass loss (%) after 4, 8 and 12 weeks for PCL-based materials.

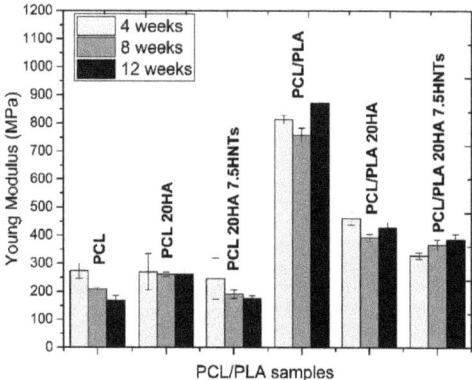

Figure 7. Results of Young modulus following 4, 8 and 12 weeks of degradation for PCL/PLA-based materials.

The Young's modulus of the PCL samples was about 275 MPa. We observed that as the incubation time and the percentage of mass loss increased, the mechanical properties gradually decreased. The neat PCL samples showed a sharp decrease in Young's modulus, but those with HA retained their Young's modulus due to the filler reinforcement. However, the samples with HA and HNT showed a gradual decrease in mechanical properties over time, since the additional threshold of the nanoloads could be exceeded as the polymer degraded, and the agglomerates could act as weak points and failure initiation points. Blending PCL with PLA resulted in the Young's modulus increasing to 800 MPa. In PCL/PLA blends with the presence of inorganic fillers, the faster degradation rate, associated with more hydrophilic properties and more amorphous phases, led to a faster decline in mechanical properties. More precisely, the PCL/PLA 20HA 7.5 HNTs blend provided a Young's modulus 60% lower than that of the PCL/PLA blend.

4. Conclusions

Studying the influence of the HA and HNTs fillers on the wettability of the selected polyesters and acrylates, we measured water contact angles, which corroborated that PCL is the most hydrophobic sample, whereas PHEMA is hydrophilic. The addition of HA and HNTs clearly improved the surface wettability of neat samples, but not of the binary samples (PCL/PLA and P(HEMA-co-EMA)).

Secondly, distribution of the fillers into the polymer matrices was studied through EDS mapping. A homogeneous distribution of HA was observed on all polymers, with the exception of PHEMA. Nonetheless, the presence of HNTs yields large aggregates on more hydrophilic polymers, derived from the hydrophobic character of the nanotubes, resulting in a good dispersion in non-polar polymers and creating agglomerations in hydrophilic polymers due to the lower interfacial adhesion. Considering HA nucleation on polymer surfaces, hydrophobic polymers did not show an efficient induction of apatite growth. Conversely, large hydroxyapatite cauliflower-shaped crystals formed on moderately hydrophilic surfaces. We discovered that PHEMA is excessively hydrophilic, so promotes HA nucleation on its surface.

Finally, the evaluation of the degradation rate of the biodegradable PCL-based samples demonstrated that both the blending PCL with PLA and the addition of HA and HNTs provide hydrophilic units, as well as decrease the overall crystallinity of PCL, thereby improving the accessibility of water molecules to ester linkages and thus the hydrolytic cleavage. Consequently, the faster degradation rate and reduced mass lead to a faster drop of mechanical properties.

Author Contributions: Conceptualization, I.D.-C. and S.C.-P.; Data curation, E.T. and I.D.-C.; Formal analysis, E.T. and S.C.-P.; Investigation, E.T. and I.D.-C.; Methodology, S.C.-P. and A.V.-L.; Project administration, A.V.-L. and V.F.;

Resources, A.V.-L.; Supervision, A.V.-L.; Visualization, V.F.; Writing—original draft, E.T. and V.F.; Writing—review & editing, V.F. All authors have read and agreed to the published version of the manuscript.

Funding: This research received no external funding.

Acknowledgments: I. Dominguez-Candela thanks the Universitat Politècnica de València for the financial support through an FPI-UPV grant (PAID-01-19).

Conflicts of Interest: The authors declare no conflict of interest.

References

1. Noyama, Y.; Miura, T.; Ishimoto, T.; Itaya, T.; Niinomi, M.; Nakano, T. Bone Loss and Reduced Bone Quality of the Human Femur after Total Hip Arthroplasty under Stress-Shielding Effects by Titanium-Based Implant. *Mater. Trans.* **2012**, *53*, 565–570. [CrossRef]
2. Temple, J.P.; Hutton, D.L.; Hung, B.P.; Huri, P.Y.; Cook, C.A.; Kondragunta, R.; Jia, X.; Grayson, W.L. Engineering anatomically shaped vascularized bone grafts with hASCs and 3D-printed PCL scaffolds. *J. Biomed. Mater. Res.* **2014**, *102*, 4317–4325. [CrossRef] [PubMed]
3. Lee, K.H.; Kim, H.Y.; Khil, M.S.; Ra, Y.M.; Lee, D.R. Characterization of nano-structured poly (epsilon-caprolactone) nonwoven mats via electrospinning. *Polymer* **2003**, *44*, 1287–1294. [CrossRef]
4. Li, X.; Cui, R.; Sun, L.; Aifantis, K.E.; Fan, Y.; Feng, Q.; Cui, F.; Watari, F. 3D-Printed biopolymers for tissue engineering application. *Int. J. Polym. Sci.* **2014**. [CrossRef]
5. Washington, K.E.; Kularatne, R.N.; Karmegam, V.; Biewer, M.C.; Stefan, M.C. Recent advances in aliphatic polyesters for drug delivery applications. *Wiley Interdiscip. Rev. Nanomed. Nanobiotechnol.* **2017**, *9*. [CrossRef]
6. Venkatesan, J.; Kim, S.K. Nano-Hydroxyapatite composite biomaterials for bone tissue engineering-a review. *J. Biomed. Nanotechnol.* **2014**, *10*, 3124–3140. [CrossRef]
7. Chen, G.Q.; Wu, Q. The application of polyhydroxyalkanoates as tissue engineering materials. *Biomaterials* **2005**, *26*, 6565–6578. [CrossRef]
8. Rezwan, K.; Chen, Q.Z.; Blaker, J.J.; Boccaccini, A.R. Biodegradable and bioactive porous polymer/inorganic composite scaffolds for bone tissue engineering. *Biomaterials* **2006**, *27*, 3413–3431. [CrossRef]
9. Lowry, K.J.; Hamson, K.R.; Bear, L.; Peng, Y.B.; Calaluce, R.; Evans, M.L.; Anglen, J.O.; Allen, W.C. Polycaprolactone/glass bioabsorbable implant in a rabbit humerus fracture model. *J. Biomed. Mater. Res.* **1997**, *36*, 536–541. [CrossRef]
10. Corden, T.J.; Jones, I.A.; Rudd, C.D.; Christian, P.; Downes, S.; McDougall, K.E. Physical and biocompatibility properties of poly-epsilon-caprolactone produced using in situ polymerisation: A novel manufacturing technique for long-fibre composite materials. *Biomaterials* **2000**, *21*, 713–724. [CrossRef]
11. Onal, L.; Cozien-Cazuc, S.; Jones, I.A.; Rudd, C.D. Water absorption properties of phosphate glass fiber-reinforced poly-epsilon-caprolactone composites for craniofacial bone repair. *J. Appl. Polym. Sci.* **2008**, *107*, 3750–3755. [CrossRef]
12. Ahmed, I.; Parsons, A.J.; Palmer, G.; Knowles, J.C.; Walkers, G.S.; Rudd, C.D. Weight loss, ion release and initial mechanical properties of a binary calcium phosphate glass fibre/PCL composite. *Acta Biomater.* **2008**, *4*, 1307–1314. [CrossRef]
13. Gough, J.E.; Christian, P.; Scotchford, C.A.; Rudd, C.D.; Jones, I.A. Synthesis, degradation, and in vitro cell responses of sodium phosphate glasses for craniofacial bone repair. *J. Biomed. Mater. Res.* **2002**, *59*, 481–489. [CrossRef] [PubMed]
14. Gough, J.E.; Christian, P.; Unsworth, J.; Evans, M.P.; Scotchford, C.A.; Jones, I.A. Controlled degradation and macrophage responses of a fluoride-treated polycaprolactone. *J. Biomed. Mater. Res.* **2004**, *69*, 17–25. [CrossRef] [PubMed]
15. Choi, W.-Y.; Kim, H.-E.; Koh, Y.-H. Production, mechanical properties and in vitro biocompatibility of highly aligned porous poly (epsilon-caprolactone) (PCL)/hydroxyapatite (HA) scaffolds. *J. Porous Mater.* **2013**, *20*, 701–708. [CrossRef]
16. Yeo, M.G.; Kim, G.H. Preparation and characterization of 3D composite scaffolds based on rapid-prototyped PCL/beta-TCP struts and electrospun pcl coated with collagen and ha for bone regeneration. *Chem. Mater.* **2012**, *24*, 903–913. [CrossRef]

17. Salerno, A.; Zeppetelli, S.; Di Maio, E.; Iannace, S.; Netti, P.A. Design of bimodal pcl and pcl-ha nanocomposite scaffolds by two step depressurization during solid-state supercritical CO2 foaming. *Macromol. Rapid Commun.* **2011**, *32*, 1150–1156. [CrossRef]
18. Jackson, I.T.; Yavuzer, R. Hydroxyapatite cement: An alternative for craniofacial skeletal contour refinements. *Br. J. Plast. Surg.* **2000**, *53*, 24–29. [CrossRef]
19. Miller, L.; Guerra, A.B.; Bidros, R.S.; Trahan, C.; Baratta, R.; Metzinger, S.E. A comparison of resistance to fracture among four commercially available forms of hydroxyapatite cement. *Ann. Plast. Surg.* **2005**, *55*, 87–92. [CrossRef]
20. Lawson, E.E.; Barry, B.W.; Williams, A.C.; Edwards, H.G.M. Biomedical applications of Raman spectroscopy. *J. Raman Spectrosc.* **1997**, *28*, 111–117. [CrossRef]
21. Loty, C.; Sautier, J.M.; Boulekbache, H.; Kokubo, T.; Kim, H.M.; Forest, N. In vitro bone formation on a bone-like apatite layer prepared by a biomimetic process on a bioactive glass-ceramic. *J. Biomed. Mater. Res.* **2000**, *49*, 423–434. [CrossRef]
22. Roach, P.; Eglin, D.; Rohde, K.; Perry, C.C. Modern biomaterials: A review-bulk properties and implications of surface modifications. *J. Mater. Sci. Mater. Med.* **2007**, *18*, 1263–1277. [CrossRef] [PubMed]
23. Torres, E.; Valles-Lluch, A.; Fombuena, V.; Napiwocki, B.; Lih-Sheng, T. Influence of the hydrophobic-hydrophilic nature of biomedical polymers and nanocomposites on in vitro biological development. *Macromol. Mater. Eng.* **2017**, *302*. [CrossRef]
24. Chen, B.Q.; Sun, K. Mechanical and dynamic viscoelastic properties of hydroxyapatite reinforced poly (epsilon-caprolactone). *Polym. Test.* **2005**, *24*, 978–982. [CrossRef]
25. Heo, S.J.; Kim, S.E.; Wei, J.; Hyun, Y.T.; Yun, H.S.; Kim, D.H.; Shin, J.W. Fabrication and characterization of novel nano- and micro-HA/PCL composite scaffolds using a modified rapid prototyping process. *J. Biomed. Mater. Res.* **2009**, *89*, 108–116. [CrossRef]
26. Lee, K.-S.; Chang, Y.-W. Thermal, mechanical, and rheological properties of poly (epsilon-caprolactone)/halloysite nanotube nanocomposites. *J. Appl. Polym. Sci.* **2013**, *128*, 2807–2816. [CrossRef]
27. Liu, M.; Guo, B.; Du, M.; Lei, Y.; Jia, D. Natural inorganic nanotubes reinforced epoxy resin nanocomposites. *J. Polym. Res.* **2008**, *15*, 205–212. [CrossRef]
28. Zhou, W.Y.; Guo, B.; Liu, M.; Liao, R.; Rabie, A.B.M.; Jia, D. Poly (vinyl alcohol)/Halloysite nanotubes bionanocomposite films: Properties and in vitro osteoblasts and fibroblasts response. *J. Biomed. Mater. Res.* **2010**, *93*, 1574–1587. [CrossRef]
29. Xue, W.; Bandyopadhyay, A.; Bose, S. Mesoporous calcium silicate for controlled release of bovine serum albumin protein. *Acta Biomater.* **2009**, *5*, 1686–1696. [CrossRef]
30. Torres, E.; Fombuena, V.; Valles-Lluch, A.; Ellingham, T. Improvement of mechanical and biological properties of Polycaprolactone loaded with Hydroxyapatite and Halloysite nanotubes. *Mater. Sci. Eng. C-Mater. Biol. Appl.* **2017**, *75*, 418–424. [CrossRef]
31. Abe, Y.; Kokubo, T.; Yamamuro, T. Apatite coating on ceramics, metals and polymers utilizing a biological process. *J. Mater. Sci. Mater. Med.* **1990**, *1*, 233–238. [CrossRef]
32. Kokubo, T.; Takadama, H. How useful is SBF in predicting in vivo bone bioactivity? *Biomaterials* **2006**, *27*, 2907–2915. [CrossRef] [PubMed]
33. Xue, L.; Greisler, H.P. Biomaterials in the development and future of vascular grafts. *J. Vasc. Surg.* **2003**, *37*, 472–480. [CrossRef] [PubMed]
34. Kim, H.M.; Kishimoto, K.; Miyaji, F.; Kokubo, T.; Yao, T.; Suetsugu, Y.; Tanaka, J.; Nakamura, T. Composition and structure of the apatite formed on PET substrates in SBF modified with various ionic activity products. *J. Biomed. Mater. Res.* **1999**, *46*, 228–235. [CrossRef]
35. Takadama, H.; Kim, H.M.; Miyaji, F.; Kokubo, T.; Nakamura, T. Mechanism of apatite formation induced by silanol groups - TEM observation. *J. Ceram. Soc. Jpn.* **2000**, *108*, 118–121. [CrossRef]
36. Lluch, A.V.; Ferrer, G.G.; Pradas, M.M. Biomimetic apatite coating on P(EMA-co-HEA)/SiO2 hybrid nanocomposites. *Polymer* **2009**, *50*, 2874–2884. [CrossRef]
37. Hutchens, S.A.; Benson, R.S.; Evans, B.R.; O'Neill, H.M.; Rawn, C.J. Biomimetic synthesis of calcium-deficient hydroxyapatite in a natural hydrogel. *Biomaterials* **2006**, *27*, 4661–4670. [CrossRef]

38. Kim, H.M.; Himeno, T.; Kawashita, M.; Kokubo, T.; Nakamura, T. The mechanism of biomineralization of bone-like apatite on synthetic hydroxyapatite: An in vitro assessment. *J. R. Soc. Interface* **2004**, *1*, 17–22. [CrossRef]
39. Azzopardi, P.V.; O'Young, J.; Lajoie, G.; Karttunen, M.; Goldberg, H.A.; Hunter, G.K. Roles of electrostatics and conformation in protein-crystal interactions. *PLoS ONE* **2010**, *5*. [CrossRef]
40. Hynes, R.O. IIntegrins-versatility, modulation, and signaling in cell-adhesion. *Cell* **1992**, *69*, 11–25. [CrossRef]
41. Wassell, D.T.H.; Hall, R.C.; Embery, G. Adsorption of bovine serum-albumin onto hydroxyapatite. *Biomaterials* **1995**, *16*, 697–702. [CrossRef]
42. Zhou, H.; Wu, T.; Dong, X.; Wang, Q.; Shen, J. Adsorption mechanism of BMP-7 on hydroxyapatite (001) surfaces. *Biochem. Biophys. Res. Commun.* **2007**, *361*, 91–96. [CrossRef] [PubMed]
43. Middleton, J.C.; Tipton, A.J. Synthetic biodegradable polymers as orthopedic devices. *Biomaterials* **2000**, *21*, 2335–2346. [CrossRef]
44. Li, H.Y.; Chen, Y.F.; Xie, Y.S. Photo-crosslinking polymerization to prepare polyanhydride/needle-like hydroxyapatite biodegradable nanocomposite for orthopedic application. *Mater. Lett.* **2003**, *57*, 2848–2854. [CrossRef]
45. Albertsson, A.C.; Varma, I.K. Aliphatic polyesters: Synthesis, properties and applications. *Degrad. Aliphatic Polyest.* **2002**, *157*, 1–40.

© 2020 by the authors. Licensee MDPI, Basel, Switzerland. This article is an open access article distributed under the terms and conditions of the Creative Commons Attribution (CC BY) license (http://creativecommons.org/licenses/by/4.0/).

Article

Effects of Magnesium Oxide (MgO) Shapes on In Vitro and In Vivo Degradation Behaviors of PLA/MgO Composites in Long Term

Yun Zhao [1,2,3], Hui Liang [1], Shiqiang Zhang [1], Shengwei Qu [1], Yue Jiang [1] and Minfang Chen [1,2,3,*]

1. School of Materials Science and Engineering, Tianjin University of Technology, Tianjin 300384, China; yun_zhaotju@163.com (Y.Z.); huiliang2014@126.com (H.L.); 18522141052@163.com (S.Z.); qsw98810@163.com (S.Q.); jiang666yue@163.com (Y.J.)
2. Key Laboratory of Display Materials and Photoelectric Device (Ministry of Education), Tianjin 300384, China
3. National Demonstration Center for Experimental Function Materials Education, Tianjin University of Technology, Tianjin 300384, China
* Correspondence: mfchentj@126.com; Tel.: +86-22-6021-6916

Received: 31 March 2020; Accepted: 28 April 2020; Published: 8 May 2020

Abstract: Biodegradable devices for medical applications should be with an appropriate degradation rate for satisfying the various requirements of bone healing. In this study, composite materials of polylactic acid (PLA)/stearic acid-modified magnesium oxide (MgO) with a 1 wt% were prepared through blending extrusion, and the effects of the MgO shapes on the composites' properties in in vitro and in vivo degradation were investigated. The results showed that the long-term degradation behaviors of the composite samples depended significantly on the filler shape. The degradation of the composites is accelerated by the increase in the water uptake rate of the PLA matrix and the composite containing the MgO nanoparticles was influenced more severely by the enhanced hydrophilicity. Furthermore, the pH value of the phosphate buffer solution (PBS) was obviously regulated by the dissolution of MgO through the neutralization of the acidic product of the PLA degradation. In addition, the improvement of the in vivo degrading process of the composite illustrated that the PLA/MgO materials can effectively regulate the degradation of the PLA matrix as well as raise its bioactivity, indicating the composites for utilization as a biomedical material matching the different requirements for bone-related repair.

Keywords: biopolymers composites; MgO nanoparticles; MgO whiskers; PLA; in vitro degradation; in vivo degradation

1. Introduction

The requirement for orthopedic implants in clinical medicine has been paid more attention by researchers in order to develop new materials to help patients avoiding the pain of secondary surgery, improve the recovery rate and reduce costs. Recently, polylactic acid (PLA) has been given more scientific research in the biomedical field, especially for bone repair [1–4], since it has the advantages of good biocompatibility, processability, biodegradability and bioabsorbability. However, some disadvantages have limited its widening applications [5–8]. For example, the mechanical properties of PLA cannot satisfy the requirements of load-bearing devices for bone fixation [5], and its poor bioactivity would significantly hinder the penetration and absorption of cells on the material surface [8]. In particular, predominant problems in its degradation have occurred. The pH variation in the degradation can accelerate the initial degradation rate of the implant and also increase the risk of adverse tissue reactions, such as the inflammatory response [9], which are not beneficial to the viability of bone cells, biodegradation properties and the requirements of bone formation.

In order to solve these problems, researchers have improved the performance of PLA by using inorganic fillers with a good biological activity and bone-similar ingredients [10–12]. Magnesium oxide (MgO), another kind of inorganic filler, is with good biocompatibility and non-toxic biological activity, and become a research hotspot of polymer modification [13–15]. It can release the Mg^{2+} ions when dissolving, which is favorable for a variety of enzymes conducive to the activation and synthesis of proteins. Moreover, its unique biological activity plays an important role in cell viability and as the antibacterial agent, to improve the survival rate of cells as well as the biological activity of composites [7,16]. Additionally, MgO also perform as an alkaline degradable material with excellent biological characteristics for the tissue engineering of regenerated bone tissue [16]. It is worth mentioning that besides the release of magnesium ions in the dissolving process, MgO, including the nanoparticles and whiskers, can impact on increasing the mechanical properties of the polymer matrix [13,17] and this is very different from the pure Mg particles as a filler in the PLA matrix [18]. It shows that since the Mg materials can produce the hydrogen during the composite degradation, it is easily apt to form the interface defects between the matrix and fillers, and meanwhile, there is no surface modification for the Mg particles prior to the preparation of the composite, which also weakens the interface bonding. In previous studies [19], it is found that the PLA/MgO composite film fabricated by the solution casting method can significantly improve the mechanical properties and biological activities of the PLA matrix, and its alkalinity has a neutralizing effect on regulating the pH of short-term degradation solutions.

It has been reported that the shape of fillers plays a crucial role in the mechanical properties and crystalline behaviors of composites. As to the materials of the PLA matrix, the whiskers usually perform the more enhancing capacity for the mechanical properties of PLA, compared with the nanoparticles [7,19]. Moreover, the difference in filler shape also affects the degrading behavior of the PLA composite during a short-time period because the dissolving rates of fillers are diverse, and this can cause the different variation in the interface bonding [18]. Luo et al. studied the degradation properties of MgO whisker/PLA composites in vitro, focusing on the long-term degradation, and since, their work has proved its better mechanical property. Nonetheless, it is necessary not only to focus on the initial property, but also to understand and investigate the long-term degradability and biocompatibility of such composites, as implants for bone repair, especially the filler shape effects on bone cells and tissues. It is worth noting that the change in the hydrophilicity and crystallization of the polymeric matrix has also been improved and influenced by the addition of fillers as well as its shape. The increasing hydrophilicity can cause the variation in the water intake in materials, and the change in the crystallization is beneficial to the growth of ordered crystal structures by affecting the degradation process of the matrix, such as the hydrolysis process of PLA, as the chain and segment are more easily apt to move in the degradation [20,21]. However, few studies have focused on the effects of the MgO shape on the degrading property of the PLA/MgO composite, and there is also no report on the in vivo degradation process of the PLA/MgO composite which provides the direct performance of its degrading property and biocompatibility to bone tissue.

In this work, MgO nanoparticles and whiskers are modified with stearic acid to obtain chemical binding through their interactions [22], as demonstrated in our previous work [17]. The PLA/MgO composites soaked in a phosphate buffer solution (PBS) under long-term degradation were studied, and particularly, the influence of the filler shape on the degradation behavior of the composites was analyzed. The in vivo degradation behavior of the composites was also evaluated.

2. Materials and Methods

2.1. Materials

PLA 2002D (NatureWorks, USA) was dried at 40 °C for 72 h before use and all other agents used were of analytical grade and did not require further treatment.

2.2. Preparations and Modifications of pMgO and wMgO

The fillers of the MgO nanoparticles (pMgO) and MgO whiskers (wMgO) were prepared in the laboratory [17,23]. Briefly, $MgCl_2 \cdot 6H_2O$ (76.24 g, CAS: 7791-18-6) was dissolved in 250 mL deionized water, and 1 mL cetyl trimethyl ammonium bromide (CTAB) (1.0 wt%, CAS: 57-09-0) was added dropwise. $C_2H_2O_4 \cdot 2H_2O$ (23.64 g, CAS: 6153-56-6) was added to this solution and the mixture was allowed to react for 20 min. The suspension was collected and centrifuged at 7000 rpm for 10 min. The pMgO precipitate was dried for 1 h in a vacuum oven (85 °C) and then sintered in a muffle furnace for 5 h at 600 °C.

The 100 mL solution of Na_2CO_3 (0.6 mol/L, CAS: 497-19-8) was added dropwise into an equal volume solution of $MgCl_2$ (0.6 mol/L, CAS: 7791-18-6) and stirred for 20 min. The mixture was aged at room temperature for 10 h and then filtered, washed, and dried at 80 °C for 3–4 h. The precursor was calcined at 750 °C for 4 h with a heating rate of 5 °C/min. Then, the resultant wMgO was obtained.

Mixtures of 0.5 g of MgO (nanoparticles or whiskers) and 50 mL of ethanol were placed in three-necked flasks, which were subjected to ultrasonic treatment in a bath to achieve a full dispersion of the mixtures. The mixtures were then heated to 45 °C under reflux condensation. Stearic acid (0.005 g, CAS: 57-11-4) was dissolved into 20 mL of ethanol, added dropwise into each MgO suspension, and allowed to react for 1 h in reflux condensation conditions. The mixtures were then centrifuged, washed, and dried. The resultant products were denoted as SpMgO or SwMgO. The morphologies of pMgO, wMgO, SpMgO and SwMgO are shown in the Supplementary Materials.

2.3. Preparation of PLA/pMgO and PLA/wMgO Composites

The PLA/pMgO and PLA/wMgO composites were manufactured by a micro twin screw extruder (Wuhan Ruiming Test Equipment, Ltd., China) in advance, with a screw speed of 40 rpm and the temperature profile varied from 165 °C at the feeding zone to 190 °C at the die. The SpMgO/SwMgO was dry-mixed with PLA and the mixtures were dried under a vacuum at 50 °C for 48 h prior to the preparing process for the PLA and composites, then followed by the pelletizing step. Detailed information on the prepared materials is presented in Table 1.

Table 1. Detailed information of the samples.

Sample	SpMgO Content/PLA Content (w/w)	SwMgO Content/PLA Content (w/w)
PLA	0	0
PPLA	1/100	0
WPLA	0	1/100

The extruded pellets were dried at 50 °C for 24 h before extrusion again by the same machine conditions to obtain the composites' wire (φ2 mm, controlled by the extrusion machine head) and then prepared the specimens' rods (φ2 × 10 mm) directly. These extrusion-molded specimens were used for the in vitro degrading experiments.

2.4. In Vitro Degradation in PBS

In vitro degradation of the pure PLA and composite rods were carried out by immersing them in 4 mL of a phosphate buffer solution (PBS) at 37 °C for 3, 4, 5, 6, 8 and 12 months. The specimens of each material were removed from the PBS at the end of these periods. After rinsing with deionized water and removing the surface water using filter paper, the samples were weighed and then dried in a vacuum at 40 °C to a constant weight.

2.5. Characterization

The morphology of the experimental samples was characterized by field-emission scanning electron microscopy (FESEM, JOEL 6700F, Japan, operating at 10 kV).

At the pre-set time points, the soaking specimens were weighed after being cleaned with deionized water and surface-dried, then they were dried to a constant weight in a vacuum at 40 °C and weighed again. The pH value of the PBS was determined by averaging the results of the 3 independent measurements, obtained from the three identical samples for each type of composite. The weight losses were calculated using the following equation:

$$W_{Loss}(\%) = \frac{m_d}{m_0} \times 100\% \tag{1}$$

where W_{Loss} is the average degrading rate, and m_0 and m_d are the initial and final weights (after removing the surface water), respectively.

The water uptake measurements were calculated using the following equation [18]:

$$W_{water}(\%) = \frac{m_w - m_d}{m_d} \times 100\% \tag{2}$$

where $W_{water}\%$ is the wateruptake rate, and m_w, m_0 and m_d are the weight of specimen after conditioning, the final weight (after removing surface water) and the initial weight, respectively.

Gel permeation chromatography (GPC) measurements were taken for the degradation samples in tetrahydrofuran (THF) (analytical purity) at a concentration of 1–2 mg/mL by a Waters 2414 system (Milford, MA) equipped with a Waters Differential Refractometer. THF was eluted at 1.0 mL/min through two Waters Styragel HT columns and a linear column. The internal and column temperatures were kept constant at 35 °C. Calibration curves were obtained based on the standard samples of the mono-dispersed polystyrene.

The thermal analysis of the samples was performed using a differential scanning calorimetry (DSC) instrument (Netzsch Co. Ltd., Freistaat, Germany). The samples, weighing approximately 5–8 mg each, were sealed in an aluminum pan, heated under a nitrogen flow from room temperature to 220 °C at a heating rate of 20 °C/min, isothermally conditioned at 220 °C for 2 min, cooled to 0 °C and then reheated to 220 °C at a heating rate of 10 °C/min. The crystallinity degree (X_c) of the samples was estimated using the following Equation:

$$X_c(\%) = \frac{\Delta H_m - \Delta H_{cc}}{\Delta H_m^0} \times 100\% \tag{3}$$

where ΔH_m (J/g) is the value of fusion, ΔH_{cc} is the cold crystallization enthalpy obtained during the DSC heating process, ΔH_m° is the fusion enthalpy of the completely crystalline PLA, and ϕ is the weight fraction of PLA in the sample. The value of PLA is selected as $\Delta H_m^0 = 93.6$ J/g [24].

2.6. In Vivo Experiment

2.6.1. Animal Models

The in vivo experiments were carried out as described by Reference [24]. Briefly, a total of 10 healthy adult (~1 year) Japanese white rabbits weighing 3 ± 0.2 kg were selected for animal testing, and they were divided in 2 groups, with 5 rabbits in one group, of which 2 rabbits were used as standby samples. One rabbit was implanted with 2 samples, where the left and right legs were implanted with one sample each (ϕ2 × 6 mm). Further, sub-cage feeding was performed for a week and no adverse reactions were found. Rabbits were anesthetized with an intramuscular injection of ketamine (0.2 mL/kg). After that, the hair on the side of the knee in a roughly 5-cm range was shaved. The iodine disinfectant was used to disinfect the knee parts of the rabbit. The anterior lateral patella of the knee was incised to about 4 cm, followed by cutting the skin, lateral support and a joint capsule. A hole with a depth of 1 cm was drilled in the femur and tibial cancellous bone of the knee. The PLA and WPLA rods were implanted into the hole separately; then, postoperative suture, iodophor disinfection and sub-cage feeding were performed. Rabbits were sacrificed with ear veins injected with air. The bone with the implanted rod was removed from the euthanized rabbits after 3, 6, 12 and 18 months, and preserved by a 10% formalin solution. For all the animal experiments, the materials and surgical instruments were radiation-disinfected in the Tianjin Jinpeng far radiation Co., Ltd. after Co60 for 24 h, with a radiation dose of 25 KGy, and the experiments were implemented by Tianjin Hospital (approval No. 2015-11155).

2.6.2. Routine Pathological Examinations

Hard histological biopsies were performed to evaluate the structure variation of the implants under long-term degradation behaviors and the tibial cancellous bone response after surgery. The surgical sites were fixed in a 10% formaldehyde solution, and then the samples were dehydrated in the order of the graded series of alcohols. Following the dehydration and decalcification, the specimens were embedded in paraffin, and the tissue sections were stained with hematoxylin and red staining.

The whole process of experiment is present in Figure 1.

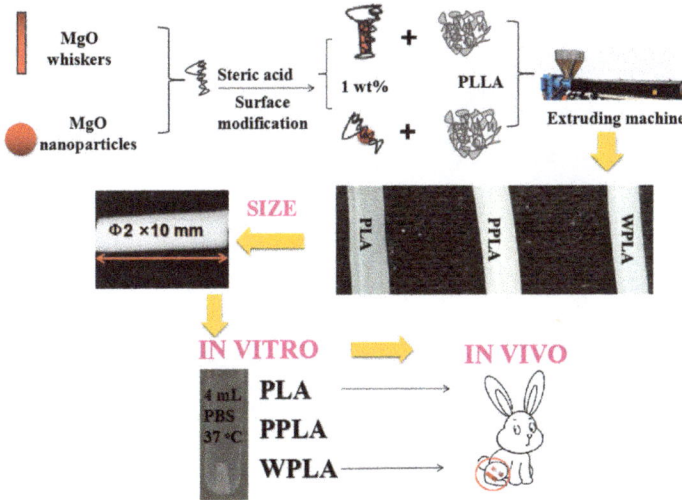

Figure 1. Schematic illustration of the experiment.

3. Results

3.1. Microstructures of In Vitro Degradation of PLA and Composites

Figure 2 shows the surface morphologies and fracture morphologies (brittle fractures treated with liquid nitrogen) of the PBS-soaked samples at the different degradation periods. It is obvious that the surface degradation of all specimens is gradually aggravated with the extension of the immersion time. Specifically, some microcracks on the surface of all the samples in the 6 months appear, and the degrading holes are also shown by the 8 and 12 months in Figure 2a, which are produced by the PLA decomposition [25]. Meanwhile, the higher number of decomposing holes of the PPLA and WPLA composites are displayed and indicate that the decomposition of composite materials is more intensive than that of PLA under the long immersion period, probably due to the presence of MgO affecting the matrix's hydrophilicity. In Figure 2b, the fracture morphologies of the composites also form the greater number of bigger holes of WPLA and PPLA observed in comparison with the contemporary PLA sample and the intensive hydrolysis behavior of the matrix over the 12 months of composition is also displayed, suggesting their accelerating degradation under the long-term immersion. However, it can be observed that the morphologies of the samples in some areas (marked by red arrows) are greatly different between WPLA and the others, particularly in the graphs of 12 months. The former performs brittle-similar fracture behavior, while the latter exhibits noticeable plastification areas, which are probably a result of the degradation product and water [20,21].

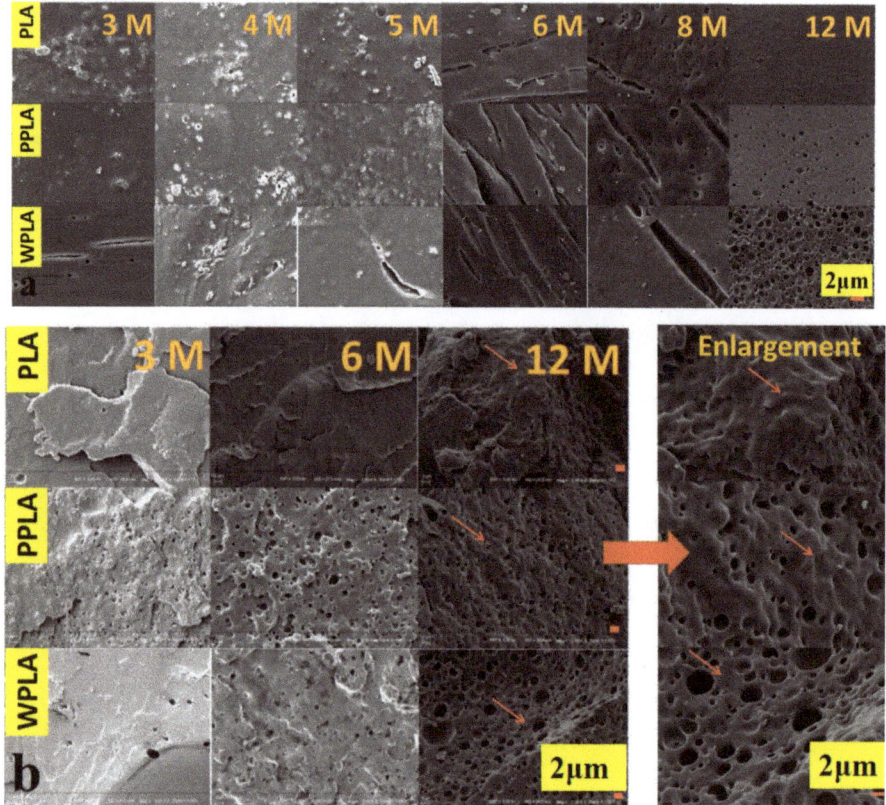

Figure 2. SEM microstructures of polylactic acid (PLA) and composites at different degradation times: (**a**) surface morphologies; (**b**) fracture morphologies.

3.2. Weight Loss, Water Intake and pH Value

The weight changes caused by the sample's degradation in the PBS are shown in Figure 3a. It can be seen that the variation in all samples is similar, namely by decreasing gradually. Obviously, the residual weights of the composites are comparatively less than that of the neat PLA over the 12 months, which is also mainly due to the MgO presence increasing the water uptake and accelerating the matrix decomposition. Initially, from the 4 months, there is an apparent increment in the degrading rate of PLA possibly because of its self-catalytic degradation [26], which accelerates its weight loss. Compared with the control PLA sample, the variation in the composites' weight loss is not greatly remarkable during the whole testing process. However, it is notable that WPLA provides the slightly more residual weight from the 5 months probably due to the hard degradation of its crystalline part leading to much more water, and there is also the increment in the water intake of WPLA over 5 months. Consequently, PPLA has the higher residual weight in comparison with WPLA in the 12 months. This is probably related to the MgO shape affecting the crystal behaviors and water uptake of the PLA matrix [18]. In Figure 3b, the composites exhibited the higher value of water absorption, which is consistent with their weight loss results, and attributed to MgO enhancing the hydrophilicity of the PLA matrix. Moreover, compared with the contemporary WPLA specimens, the samples of PPLA perform much a stronger hydrophilic capability during the experimental period and this is probably caused by the higher quantity of nanoparticles compared with the same weight whiskers. Undoubtedly, the density of the crystal areas of PPLA can also be increased more notably by nanoparticle nucleation [23] but these forming areas are probably smaller due to the inhibiting growth effect between each other. Moreover, there may be much more parts of the imperfect crystal and amorphous region present, which can easily occur at the beginning of the degradation and lead to the acceleration of the degrading process, since metal oxides can effectively accelerate the decomposition of the PLA matrix, especially in a comparatively short-term degradation [27]. As to the variation in the pH value shown in Figure 3c, the pure PLA has a significantly lower value than that of the composites. With the addition of MgO, the pH values of the degrading solution are higher, in spite of decreasing gradually, especially at the final stage. The delay in the pH reduction in the composite solution may be mainly due to the neutralization of the alkaline magnesium ions released by MgO in the degradation medium [28]. Additionally, it is noted that PPLA also performs slightly better in controlling the solution's pH value, although the pH value of the later period was around at 6.4.

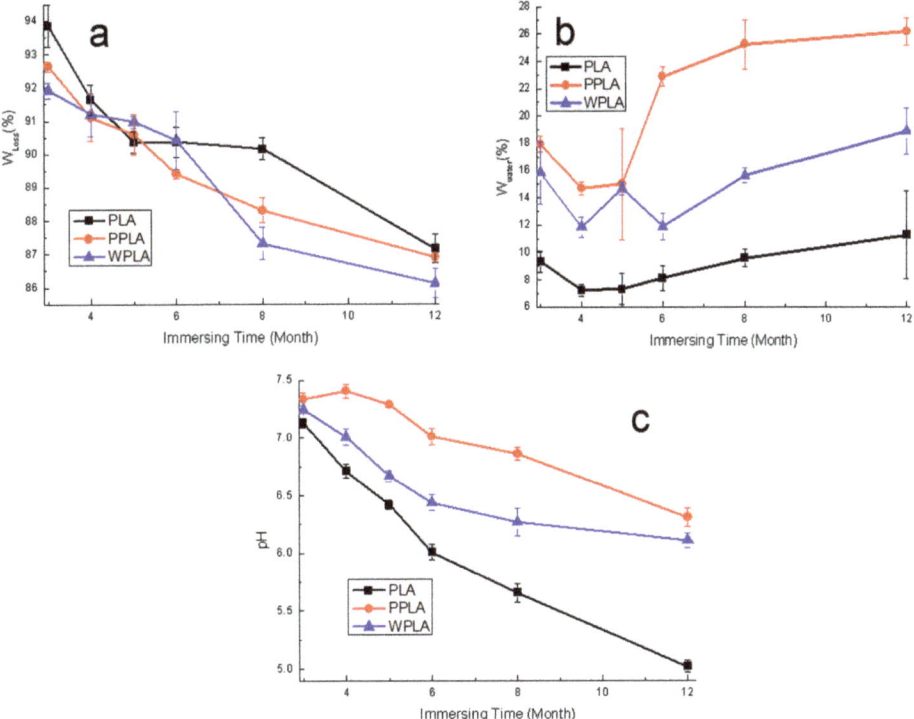

Figure 3. The (**a**) weight loss, (**b**) water intake and (**c**) pH value as a function of the degradation time of the neat PLA and composite immersed in the phosphate buffer solution (PBS).

3.3. GPC and DSC Results

In Table 2, the molecular weight data obtained from the GPC analysis are included. It is observed that the M_w of PLA and the composites decrease as the time is prolonged, as a consequence of the hydrolytic process. Initially, there is no intensively significant difference between the composites and PLA. The M_w of WPLA over 3 months is a little higher and this tendency is maintained until the 12-months of the degradation. As to the variation in PI (dispersity indexes), initially, the neat PLA performs the lower value compared with the composites for the samples of 3 and 4 months, while with the time prolonging, there is an increasing tendency of PLA PI. This change is probably due to the hydrolysis of the matrix and the formation of the degradation product with a lower M_w from the amorphous areas in the PLA sample. It is noticeable that the variation in the M_w of PPLA is with the lowest value from 3 months to 6 months, and this is consistent with the previous study's report that metal oxides can enhance the degradation of the PLA matrix before 3 months [29]. However, with the extension of the immersing time, PPLA has the higher M_w value than that of PLA in 8 and 12 months, indicating the delaying degradation of PPLA. This is probably due to the more crystalline structure formed by the nanoparticles and the slowing degradation of the crystalline part occurring in the polymeric matrix. The MgO, especially the whiskers, is favorable for forming the more crystal regions, resulting in it effectively inhibiting and alleviating the degradation of the specimen [7], in spite of increasing the water uptake and hydrophilicity of the PLA matrix. In addition, there is a prominent difference between the M_w of PPLA and WPLA in the 12 months, and the higher M_w value of WPLA observed is possibly ascribed to the presence of bigger and more complete crystal regions, which lead to a higher crystallinity and is formed by the tighter arrangement of the PLA molecule induced by the whiskers [30]. This also indicates that the large amount of water entering the matrix is apt to accelerate the hydrolyzing of the amorphous material, but it slightly impacts on the crystal area even for the long-term degradation.

Table 2. Gel permeation chromatography (GPC) data corresponding to the samples.

Time/Months	PLA			PPLA			WPLA		
	M_w	M_n	PI	M_w	M_n	PI	M_w	M_n	PI
3	135,777	85,630	1.59	124,969	79,492	1.57	142,721	87,717	1.63
4	122,638	78,164	1.57	119,200	74,087	1.61	128,695	78,720	1.63
5	113,343	70,064	1.62	105,310	65,020	1.62	111,342	75,066	1.48
6	84,728	55,543	1.53	77,485	50,884	1.52	103,873	67,122	1.55
8	51,364	28,036	1.83	64,290	41,053	1.57	68,602	44,645	1.54
12	4017	3365	1.22	9798	16,703	1.70	21,573	13,011	1.66

Weight averaged molecular weight (M_w), number averaged molecular weight (M_n) and dispersity indexes (PI = M_w/M_n) of the specimens as determined by GPC.

The DSC data for the pure PLA and the composites in the experiment are displayed in Figure 4. During the degradation process, the Tg peaks observed for the samples is gradually shifted to the lower value, and the same variation is also observed for Tm. There are also two endothermic peaks for some curves and this is a common phenomenon for PLA degradation due to the recrystallization process of the defective crystals as decomposed products [7]. The whole degrading process starts with an amorphous degradation and then gradually combines with the destruction of the crystalline structure in a long-term degradation, which is also with the similar decline process of the molecular weight proved by the GPC results. It is found that there is only one endothermic peak in the PLA sample in 12 months with the lower value of Tm, and according to the GPC results in the corresponding period, it probably implies the formation of a decomposed product with a low molecular weight during the long-term degradation, which is distinguished from the crystalline region in PPLA and WPLA, as observed in the contemporary comparison of the curves. For the samples in 3 months, the curve of the PPLA specimen has double the endothermic peaks, possibly due to its accelerating degradation caused by the MgO nanoparticles [29] at the former 3-months stage. Moreover, in comparison with the curves of PPLA, the variation in the WPLA DSC line implies the lower degrading rate of WPLA, and this can be ascribed to its high crystallinity by the whiskers, demonstrated by the results of Xc in Table 3. Additionally, there is also no obvious cold crystallization phenomenon of WPLA in 12 months, differing from the PPLA performance, which also verifies the high crystallization of the WPLA composite. It is also greatly remarkable that the whole variation of PPLA in the DSC from 3 months to 12 months is not similar to that of WPLA, and the double endothermic peaks are maintained in all the curves of PPLA. It is probably due to the relatively denser or more compact crystalline network structure in the

PPLA matrix after the short-term degradation, which may be helpful to delay its degradation, in spite of its lower crystallinity.

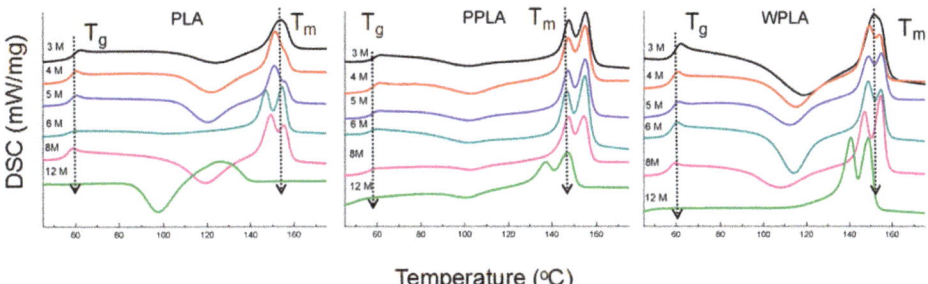

Figure 4. The differential scanning calorimetry (DSC) curves of the secondary heating curves obtained for the degraded samples.

Table 3. DSC data corresponding to the samples.

Time(Months)/Samples		T_{m1} (°C)	T_{m2} (°C)	ΔH_{cc} (J/g)	ΔH_m (J/g)	X_c (%)
3	PLA	153.9	-	−12.8	25.5	40.8
	PPLA	147.7	155.2	−8.37	35.3	46.6
	WPLA	151.7	154.5	−23.6	37.7	65.4
4	PLA	151.5	156.6	−21.0	33.1	57.8
	PPLA	147.6	155.2	−10.9	37.5	51.6
	WPLA	149.4	154.2	−22.1	34.3	60.2
5	PLA	151.4	156.8	−25.0	29.6	58.4
	PPLA	147.3	155.1	−6.3	38.9	48.4
	WPLA	149.1	155.1	−22.1	30.6	56.2
6	PLA	147.2	154.7	−5.0	39.6	47.6
	PPLA	146.8	154.9	−5.9	43.0	52.2
	WPLA	148.8	154.8	−30.3	32.9	67.5
8	PLA	149.4	155.6	−26.5	45.6	77.1
	PPLA	147.7	154.6	−4.8	35.8	43.3
	WPLA	147.1	154.9	−16.8	43.0	63.8
12	PLA	126.9	-	−33.0	24.9	61.9
	PPLA	137.5	147.2	−6.8	28.7	38.0
	WPLA	140.6	148.9	−4.0	43.1	50.3

3.4. Dying of In Vivo Degradation

According to the degradation results in vitro, the in vivo degradation of PLA and WPLA were carried out for the comparatively long-term implanting. More attention was given to the changes in the materials in vivo with the time of 3 months, 6 months, 12 months and 18 months, and the graphs of the histological examinations stained by hematoxylin and red staining are displayed in Figure 5. After 3 months, the edge of the implanted WPLA composite was not flat, while the control PLA was still relatively smooth, as shown. Combined with the results of the water intake and molecular weight of the materials, it indicates that the swelling stage of the decomposing process is more intensive for WPLA, due to its enhanced hydrophilicity. Meanwhile, after the 6-month implantation, it is noteworthy that the implants of PLA and WPLA exhibited a large number of decomposition cracks, and the apparent destruction of the implants was also observed (12 months of Figure 5), but the cracks for both samples are not similar. Specifically, there are much more and parallel cracks displayed in the control PLA in spite of the insignificant swelling, nevertheless the specimen of WPLA seems to swell significantly and break into some block regions by less random cracks. The results illustrate that the water intake rate of WPLA with the MgO whiskers is faster than that of the pure PLA, but the degradation rate of the pristine PLA may be more prominent, suggesting that the degradation process of the matrix in the control and composites were different from each other. Moreover, after the 18 months, the implant of WPLA is basically decomposed, the partial degrading matrix enter into the newly-formed bone trabecula with a tight touch (marked by red arrows in 18 months of Figure 5) and PLA is still in the comparatively complete rod shape with a slight expansion, implying the accelerating decomposition of the swelled WPLA in the final degrading stage. This illustrates that MgO has a significant effect on the degradation process of the PLA matrix. The swelling rate and process are more accelerated and more intensively impacted on the degradation, along with the matrix decomposition. Additionally, from the parts noted by the red lines in the graph, it can be seen that the presence of MgO is favorable for inducing the tight connect between the matrix and bone trabeculae, due to both the better hydrophilicity and biological activity of the significant effect of Mg^{2+} on bone formation and healing [28].

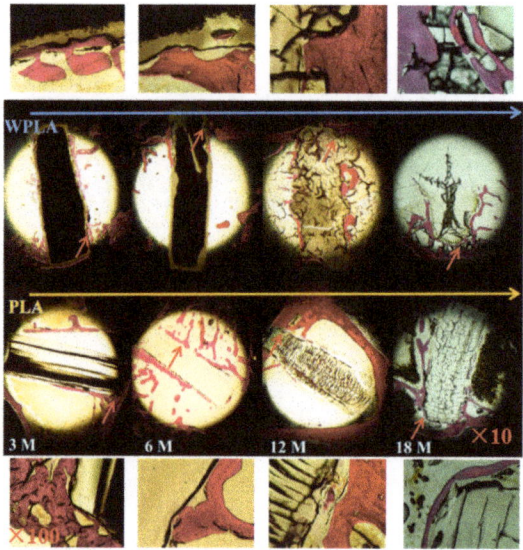

Figure 5. Histological morphologies of the implanted PLA and WPLA after 3-, 6-, 12- and 18-months duration of implantation.

4. Discussions

The nucleation abilities of the nanoparticles and whiskers for the polymeric composites are different, leading to the variation in their crystallinization. Undoubtedly, with the equal quality, the quantity of the nanoparticles is much more than that of the whiskers, and the hydrophilicity improvements of PPLA are more prominent, as shown in Figure 3b. The hydrolysis process of the composites is mainly impacted by water absorption. The SEM graphs reveal that the degradation behaviors of the composites PPLA and WPLA in the long-term degradation process are similar to that of the pure PLA, and there is the beginning of the amorphous region's decomposition for the polymeric matrix. However, the decomposition of the PLA amorphous area is obviously faster, which shows the "tough fracture", although no obvious hole is shown in the 12-months sample of Figure 1. It is probably due to the amorphous status from the degradation product of a low molecular weight [31]. Meanwhile, it also seems that compared with PPLA, the crystalline region of WPLA is more difficult to destroy, possibly due to two aspects. One is that the whiskers in WPLA induce the PLA molecular chain to form a crystal structure along the whiskers, which enhance their interface bonding and increase the crystallinity [30]. The other is that the whiskers have the ability to promote the ordering of the hydrolyzed PLA molecular chains, and its looser percolation network structure is more conducive to the growth of the PLA chip [31], which is also confirmed by the results of a higher value of crystallinity in the DSC and molecular weight in the GPC of the 12-months degradation. Additionally, it should be noted that although PPLA is easier to lose the amorphous area and perform a fast degradation during a short-time immersion, the nanoparticles with a smaller size are apt to fabricate the relatively denser or more compact crystalline structure and this is helpful to delay the PPLA degradation in the long-term degradation, which can also maintain the pH stability, as shown in Figure 2.

As to the in vivo degradation, the variation in the decomposing process of the WPLA composite affected by the whiskers is also distinct from that of the control PLA. The long-term degradation results show that the composites swell more prominently and faster due to the enhanced hydrophilicity which increases the water intake, and the swelling rate of the composite is also much faster than the degradation rate of the polymeric matrix. Meanwhile, the degraded status of the WPLA implant observed directly from the histological morphologies differs from the sample without whiskers. Due to the swelling effect, the composite is decomposed into the "block" shape, although it still degraded through bulk erosion [32], which is identical to the degrading mechanism of the pristine PLA. However, with the higher water uptake, it indicates that the degradation of WPLA is greatly controlled by the diffusion rate of the water in the PLA matrix due to the significant enhancement of hydrophilicity. Moreover, the biological activity of the PLA matrix is obviously improved, the specimen is tightly surrounded by bone tissue and the materials and newly formed bone trabecula are integrated into each other.

5. Conclusions

The in vitro and in vivo degradation of the biodegradable composites prepared by the PLA matrix blending with the 1% wt. MgO were investigated to determine the effect of the MgO shape (nanoparticles and whiskers) on the degradation process of PLA/MgO composites. We have demonstrated that the addition of MgO can accelerate the water uptake rate of the PLA matrix, and the degrading procedure of composites begins with the loss of the non-crystalline region prior to the destruction of the crystalline parts, demonstrated by the DSC results, but the water uptake rate of the composites, especially with MgO whiskers, is obviously faster than its bulk degradation in the degrading procedure, caused by the presence of MgO. The in vivo results of the histological morphologies suggest that the PLA matrix of the composite has the prominently enhanced bioactivity and that this PLA/MgO biocomposite can be potentially utilized for bone repair with better performance.

Supplementary Materials: The following are available online at http://www.mdpi.com/2073-4360/12/5/1074/s1. Figure S1. The SEM graphs of the MgO nanoparticles and whiskers.

Author Contributions: Conceptualization, M.C.; Data curation, Y.Z., H.L. and S.Q.; Formal analysis, Y.Z.; Funding acquisition, M.C.; Investigation, H.L.; Methodology, S.Z.; Project administration, M.C.; Resources, Y.Z., H.L., S.Z. and Y.J.; Supervision, M.C.; Validation, Y.Z.; Writing—original draft, Y.Z.; Writing—review & editing, Y.Z. All authors have read and agreed to the published version of the manuscript.

Funding: The work was funded by National Nature Science Foundation of China (No. 51801137, No. 51871166); the Tianjin Natural Science Foundation (No. 17JCQNJC03100); the Joint Foundation of National Natural Science Foundation of China (Grant No. U1764254).

Conflicts of Interest: The authors do not have any financial/personal conflict of interest that could affect their objectivity, or inappropriately influence their actions.

References

1. Alsberg, E.; Kong, H.J.; Hirano, Y.; Smith, M.K.; Albeiruti, A.; Mooney, D.J. Regulating bone formation via controlled scaffold degradation. *J. Dent. Res.* **2003**, *82*, 903–908. [CrossRef] [PubMed]
2. Kakinoki, S.; Uchida, S.; Ehashi, T.; Murakami, A.; Yamaoka, T. Surface Modification of Poly (L-lactic acid) Nanofiber with Oligo (D-lactic acid) Bioactive-Peptide Conjugates for Peripheral Nerve Regeneration. *Polymers* **2011**, *3*, 820–832. [CrossRef]
3. Rogina, A.; Pribolšan, L.; Hanžek, A.; Gómez-Estrada, L.; Ferrer, G.G.; Marijanovi´c, I.; Ivankovi´c, M.; Ivankovi´c, H. Macroporous poly (lactic acid) construct supporting the osteoinductive porous chitosan-based hydrogel for bone tissue engineering. *Polymer* **2016**, *98*, 172–181. [CrossRef]
4. Wang, X.J.; Song, G.J.; Lou, T. Fabrication and characterization of nano composite scaffold of poly (L-lactic acid)/hydroxyapatite. *J. Mater. Sci.* **2010**, *21*, 183–188. [CrossRef]
5. Park, J.E.; Todo, M. Development and characterization of reinforced poly(L-lactide) scaffolds for bone tissue engineering. *J. Mater. Sci. Mater. Med.* **2011**, *22*, 1171–1182. [CrossRef]
6. Li, X.; Chu, C.L.; Liu, L.; Liu, X.K.; Bai, J.; Guo, C.; Xue, F.; Lin, P.H.; Chu, P.K. Biodegradable poly-lactic acidbased-composite reinforced unidirectionally with high-strength magnesium alloy wires. *Biomaterials* **2015**, *49*, 135–144. [CrossRef]
7. Wen, W.; Zou, Z.P.; Luo, B.H.; Zhou, C.R. In vitro degradation and cytocompatibility of g-MgO whiskers/PLLA composites. *J. Mater. Sci.* **2017**, *52*, 2329–2344. [CrossRef]
8. Nampoothiri, K.M.; Nair, N.R.; John, R.P. An overview of the recent developments in polylactide (PLA) research. *Bioresour. Technol.* **2010**, *101*, 8493–8501. [CrossRef]
9. Böstman, O.M.; Pihlajamäki, H.K. Adverse tissue reactions to bioabsorbable fixation devices. *Clin. Orthop. Relat. Res.* **2000**, *371*, 216–227. [CrossRef]
10. Jiang, L.Y.; Xiong, C.D.; Jiang, L.X.; Xu, L.J. Effect of hydroxyapatite with different morphology on the crystallization behavior, mechanical property and in vitro degradation of hydroxyapatite/poly (lactic-co-glycolic) composite. *Compos. Sci. Technol.* **2014**, *93*, 61–67.
11. Shalumon, K.T.; Sheu, C.; Fong, Y.T.; Liao, H.T.; Chen, J.P. Microsphere-Based Hierarchically Juxtapositioned Biphasic Scaffolds Prepared from Poly(Lactic-co-Glycolic Acid) and Nanohydroxyapatite for Osteochondral Tissue Engineering. *Polymers* **2016**, *8*, 429. [CrossRef]
12. Kothapalli, C.R.; Shaw, M.T.; Wei, M. Biodegradable HA-PLA 3-D porous scaffolds: Effect of nano-sized filler content on scaffold properties. *Acta Biomater.* **2005**, *1*, 653–662. [CrossRef]
13. Ma, F.Q.; Lu, X.L.; Wang, Z.M.; Sun, Z.J.; Zhang, F.F.; Zheng, Y.F. Nanocomposites of poly(L-lactide) and surface modified magnesia nanoparticles: Fabrication, mechanical property and biodegradability. *J. Phys. Chem. Solids* **2011**, *72*, 111–116. [CrossRef]
14. Kum, C.H.; Cho, Y.; Seo, S.H.; Joung, Y.K.; Ahn, D.J.; Han, D.K. A poly (lactide) stereocomplex structure with modified magnesium oxide and its effects in enhancing the mechanical properties and suppressing inflammation. *Small* **2014**, *10*, 3783–3794. [CrossRef]
15. Yang, J.J.; Cao, X.X.; Zhao, Y.; Wang, L.; Liu, B.; Jia, J.P.; Liang, H.; Chen, M.F. Enhanced pH stability, cell viability and reduced degradation rate of poly(L-lactide)-based composite in vitro: Effect of modified magnesium oxide nanoparticles. *J. Biomater. Sci. Polym. Ed.* **2017**, *28*, 486–503. [CrossRef]
16. Wetteland, C.L.; Nguyen, N.T.; Liu, H.N. Concentration-dependent behaviors of bone marrow derived mesenchymal stem cells and infectious bacteria toward magnesium oxide nanoparticles. *Acta Biomater.* **2016**, *35*, 341–356. [CrossRef]
17. Zhao, Y.; Liu, B.; You, C.; Chen, M.F. Effects of MgO whiskers on mechanical properties and crystallization behavior of PLLA/MgO composites. *Mater. Des.* **2016**, *89*, 573–581. [CrossRef]

18. Cifuentes, S.C.; Gavilán, R.; Lieblich, M.; Benavente, R.; González-Carrasco, J.L. In vitro degradation of biodegradable polylactic acid/magnesium composites: Relevance of Mg particle shape. *Acta Biomater.* **2016**, *32*, 348–357. [CrossRef]
19. Wen, W.; Luo, B.; Qin, X.; Li, C.; Liu, M.; Ding, S.; Zhou, C. Strengthening and toughening of poly(L-lactide) composites by surface modified MgO whiskers. *Appl. Surf. Sci.* **2015**, *332*, 215–223. [CrossRef]
20. Chen, H.M.; Shen, Y.; Yang, J.H.; Huang, T.; Zhang, N.; Wang, Y.; Zhou, Z.W. Molecular ordering and α'-form formation of poly(L-lactide) during the hydrolytic degradation. *Polymer* **2013**, *54*, 6644–6653. [CrossRef]
21. Chen, H.M.; Feng, C.X.; Zhang, W.B.; Yang, J.H.; Huang, T.; Zhang, N.; Wang, Y. Hydrolytic degradation behavior of poly(L-lactide)/carbon nanotubes nanocomposites. *Polym. Degrad. Stab.* **2013**, *98*, 198–208. [CrossRef]
22. Zhao, Y.; Liu, B.; Yang, J.J.; Jia, J.J.; You, C.; Chen, M.F. Effects of modifying agents on surface modifications of magnesium oxide whiskers. *Appl. Surf. Sci.* **2016**, *388A*, 370–375. [CrossRef]
23. Jia, J.P.; Yang, J.J.; Zhao, Y.; Liang, H.; Chen, M.F. The crystallization behaviors and mechanical properties of poly(L-lactic acid)/magnesium oxide nanoparticle composites. *RSC Adv.* **2016**, *6*, 43855–43863. [CrossRef]
24. Liu, Y.L.; Shao, J.; Sun, J.R.; Bian, X.C.; Li, D.; Xiang, F.S.; Sun, B.; Chen, Z.; Li, G.; Chen, X.S. Improved mechanical and thermal properties of PLLA by solvent blending with PDLA-b-PEG-b-PDLA. *Polym. Degrad. Stab.* **2014**, *101*, 10–17. [CrossRef]
25. Kumar, S.; Singh, S.; Senapati, S.; Singh, A.P.; Ray, B.; Maiti, P. Controlled drug release through regulated biodegradation of poly(lactic acid) using inorganic salts. *Int. J. Biol. Macromol.* **2017**, *104*, 487–497. [CrossRef]
26. Larrañaga, A.; Aldazabal, P.; Martin, F.J.; Sarasua, J.R. Hydrolytic degradation and bioactivity of lactide and caprolactone based sponge-like scaffolds loaded with bioactive glass particles. *Polym. Degrad. Stab.* **2014**, *110*, 121–128. [CrossRef]
27. Lizundia, E.; Mateos, P.; Vilas, J.L. Tuneable hydrolytic degradation of poly(L-lactide) scaffolds triggered by ZnO nanoparticles. *Mat. Sci. Eng. C* **2017**, *75*, 714–720. [CrossRef]
28. Wu, Y.H.; Li, N.; Cheng, Y.; Zheng, Y.F.; Han, Y. In vitro study on biodegradable AZ31 magnesium alloy fibers reinforced PLGA composite. *J. Mater. Sci. Technol.* **2013**, *29*, 545–550. [CrossRef]
29. Qu, M.; Tu, H.L.; Amarante, M.; Song, Y.Q.; Zhu, S.S. Zinc Oxide Nanoparticles Catalyze Rapid Hydrolysis of Poly(lactic acid) at Low Temperatures. *J. Appl. Polym. Sci.* **2014**, *131*, 40287. [CrossRef]
30. Ning, N.Y.; Fu, S.R.; Zhang, W.; Chen, F.; Wang, K.; Deng, H.; Zhang, Q.; Fu, Q. Realizing the enhancement of interfacial interaction in semicrystalline polymer/filler composites via interfacial crystallization. *Prog. Polym. Sci.* **2012**, *37*, 1425–1455. [CrossRef]
31. Chen, H.M.; Chen, J.; Shao, L.N.; Yang, J.H.; Huang, T.; Zhang, N.; Wang, Y. Comparative Study of Poly(L-lactide) Nanocomposites with Organic Montmorillonite and Carbon Nanotubes. *J. Polym. Sci. Pol. Phys.* **2013**, *51*, 183–196. [CrossRef]
32. Von Burkersroda, F.; Schedl, L.; Göpferich, A. Why degradable polymers undergo surface erosion or bulk erosion. *Biomaterials* **2002**, *23*, 4221–4231. [CrossRef]

© 2020 by the authors. Licensee MDPI, Basel, Switzerland. This article is an open access article distributed under the terms and conditions of the Creative Commons Attribution (CC BY) license (http://creativecommons.org/licenses/by/4.0/).

Article

Manufacturing and Characterization of Functionalized Aliphatic Polyester from Poly(lactic acid) with Halloysite Nanotubes

Sergi Montava-Jorda [1], Victor Chacon [2], Diego Lascano [2,3,*], Lourdes Sanchez-Nacher [2] and Nestor Montanes [2]

1. Department of Mechanical and Materials Engineering, Universitat Politècnica de València (UPV), Plaza Ferrándiz y Carbonell 1, 03801 Alcoy, Spain
2. Technological Institute of Materials (ITM), Universitat Politècnica de València (UPV), Plaza Ferrándiz y Carbonell 1, 03801 Alcoy, Spain
3. Escuela Politécnica Nacional, 17-01-2759 Quito, Ecuador
* Correspondence: dielas@epsa.upv.es; Tel.: +34-966-528-433

Received: 12 July 2019; Accepted: 3 August 2019; Published: 6 August 2019

Abstract: This work reports the potential of poly(lactic acid)—PLA composites with different halloysite nanotube (HNTs) loading (3, 6 and 9 wt%) for further uses in advanced applications as HNTs could be used as carriers for active compounds for medicine, packaging and other sectors. This work focuses on the effect of HNTs on mechanical, thermal, thermomechanical and degradation of PLA composites with HNTs. These composites can be manufactured by conventional extrusion-compounding followed by injection molding. The obtained results indicate a slight decrease in tensile and flexural strength as well as in elongation at break, both properties related to material cohesion. On the contrary, the stiffness increases with the HNTs content. The tensile strength and modulus change from 64.6 MPa/2.1 GPa (neat PLA) to 57.7/2.3 GPa MPa for the composite with 9 wt% HNTs. The elongation at break decreases from 6.1% (neat PLA) down to a half for composites with 9 wt% HNTs. Regarding flexural properties, the flexural strength and modulus change from 116.1 MPa and 3.6 GPa respectively for neat PLA to values of 107.6 MPa and 3.9 GPa for the composite with 9 wt% HNTs. HNTs do not affect the glass transition temperature with invariable values of about 64 °C, or the melt peak temperature, while they move the cold crystallization process towards lower values, from 112.4 °C for neat PLA down to 105.4 °C for the composite containing 9 wt% HNTs. The water uptake has been assessed to study the influence of HNTs on the water saturation. HNTs contribute to increased hydrophilicity with a change in the asymptotic water uptake from 0.95% (neat PLA) up to 1.67% (PLA with 9 wt % HNTs) and the effect of HNTs on disintegration in controlled compost soil has been carried out to see the influence of HNTs on this process, which is a slight delay on it. These PLA-HNT composites show good balanced properties and could represent an interesting solution to develop active materials.

Keywords: poly(lactic acid), halloysite nanotubes; mechanical characterization; morphology; thermal characterization

1. Introduction

In the last decade, the polymer industry has faced important challenges related to new regulations, increasing concern about environment, petroleum depletion and others. Sustainability has consolidated as a leading force in the development of new high environmentally friendly materials [1–3]. Petroleum-derived polymers have, in general, a remarkable effect on the overall carbon footprint, so that many researches have focused on the development of new polymers (thermoplastics, thermosetting and elastomers) with a positive effect on their carbon footprint [4–7]. These environmentally friendly polymers could be classified into three different groups with different environmental connotations [8]. One group is composed of biodegradable petroleum-derived polymers such as most aliphatic polyesters, i.e., poly(ε-caprolactone)—PCL, poly(butylene succinate)—PBS, poly(glycolic acid)—PGA and so on [9–11]. Although they are petroleum-derived polymers, they show interest from an environmental standpoint as they can be disintegrated in controlled compost soil conditions. Another interesting and growing group is that of bio-based and non-biodegradable polymers which include polymers such as poly(ethylene)—PE, poly(ethylene terephthalate)—PET, poly(amides)—PAs and so on, which are obtained from renewable resources but are not biodegradable [12–16]. In fact, they show almost identical properties to their corresponding petroleum-derived counterparts. Their interests in that they can contribute to reduce the carbon footprint as they are obtained from plants that are able to fix CO_2. Finally, there is an increasing interest on a group of polymers that are both bio-based and biodegradable. This group includes polysaccharides and derivatives, e.g., cellulose, chitin, chitosan, starch and derivatives, among others [17–19]. Protein-based polymers, e.g., gluten, soybean and collagen, among others are also included in this group [20–23]. Finally, bacterial polyesters or poly(hydroxyalkanoates)—PHAs, which derive from bacterial fermentation are gaining interest but up today, their synthesis process and purification is still an expensive process but these materials are thought to be a real alternative to a wide variety of polymers due the high number of poly(esters) that can be synthesized by different bacteria. Among all poly(hydroxyalkanoates), it is worthy to note the key role that poly(hydroxybutyrate)—PHB and its copolymers could acquire in the future [24–28].

Above all these polymers, it is worthy to note the current interest of poly(lactic acid)—PLA, which can be obtained by starch fermentation and, subsequently, polymerization. PLA and its blends have gained a privileged position in the polymer industry as it shows balanced properties (mechanical, thermal, barrier, physical and so on) with an everyday most competitive price [29–31]. Today it is possible to find PLA in a wide variety of industrial applications that include automotive, construction and building, packaging, houseware and so on [32–36], Behera et al. [37] proposed that by controlling the manufacturing process, the properties of PLA could be tailored e.g., by lowering M_w and the glass transition temperature (T_g). On the other hand, PLA is one of the most widely used material employed for fused deposition modeling (FDM) for 3D printed parts for whatever industry [38–41]. In addition to its biodegradability, PLA is resorbable and therefore it finds increasing use in the medicine industry (stents, screws, fixation plates, scaffolds for tissue engineering, surgical suture and so on) [42–46]. In these applications, PLA could also act as a drug carrier to provide controlled drug release [47,48]. Micro and nanoencapsulation have given interesting results for different purposes. This can be accomplished by loading the PLA matrix with encapsulated micro- or nanoparticles. In the recent years, nanotubes (NTs) have gained interest in both pharmacy and medicine applications as they can be used as carriers to deliver different active compounds such as antioxidants, antibiotics, antimicrobials and wound healing, among others [49–51]. Different nanotubes have been proposed with this aim and different possibilities. Despite carbon nanotubes (CNTs) have attracted most interest, other nanotubes are being studied as potential drug carriers in medicine and pharmacy [52–55], e.g., titanate and titania nanotubes (TiNTs) [56,57], hydroxyapatite (HApNTs) [58] and zinc oxide (ZnONTs) [59]. One important drawback of nanotubes is their synthesis process, which is, usually, complex and expensive. Nevertheless, aluminosilicates offer interesting properties as potential carriers. Halloysite nanotubes (HNTs) are naturally occurring nanotubes derived from a particular aluminosilicate structure, composed of an external silicate layer and an internal alumina layer with a different specific volume. This particular structure allows growing nanotubes in the form of rolled silica-alumina layers. They are worldwide available at a competitive price and they can be selectively etched to increase their carrying capacity and above all, they are completely biocompatible [60,61]. HNTs have been loaded into different polymer matrices such as poly(lactic acid), N,N'-ethylenebis(stearamide) (EBS), poly(propylene), poly(amide) 11, poly(ethylene-co-vinyl acetate) (EVA), high impact poly(styrene) (HIPS) [62–64] and different loaded drugs such as insulin, thymol and PEG-hirudin [65,66]. In addition to the bioactivity of these composites, they must fulfill some mechanical, thermal, chemical and physical properties to be used in a particular application [67].

This work explores the effect of halloysite nanotubes (HNTs) on mechanical, thermal and thermomechanical properties of a poly(lactic acid) matrix for potential uses in engineering applications. The effect of the loading amount comprised between 3 and 9 wt% is studied.

2. Experimental

2.1. Materials

A poly(lactic acid) commercial grade Ingeo 6201D in pellet form was supplied by NatureWorks LLC (Minnesota, USA). This grade possesses a density of 1.24 g cm^{-3} and a melt flow index comprised between 15–30 g/10 min measured at a temperature of 210 °C. Despite this commercial grade is intended for melt spinning of fibers, this melt flow index is also suitable for injection molding and this grade has been previously used as a base material for plasticization, blending and manufacturing of wood plastic composites. Halloysite nanotubes (HNTs) with a chemical composition of $Al_2Si_2O_5(OH)_4 \cdot 2H_2O$ and CAS 1332-58-7, were purchased from Sigma Aldrich (Madrid, Spain). HNTs had an average molecular weight of 294.19 g mol^{-1}. The average length was comprised between 1 and 3 µm while the external diameter varied from 30 to 70 nm. The typical morphology of these HNTs is shown in Figure 1 where the typical tubular shape can be detected and even the lumen diameter can be seen by transmission electron microscopy technique, it was performed in a Philips microscope model CM10 (Eindhoven, the Netherlands), the acceleration voltage was set at 100 kV. The HNTs samples were spread in acetone then immersed in an ultrasound bath; afterward, a small drop of this solution was poured onto a carbon grid, it was subjected to solvent evaporation at 25 °C. From the TEM images, the lumen size of HNTs was obtained. At least 50 measurements were performed to obtain the size distribution. As reported in a previous work by Garcia-Garcia et al. [60], the composition of HNTs is mainly SiO_2 (53.75%) and Al_2O_3 (44.57%) and, in a less extent, some additional oxides at less than 1% (P_2O_5, Fe_2O_3, SO_3 and CaO, among others). As it can be seen in Figure 1, HNTs were also characterized by field emission electron microscopy (FESEM) operated at an acceleration voltage of 2 kV. Due to the high hydrophilicity of HNTs, they tended to form aggregates (Figure 1c), which shows an aggregate of dimensions 5.1 × 71 µm^2. The tubular shape could also be observed by FESEM as it is shown in Figure 1d, in which a single HNT can be observed with an external diameter of 105 nm and a length of about 0.5 µm. Nevertheless, as indicated by Garcia-Garcia et al. [60] the size distribution of HNTs is rather heterogeneous.

Both PLA pellets and HNTs were dried at 60 °C for 24 h in an air-circulating oven to remove residual moisture. Different formulations were prepared containing different HNTs loading as summarized in Table 1. Poly(vinyl acetate)—PVAc was obtained from Sigma-Aldrich (Madrid, Spain), it was used as a compatibilizer due to its high hydrophilicity.

Table 1. Labeling and composition of poly(lactic acid) composites with different HNTs loadings with poly(vinyl acetate) compatibilizer.

Code	PLA (wt%)	HNTs (wt%)	PVAc (phr) *
PLA	100	-	-
PLA-3HNT	97	3	0.3
PLA-6HNT	94	6	0.6
PLA-9HNT	91	9	0.9

* phr represents the weight parts of additive per one hundred weight parts of the PLA/HNT composite material.

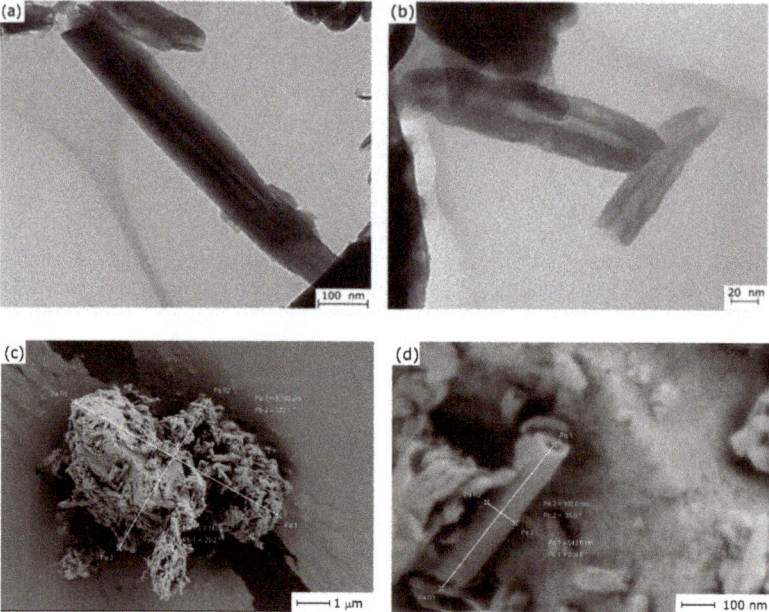

Figure 1. Different images showing the structure of halloysite nanotubes (HNTs). Transmission electron microscopy (TEM) images of halloysite nanotubes at different magnifications (**a**) 30,000×, (**b**) 80,000×. Field emission scanning electron microscopy (FESEM) images of (**c**) HNT aggregate at 10,000× and (**b**) isolated HNTs showing dimensions and the tubular structure at 100,000×.

2.2. Manufacturing of PLA/HNTs Composites

These formulations were placed in a zipper bag and subjected to an initial homogenization process. The mixtures were fed into a twin-screw co-rotating extruder from DUPRA S.L. (Alicante, Spain) with the following temperature profile: 165 °C (hopper), 170 °C, 175 °C and 180 °C (dye). The rotating speed of the screw was set to 40 rpm and the maximum extruder capacity was set at 60%. The obtained compounds, were pelletized for further processing by injection moulding in a Meteor 270/75 from Mateu & Solé (Barcelona, Spain) with the following temperature profile (from the hopper to the injection nozzle): 170 °C, 180 °C, 190 °C and 200 °C. After this, standard samples for characterization were obtained as can be seeing in Figure 2. As it can be seen, the addition of HNTs contribute to a remarkable change in color from white (neat PLA) to dark brown for the composite with 9 wt% HNTs. Therias et al. [68] have reported identical change in color with increasing HNTs loading.

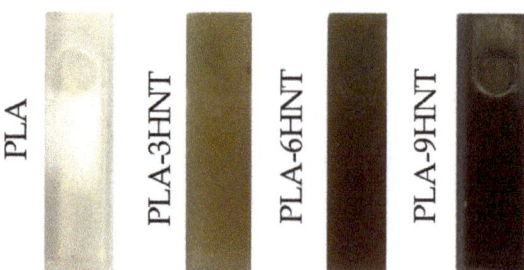

Figure 2. Digital images of PLA-HNT composites with different HNTs loading.

2.3. Thermal Characterization

Thermal characterization was carried out by differential scanning calorimetry (DSC) in a DSC 821 from Mettler-Toledo Inc. (Schwerzenbach, Switzerland). The selected thermal cycled was divided into three different stages: A first heating from 30 °C to 200 °C was followed by a cooling down to 0 °C and, finally, a second heating stage from 0 °C up to 350 °C was scheduled. All the stages were run at a rate of 10 °C min^{-1} in nitrogen atmosphere (66 mL min^{-1}). Different parameters were obtained from the second heating cycle, namely, the glass transition temperature (T_g), the cold crystallization peak temperature (T_{cc}) and enthalpy (ΔH_{cc}), the melt peak temperature (T_m) and enthalpy (ΔH_m) and the degradation temperature (T_d). The degree of crystallinity (χ_c%) was calculated following Equation (1).

$$\chi_{cPLA}\ (\%) = \left[\frac{|\Delta H_m| - |\Delta H_{cc}|}{|\Delta H_{100\%}| \cdot w_{PLA}}\right] \cdot 100 \tag{1}$$

where $\Delta H_{100\%}$ is a theoretical value that represents the estimated melt enthalpy of a fully crystalline PLA polymer, i.e., 93.7 J g^{-1} as reported in literature [69]. Finally, the term w$_{PLA}$ represents the weight fraction of PLA on composites [70–72].

In addition to this thermal characterization, samples were analyzed by thermomechanical analysis (TMA) to obtain the dimensional stability of the obtained composites. In particular, the coefficient of linear thermal expansion (CLTE) was calculated from the slope of the characteristic TMA curves. To this, a thermomechanical analyzer TA Q400 (DE, USA) was used with a constant force of 20 mN. The heating program was set from −80 °C up to 100 °C at a heating rate of 2 °C min^{-1}.

To better understand the thermomechanical behavior of the samples, they underwent dynamic mechanical thermal analysis (DMTA) in an oscillatory rheometer AR-G2 supplied by TA Instruments (New Castle, USA). A special clamp system designed for solid samples (40 × 10 × 4 mm^3) were used, it works in torsion-shear conditions. The heating program was a temperature sweep from 30 °C up to 140 °C at a constant heating rate of 2 °C min^{-1}. The maximum shear/torsion deformation (γ), and the oscillations frequency were defined as a percentage of 0.1% and 1 Hz, respectively.

2.4. Mechanical Properties

The most relevant information about mechanical properties was obtained by tensile, flexural, impact and hardness tests. Tensile and flexural tests were carried out as indicated in ISO 527 and ISO 178 respectively in a universal test machine ELIB 30 from S.A.E. Ibertest (Madrid, Spain). A load cell of 5 kN was used for both tests and the crosshead rate was set to 5 mm min^{-1}. With regard to the impact properties, a Charpy pendulum with a total energy of 6 J from Metrotec S.A. (San Sebastián, Spain) was used on unnotched samples. The hardness was estimated using a Shore durometer with the "D" scale, model 673-D from J. Bot S.A. (Barcelona, Spain). All mechanical tests were run on five different samples and the average values of the corresponding properties were obtained.

2.5. Morphology Characterization

The morphology of the fractured samples (after failure in the impact test) was analyzed by field emission scanning electron microscopy (FESEM). A Zeiss Ultra55 FESEM microscope from Oxford Instruments (Oxfordshire, UK) working at an acceleration voltage of 2 kV was used. As PLA/HNTs composites are not electrically conducting, samples were subjected to a sputtering process with gold-palladium in a high vacuum sputter coater EM MED020 from Leica Microsystems (Milton Keynes, UK).

2.6. Water Uptake of PLA/HNTs

Water uptake of PLA/HNTs composites was carried out as indicated in ISO 62:2008 with distilled water at 30 ± 1 °C for a period of 98 days. Rectangular samples with dimensions 80 × 10 × 4 mm^3 were initially dried at 60 °C for 24 h to remove residual moisture in an air circulating oven 2001245

DIGIHEAT-TFT from J.P. Selecta, S.A. (Barcelona, Spain). After this drying process, samples were submerged in distilled water and removed after planned times. These samples are dried with a cotton cloth and subsequently, are weighed in an analytic balance AG245 from Mettler-Toledo Inc. (Schwerzenbach, Switzerland); after weighting, samples were submerged aging in distilled water. The amount of absorbed water during the water uptake was calculated from Equation (2).

$$\Delta m_t(\%) = \left(\frac{W_t - W_0}{W_0}\right) \times 100, \quad (2)$$

where W_t is the mass after a time t while W_0 represents the dry weight of PLA/HNTs composites before the water uptake process. After a particular immersion time (saturation time) changes in water absorption can be neglected as they are extremely low. Then, it is possible to obtain the saturation mass that is denoted as Δmass.

2.7. Disintegration in Controlled Compost Soil

Biodegradation test, or more correctly, disintegration in the controlled compost soil test, was carried out following ISO 20200. The disintegration process was carried out at a temperature of 58 °C and a relative humidity on soil of 55%. Squared samples (20 × 20 mm^2) and a thickness of 1 mm were buried into a biodegradation reactor, prepared as indicated in the corresponding standard. After some periods, samples were unburied, washed with distilled water and dried at 50 °C and, finally, weighed in an analytic thermobalance. The weight loss during the biodegradation process was calculated from Equation (3).

$$\text{Weight loss }(\%) = \left(\frac{W_0 - W_t}{W_0}\right) \times 100. \quad (3)$$

In this equation, W_t represents the mass after a degradation time t and W_0 stands for the dry weight of PLA/HNTs composites before disintegration.

3. Results and Discussion

3.1. Influence of HNTs Content on Mechanical Performance of PLA/HNTs Biocomposites

Table 2 shows a summary of the main mechanical parameters obtained from different tests, i.e., tensile, flexural, impact and hardness, as a function of increasing HNTs content. The first thing one can realize is that the tensile strength (σ_t) was not highly affected by the presence of HNTs. It is true that we could observe a slight decrease in σ_t with increasing HNTs content. In particular, neat PLA showed a σ_t of 64.6 MPa while all composites with HNTs showed a σ_t comprised between 58–59 MPa, which represents a maximum percentage decrease of about 10%. On the other hand, the effect of HNTs on ductile properties such as elongation at break (%ε_b) was much more pronounced. PLA is an intrinsically fragile polymer with a restricted elongation at break of only 6.1%. As we can see in Table 2, HNTs produced an embrittlement effect on PLA, leading to %ε_b lower than 4%. The maximum decrease corresponded to the PLA composite with 9 wt % HNTs with an elongation at break of 3.3% (which represents a percentage decrease of about 45%). HNTs were dispersed into the PLA polymer matrix and promoted a loss on material cohesion. Therefore, the stress transfer from the embedded particles to the surrounding matrix was restricted, and this had a negative effect on elongation at break. Similar findings have been reported in literature with PLA matrices [73,74]. As expected, the tensile modulus increased with increasing HNTs content. It is important to remember that the tensile modulus (E_t) represents the ratio between the applied stress (σ) and the obtained elongation (ε) in the linear region of a stress-strain curve. As above-mentioned, both tensile strength and elongation at break decrease with increasing HNTs content, but the decrease in %ε_b was much more pronounced than that observed for σ_t. As the strain was in the denominator, as the denominator (ε) decreased more than the numerator (σ), the overall effect was an increase in E_t. While neat PLA was characterized by an

E_t value of 2086 MPa and the composite with 9 wt% HNTs was more brittle, with a tensile modulus of 2311 MPa (>10% increase). These results are in accordance to those reported by Chen et al. [45]. This embrittlement could be related to an increase in the structural stiffness of PLA chains due to HNTs-PLA interactions [75,76]. These low decrease in tensile properties is related to the use of the PVAc compatibilizer, which is highly hydrophilic that can establish interactions with both hydroxyl groups (–OH) contained in PLA (end chains) and HNTs surface, thus leading to good embedding of the HNTs into the PLA matrix with a positive effect on mechanical performance. Pracella et al. [77] reported the interesting compatibilizing effect of poly(vinyl acetate), PVAc to enhance compatibilization in PLA/cellulose nanocrystals, as it provides increased adhesion between these two components.

Table 2. Mechanical properties of PLA/HNTs composites obtained from tensile tests (tensile modulus—E_t, tensile strength—σ_t and elongation at break—%ε_b), flexural tests (flexural modulus—E_f and flexural strength—σ_f), hardness (Shore D) and impact (impact strength from the Charpy test).

Code	Tensile			Flexural		Shore D Hardness	Impact Strength (kJ m^{-2})
	Modulus, E_t (MPa)	Strength, σ_t (MPa)	Elongation at Break (ε_b%)	Modulus, E_f (MPa)	Strength, σ_f (MPa)		
PLA	2086 ± 82	64.6 ± 1.6	6.1 ± 0.8	3570 ± 39	116.1 ± 2.1	81.7 ± 1.0	1.46 ± 0.5
PLA-3HNT	2097 ± 73	59.2 ± 2.6	3.7 ± 0.5	3701 ± 100	105.5 ± 3.9	81.0 ± 2.0	1.18 ± 0.2
PLA-6HNT	2160 ± 78	58.7 ± 0.7	3.9 ± 0.1	3918 ± 252	110.4 ± 3.4	82.0 ± 2.0	1.03 ± 0.2
PLA-9HNT	2311 ± 132	57.7 ± 2.4	3.3 ± 0.2	3927 ± 128	107.6 ± 6.8	83.2 ± 2.7	0.71 ± 0.2

With regard to flexural properties, the behavior of PLA-HNT composites followed the same tendency observed for tensile tests. All the herein developed composites with PLA and HNTs show increased flexural modulus (E_f) compared to neat PLA. It is worthy to remark that PLA composites with 9 wt% HNTs offered a E_f value of 3927 MPa, which was noticeably higher than neat PLA (3570 MPa). As the above-mentioned, addition of HNTs provided increased structural stiffness on PLA polymer chains [75,76,78]. On the contrary, the flexural strength (σ_f) decreased with increasing HNT content. At this point it is worthy to remark that deformation and strength were highly sensitive to material cohesion. Although some HNTs-PLA interactions are expected, presence of HNTs aggregates breaks cohesion and this has a negative effect on both properties in tensile or flexural conditions. Therefore, the flexural strength (sf) of neat was 116 MPa while this was reduced down to values of 107 MPa for PLA composites containing 9 wt % HNTs (around 7.7% decrease). Although it has been reported the reinforcing properties of HNTs on a PLA matrix, Therias et al. [68] have reported a slight decrease (almost negligible) of tensile strength with increasing HNTs content (up to 12 wt %), the most noticeable effect is an increase in stiffness. It is important to remark that the modulus represents the stress to strain ratio in the linear region of a tensile or flexural diagram. As the tensile strength remains with high values and elongation at break is remarkably reduced, then this ratio increases. Therefore, the tensile and flexural moduli increase with increasing HNTs content. In contrast, Chen et al. [79] have reported alternating values of tensile strength (increase and decrease) with different HNTs loading, but, in general, the stiffness is always increased with HNTs loading. It is worthy to remark the work by Guo et al. [80]. In their work they report the effect of pristine HNTs and alkali-modified HNTs. The show a decrease in tensile strength and elongation at break by increasing the loading of pristine HNTs but, in contrast, alkali-modified HNTs contribute to an increase in tensile strength, which is attributed to improved PLA/HNTs interactions due to the alkali treatment, which enhances more available hydroxyl groups in the outer surface.

Table 2 also summarizes Shore D hardness values for all composites, compared with neat PLA. All values were close to 82 and there was a very slight increase with the presence of HNTs. Nevertheless, this slight change was included in the deviation itself and, therefore, it was not possible to conclude a clear tendency.

Table 2 also includes the impact strength values obtained from the Charpy test. As expected, addition of HNTs provided an embrittlement effect as indicated by the modulus (in both tensile and flexural conditions). This embrittlement was also evident from the impact strength values. Neat PLA offered an impact strength of 1.46 kJ m^{-2} and this was reduced down to values close to half (0.71 kJ m^{-2}) in composites with 9 wt% HNTs. Impact strength was also related to material cohesion and, as indicated previously, presence of HNTs aggregates led to poor material cohesion and this had a negative effect on both resistance (maximum stress that the material can withstand) and deformation, both parameters playing a key role in impact strength. The embedded HNTs particles acted as starting points for crack initiation and growth thus leading to a clear embrittlement, which can be observed by morphology analysis.

The morphologies of the fractured samples after impact tests are gathered in Figure 3 at different magnifications. Neat PLA (Figure 3a,b) showed a typical brittle fracture surface, with low roughness and a smooth fracture surface. As the HNTs loading increased, it was possible to detect an increase roughness, which was attributed to the presence of HNTs. HNTs promoted an embrittlement against impact, which favored crack formation and growth. This is because HNTs are highly hydrophilic and do not establish strong interactions with PLA matrix. At higher magnification (5000×) it was possible to see a fine particle dispersion of HNTs embedded into the PLA polymer matrix (Figure 3d,f,h). It is possible to observe the presence of HNTs aggregates as well as some individual HNTs embedded in the PLA matrix. As the HNTs content increased the density of these aggregates/individual HNTs increased. Although it seems these particles were fully embedded into the PLA matrix, there was poor interaction between them and the surrounding matrix even with the presence of PVA compatibilizer. This phenomenon provided poor material cohesion and, therefore, stress concentration phenomena could take place. Some research works indicate that HNTs usually give good particle dispersion due to the tubular structure if HNTs, which is responsible for a weakening of nanotube–nanotube interactions. In addition, weak interactions can be obtained between the PLA matrix and the embedded HNTs particles by hydrogen bonds between the carbonyl groups in PLA and the hydroxyl groups in halloysite [81,82].

Magnified FESEM images taken at 10,000× (Figure 4) showed in a clearer way, the presence of these aggregates and individual HNTs.

As it can be seen in Figure 4a, good HNT dispersion is achieved for this relatively low HNT loading of 3 wt%. Individual HNTs can be detected as well as some aggregates (in de upper-right side). This aggregate formation is also detectable in PLA/HNTs composites containing 6 wt% HNTs (Figure 4b) located at the left side; nevertheless, the presence of individual well-dispersed HNTs can also be observed. Finally, Figure 4c shows the fracture surface corresponding to the PLA/HNT composite with 9 wt% HNTs. It is clearly detectable the presence of larger aggregates and some individual HNTs embedded in the PLA matrix. To check this dispersion Figure 5 gathers the FESEM image of a PLA/HNT composite (9 wt% HNT) and their corresponding energy-dispersive X-ray spectroscopy mapping images for oxygen carbon (C Kα1_2), aluminum (Al Kα1; O Kα1) and silicon (Si Kα1). As the above-mentioned, HNTs are aluminosilicate structures and, therefore, the presence of Al_2O_3 (Al Kα1) and SiO_2 (Si Kα1 and O Kα1) was evidenced by the EDX analysis. In addition, these elements mapping suggest the presence of HNTs aggregates as the size was of 2.5×5 μm^2 (see upper-left side of O Kα1, Al Kα1 and Si Kα1 EDX mapping images), as well as some individual points related to individual HNTs as observed by the FESEM analysis.

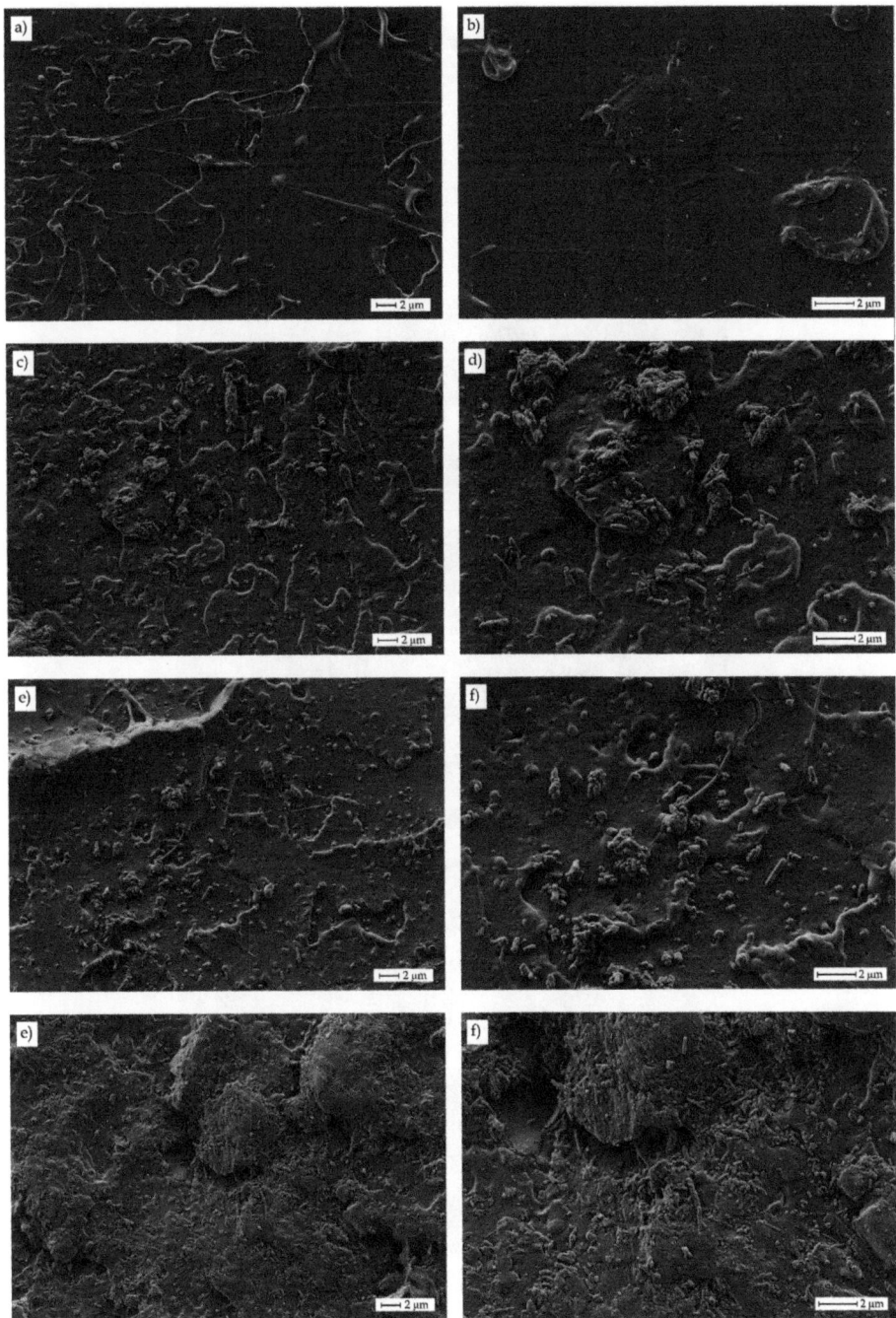

Figure 3. Field emission scanning electron microscopy (FESEM) images at different magnifications (2500×: left column; 5000×: right column) of fractured samples from impact tests corresponding to PLA-HNT composites with different HNTs loading. (**a**,**b**) neat PLA, (**c**) and (**d**) 3 wt% HNTs, (**e**) and (**f**) 6 wt% HNTs and (**g**) and (**h**) 9 wt% HNTs.

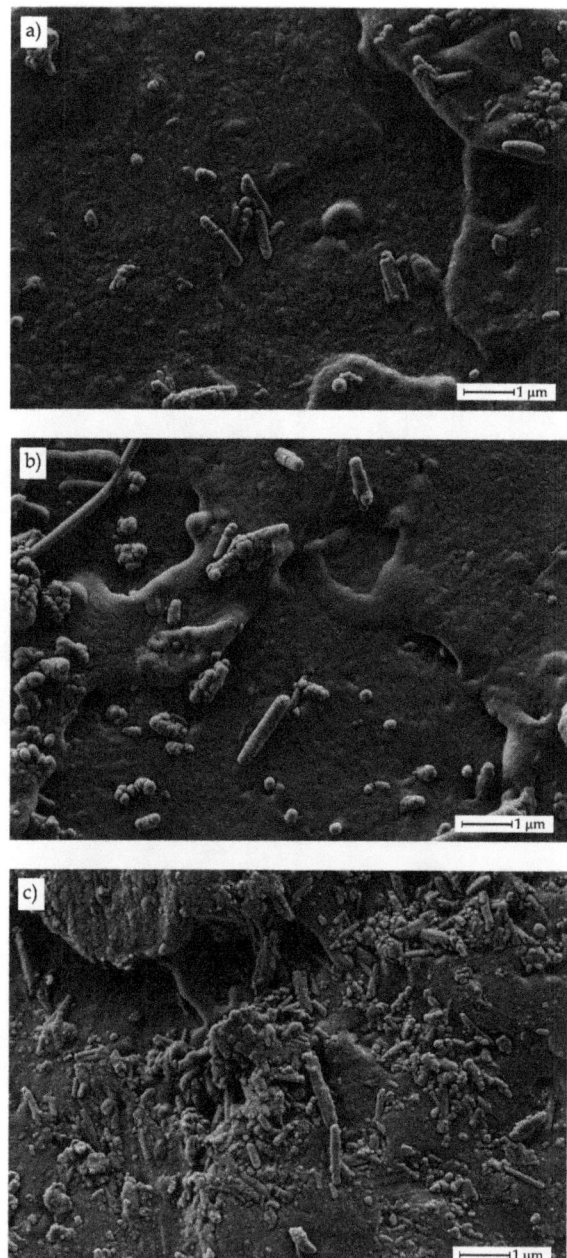

Figure 4. Field emission scanning electron microscopy (FESEM) images at 10,000× of fractured samples from impact tests corresponding to PLA-HNT composites with different HNTs loading. (**a**) 3 wt% HNTs, (**b**) 6 wt% HNTs and (**c**) 9 wt% HNTs.

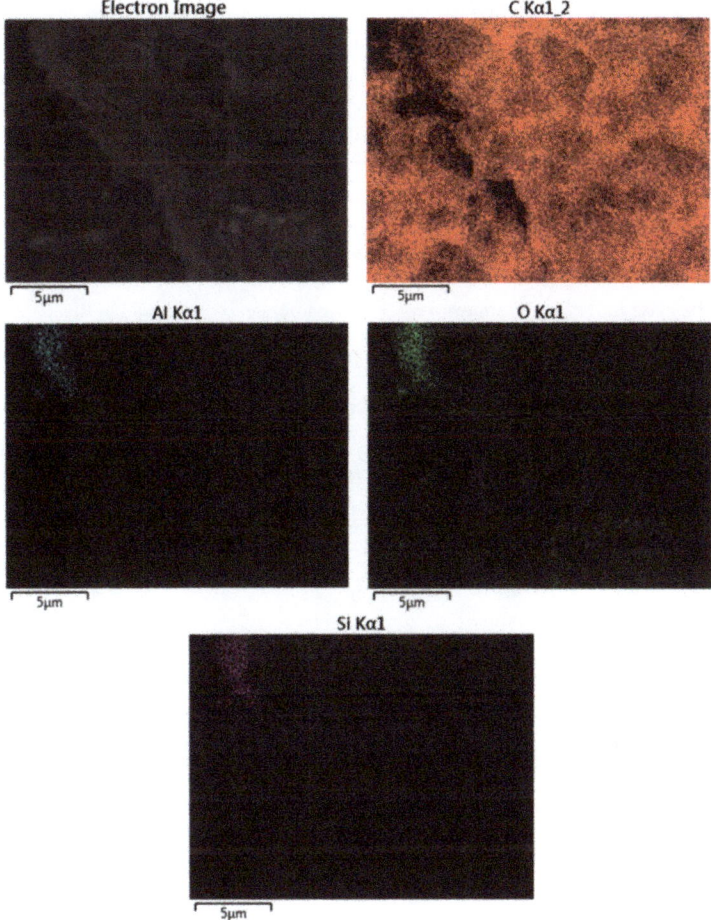

Figure 5. Field emission scanning electron microscopy (FESEM) image of a PLA/HNT composite with 9 wt% HNTs at 5000× and EDX mapping corresponding to carbon (C Kα1_2), aluminum (Al Kα1), oxygen (O Kα1) and silicon (Si Kα1).

3.2. Influence of HNTs Content on Thermal Behaviour of PLA/HNTs Biocomposites

The effect of HNTs on thermal properties was assessed by differential scanning calorimetry (DSC). Figure 6 and Table 3 gathers the main parameters from DSC analysis for neat PLA and PLA-HNTs composites with increasing HNTs loading. With regard to the glass transition temperature (T_g), the addition of HNTs did not provide any remarkable change in this parameter, which indicates poor HNTs-PLA interactions. In fact, the T_g values remained almost constant with values around 64 °C. Regarding the cold crystallization process, it is possible to observe a slight decrease in the cold crystallization peak temperatures (T_{cc}) as observed by Prashanta et al. [74]. This is directly related to the nucleant effect that HNTs can exert in PLA chains, thus allowing PLA chains to fold in a packed way to form crystallites. The melt peak temperature (T_m) was not affected by the presence of whatever loading of HNTs, remaining at values of about 172–173 °C. Regarding the degradation onset temperatures (T_d), no clear effect of HNTs could be detected as all values were comprised between 322 and 327 °C [83]. With the obtained values of the melt and cold crystallization enthalpies (ΔH_m

and ΔH_{cc}) respectively, the percentage degree of crystallinity (%χ_c) was calculated and it is shown in Table 3. As observed by Murariu et al. [84], addition of halloysite provides slightly lower crystallinity values. Table 3 shows two different crystallinity values, one is %χ_{c_s}, which represents the degree of crystallinity of the injection molded sample without any removal of the thermal history (this takes into account the difference between the melt and cold crystallization enthalpies) and a second %χ_{c_max}, which stands for the maximum crystallinity PLA can reach in these composites (this only considers the melt enthalpy). As it can be seen, this maximum crystallinity followed the same tendency, it means, a decrease with increasing HNT loading. This could be related to formation of less perfect crystals due to the presence of HNTs as reported by Chen et al. [79] with similar crystallinities for neat PLA of about 37% and lower values of 33–34% for a HNT loading of 10–15%. They also reported a clear decrease in the cold crystallization process as observed in this work, while very slight changes were observed for the glass transition temperature, T_g, in accordance to the results in this research.

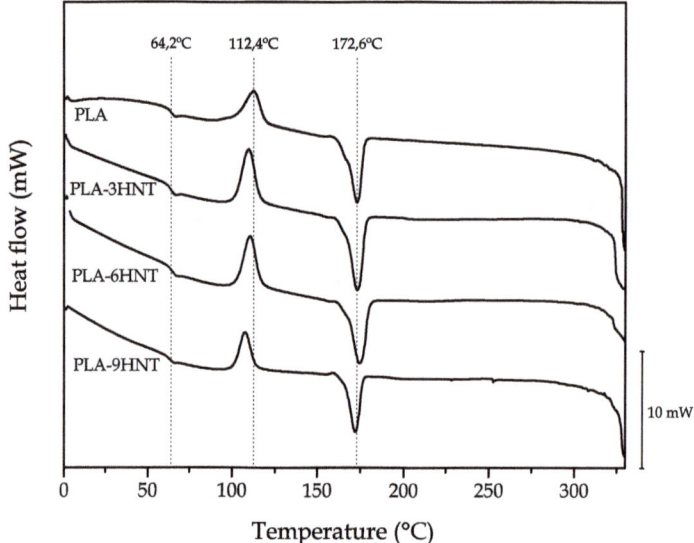

Figure 6. Comparative plot of the differential scanning calorimetry (DSC) thermograms corresponding to PLA-HNT composites with different HNTs loading.

Table 3. Main thermal parameters of PLA-HNT composites with different HNTs loading obtained by differential scanning calorimetry (DSC) analysis.

Code	T_g (°C)	T_{cc} (°C)	ΔH_{cc} (J g^{-1})	T_m (°C)	ΔH_m (J g^{-1})	T_d (°C)	χ_{c_s} * (%)	χ_{c_max} ** (%)
PLA	64.2 ± 3.2	112.4 ± 7.9	25.1 ± 2.6	172.6 ± 13.8	−33.7 ± 0.6	326.2 ± 6.9	9.2 ± 0.2	36.0 ± 0.8
PLA-3HNT	64.0 ± 2.7	109.6 ± 6.7	24.8 ± 3.1	172.6 ± 12.0	−32.3 ± 1.2	321.0 ± 18.5	8.2 ± 0.3	35.6 ± 1.1
PLA-6HNT	64.6 ± 4.5	106.2 ± 10.8	23.4 ± 1.2	174.0 ± 12.8	−29.2 ± 1.8	323.0 ± 23.7	6.6 ± 0.2	33.4 ± 0.9
PLA-9HNT	64.6 ± 4.0	105.4 ± 4.5	22.7 ± 2.6	171.6 ± 6.0	−28.6 ± 0.7	327.1 ± 20.4	6.9 ± 0.2	33.8 ± 1.0

* χ_{c_s} represents the crystallinity of the injection-molded materials without removing the thermal history. ** χ_{c_max} stands for the maximum crystallinity that can be obtained in PLA/HNT composites.

In addition to DSC characterization, the dimensional stability was assessed by thermomechanical analysis (TMA) through determining the coefficient of linear thermal expansion (CLTE) below and above the glass transition temperature (T_g). Obviously, the CLTE was much lower at temperatures below the T_g compared to temperatures above T_g. This is due to the glassy behavior of the material below the T_g, as expected. On the contrary, above T_g the material becomes more plastic and this allows higher expansion. Nevertheless, the effect of HNTs on overall CLTE can be neglected as all the CLTE values for neat PLA and all its composites with HNTs changed in a very narrow range comprised between 76.1 and 78.7 µm m^{-1} °C^{-1}. This identical tendency can be observed for the CLTE above the T_g with values comprised in the 130.5–133.7 µm m^{-1} °C^{-1} range. It has been reported the positive effect of HNTs on dimensional stability of [85] polyimide (PI) composites up 40–50 wt% loading of HNTs and polyamic acid (PAA). This high loading contributes to lowering the CLTE of polyimide (PI) films to be compatible with most metals, thus leading to polymer matrix composites with application as film capacitors. The obtained results in the present work suggest a slight improvement on the dimensional stability but it is not highly pronounced due to the relatively low HNTs loading compared to other works.

Dynamic-mechanical thermal analysis (DMTA) is very helpful to assess the effect of HNTs on both thermal and mechanical properties. Figure 7 shows the plot evolution of the storage modulus (G'; Figure 5a) and the dynamic damping factor (tan δ; Figure 7b). Regarding to neat PLA, its characteristic DMTA curve gave interesting information. Below 50 °C, G' remained almost constant and in the temperature, range comprised between 55–70 °C, a threefold decrease could be detected. This dramatic decrease in the storage modulus, G' was directly related to the temperature range in which the glass transition occurs. Another important process that could be observed by DMTA in neat PLA was the cold crystallization process. Crystallization is related to the formation of a packed-ordered structure, which involves an increase in the storage modulus, G'. Therefore, as we can see, G' increased in the temperature range of 80–100 °C. Regarding the effect of HNTs on DMTA behaviour, it is possible to see that all PLA-HNTs DMTA curves showed the same shape of that of neat PLA but with slight changes. On one hand, all G' curves were shifted to higher values thus indicating an increase in the storage modulus, G' which is directly related to the tensile and flexural modulus as described previously. Similar findings have been reported by Prashanta et al. [74]. Regarding the glass transition process, it seems that all curves were overlapped, which indicates slight or no changes in T_g as observed by DSC. On the other hand, regarding the cold crystallization process, DSC revealed a decrease in the peak temperature, and this is in total agreement with the DMTA curves of neat PLA and PLA-HNTs composites. It is possible to see in Figure 7a that the characteristic curves were moved to lower temperatures as the HNTs content increased. On the other hand, Figure 7b shows the evolution of the dynamic damping factor (tan δ) with the temperature. Despite that there are several methods to obtain the T_g from the DMTA analysis, one of the most accepted methods considers T_g as the peak maximum of the dynamic damping factor. The peak was located at 65 °C, which is in total agreement with the values obtained by DSC and there were no detectable changes in T_g. It is possible to find a correlation with the maximum values of the dynamic damping factor with increasing HNTs loading since tan δ represents the ratio between the loss modulus (G'') and the storage modulus (G'). As indicated, G' increased with HNTs loading and as it is in the denominator of the dynamic damping factor, it led to a decrease in tan δ as it can be observed in Figure 4b. Table 4 shows a summary of some DMTA parameters taken at different temperatures, so that one can see in a clear way how the storage modulus of the samples behaved at different temperatures.

Table 4. Summary of some dynamic mechanical thermal properties of PLA-HNTs composites with different HNTs loading obtained by DMTA.

Code	G' at 30 °C	G' at 50 °C	G' at 80 °C	G' at 120 °C	Max tan δ
PLA	1599	1455	2.10	95.26	2.621
PLA-3 HNT	1678	1549	2.24	102.5	2.582
PLA-6 HNT	1731	1650	2.31	109.1	2.543
PLA-9 HNT	1843	1756	2.70	123.7	2.382

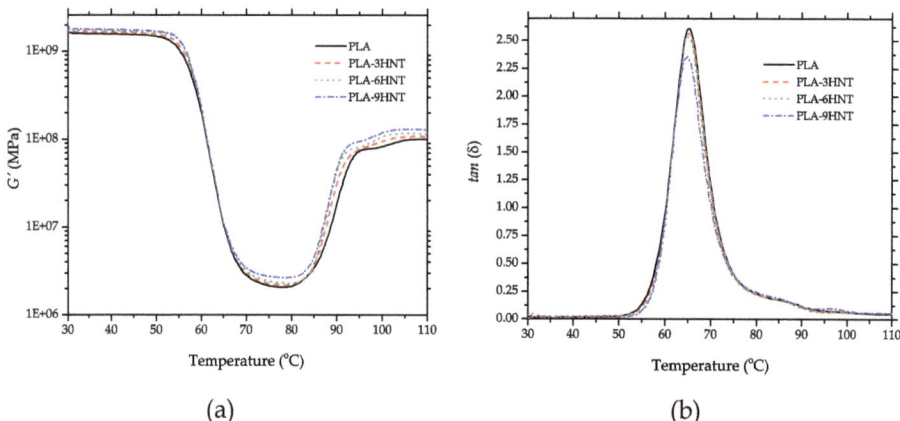

Figure 7. Plot evolution of the dynamic mechanical thermal properties (DMTA) for PLA-HNTs composites with different HNTs loading (**a**) storage modulus, G' and (**b**) dynamic damping factor, $\tan \delta$.

3.3. Study of the Water Uptake of PLA/HNTs Biocomposites

The evolution of the water uptake of PLA-HNTs composites is shown in Figure 8. As it can be seen, there was an increase in the absorbed water as a function of the immersion time and followed the typical Fick's Law. An initial stage with high slope related to water absorption (Δmass) could be seen. In a second stage this increase was less pronounced and finally, a stationary mass was obtained (equilibrium water uptake), denoted as ($\Delta mass_\infty$).

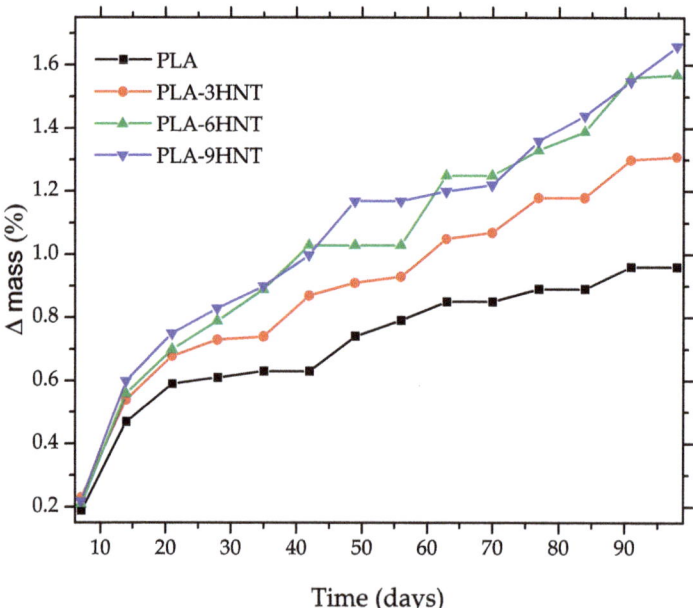

Figure 8. Evolution of the water uptake process in PLA-HNTs composites with different HNTs loading.

The lowest water uptake values were observed for neat PLA with a saturation water content of 0.95% reached after 91 days. Halloysite nanotubes are highly hydrophilic materials due to the aluminosilicate structure and, consequently, as the HNTs loading increases, the saturation water increases as observed by Russo et al. [73]. In particular, the saturation for the PLA composite with 3 wt% HNTs was 1.31% while this was still higher for the composite containing 9 wt% HNTs (1.67%). The hydrophilic nature of HNTs was responsible for this increase. HNTs offer a high surface area with a high number of hydroxyl groups that contribute to the water uptake rate and equilibrium water. Therefore, as the HNTs loading increases, the $\Delta mass_\infty$ also increases.

3.4. Disintegration in Controlled Compost Soil of PLA/HNTs Biocomposites

The weight loss of PLA-HNTs composites during the biodegradation (disintegration in controlled compost soil) is shown in Figure 9. All materials showed an incubation time of about nine days in which very slight weight changes took place. Above this induction time, all materials started a slow weight loss of up to 30 days and above this, the disintegration rate increased. Disintegration of PLA and other aliphatic polyesters was directly related to hydrolytic degradation of the ester groups; for this reason, it is necessary to maintain a relative humidity of 55% at a moderate temperature of 58 ± 2 °C to speed up the process. Neat PLA is the one with the highest disintegration rate as reported in literature [86,87]. After 35 days the weight loss was close to 30% and a 90% weight loss was achieved after 49 days. Above 49 days, PLA was fully disintegrated, and it was not possible to recover any piece to weight it. The presence of HNTs delayed the disintegration process. At 48 days, neat PLA was fully disintegrated while all PLA-HNTs composites showed a weight loss of about 80%. The overall effect of HNTs on the disintegration was a delay in the rate and lower weight loss after the same period compared to neat PLA as halloysite did not degrade due to its inorganic nature.

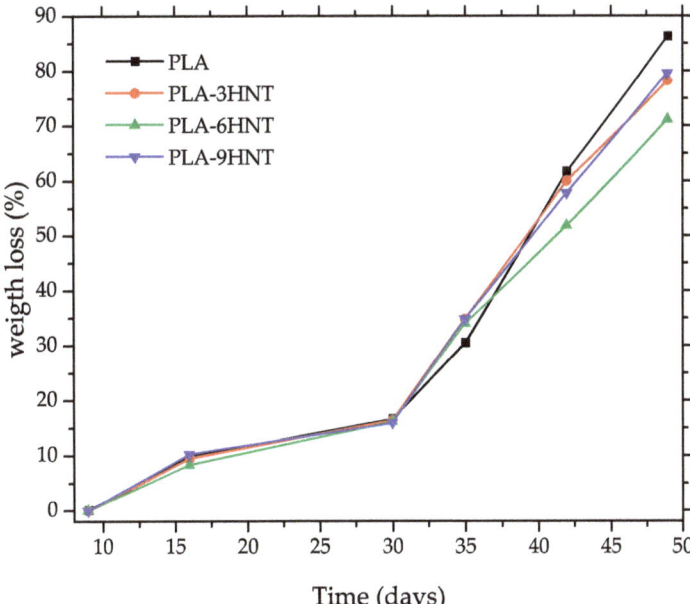

Figure 9. Follow up of the disintegration process in controlled compost soil of PLA-HNTs composites with different HNTs loading.

Figure 10 gathers some optical images of the disintegration process of neat PLA and PLA-HNTs composites with varying HNTs loading. Above two weeks, it was possible to appreciate a clear change in surface in almost all materials, which is representative of embrittlement (crack formation and growth), as Aguero et al. [88] reported, during the first two and three weeks, neat PLA is in the induction period. After four weeks all materials show some disintegration level. Neat PLA shows a less fragmented structure at this time. As it has been detected previously with the weight loss, presence of HNTs delay the disintegration process. After six weeks, the consistency of all materials has been lost and there is not a clear effect of the HNTs content, Paul et al. [89] have investigated the effect of organo-nanomodifiers in PLA blends in which it can be seen that they tend to degrade faster than neat PLA, but it is expectable that the final weight loss is directly related to the HNT content as this inorganic component does not degrade in these conditions, unlike this, in other researches [68], halloysite nanotubes have an effect that promotes degradation in photooxidation media.

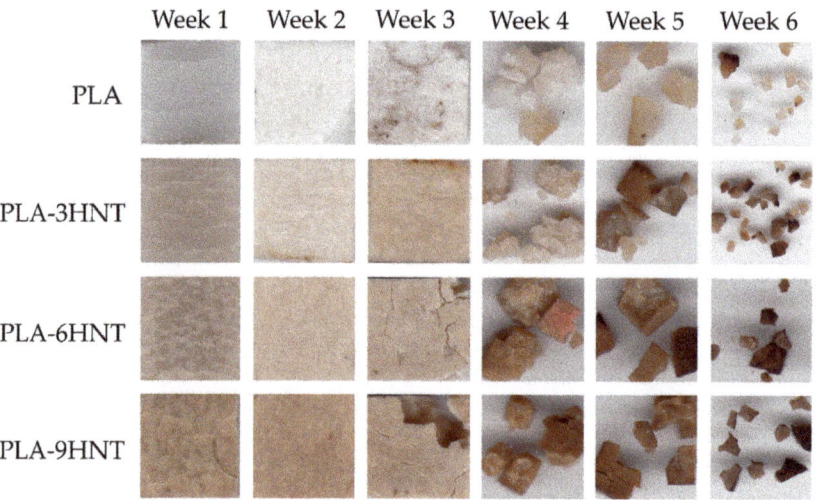

Figure 10. Optical images of the disintegration in controlled compost soil of PLA-HNT composites with different HNTs loading.

4. Conclusions

This work assessed the technical viability of PLA composites with different halloysite nanotubes content (HNTs) for further uses. In particular, this work focused on the effect of HNTs loading (in the 3–9 wt% range) on mechanical, thermal, thermomechanical and disintegration properties of PLA-HNT composites. In general, the mechanical properties were slightly lower than those of neat PLA but the decrease in tensile or flexural strength was less than 7%, which was a positive effect. On the contrary, the elongation at break was reduced to a half with the presence of HNTs. Nevertheless, it is worthy to note that PLA itself was extremely brittle. With regard to thermal properties, presence of HNTs led to lowering the cold crystallization process but the glass transition temperature (T_g) did not change. Identical behavior was obtained with differential scanning calorimetry (DSC) and dynamic-mechanical thermal analysis (DMTA) thus showing the consistency of the obtained results. Regarding the water uptake, the presence of highly hydrophilic HNTs contributes to increased water uptake (which is a negative effect on most industrial applications, but could be a positive effect on medical applications). Finally, the disintegration at PLA with HNTs was slightly delayed but, in general, all composites were almost disintegrated in a reasonable time. This work opens new possibilities to these composites for further applications in medicine as the overall properties are maintained and HNTs could be used as carriers for controlled drug delivery.

Author Contributions: Conceptualization, N.M. and S.M.-J.; methodology, L.S.-N. and D.L; validation N.M., L.S.-N. and S.M.-J.; data analysis, V.C. and D.L.; original draft preparation, V.C. and D.L.; review and editing, S.M.-J. and L.S.-N.; supervision, L.S.-N.; project administration, N.M.

Acknowledgments: This research was supported by the Ministry of Science, Innovation, and Universities (MICIU) through the MAT2017-84909-C2-2-R program number. D. Lascano wants to thank UPV for the grant received though the PAID-01-18 program. Microscopy services at UPV are acknowledged for their help in collecting and analyzing FESEM images.

Conflicts of Interest: The authors declare no conflict of interest.

References

1. Andreessen, C.; Steinbuechel, A. Recent developments in non-biodegradable biopolymers: Precursors, production processes, and future perspectives. *Appl. Microbiol. Biotechnol.* **2019**, *103*, 143–157. [CrossRef] [PubMed]
2. Djukic-Vukovic, A.; Mladenovic, D.; Ivanovic, J.; Pejin, J.; Mojovic, L. Towards sustainability of lactic acid and poly-lactic acid polymers production. *Renew. Sustain. Energy Rev.* **2019**, *108*, 238–252. [CrossRef]
3. Matson, J.B.; Baker, M.B. Polymers for biology, medicine and sustainability. *Polym. Int.* **2019**, *68*, 1219. [CrossRef]
4. Fombuena, V.; Sanchez-Nacher, L.; Samper, M.D.; Juarez, D.; Balart, R. Study of the properties of thermoset materials derived from epoxidized soybean oil and protein fillers. *J. Am. Oil Chem. Soc.* **2013**, *90*, 449–457. [CrossRef]
5. Carbonell-Verdu, A.; Bernardi, L.; Garcia-Garcia, D.; Sanchez-Nacher, L.; Balart, R. Development of environmentally friendly composite matrices from epoxidized cottonseed oil. *Eur. Polym. J.* **2015**, *63*, 1–10. [CrossRef]
6. Espana, J.M.; Samper, M.D.; Fages, E.; Sanchez-Nacher, L.; Balart, R. Investigation of the effect of different silane coupling agents on mechanical performance of basalt fiber composite laminates with biobased epoxy matrices. *Polym. Compos.* **2013**, *34*, 376–381. [CrossRef]
7. Scaffaro, R.; Maio, A.; Sutera, F.; Gulino, E.F.; Morreale, M. Degradation and recycling of films based on biodegradable polymers: A short review. *Polymers* **2019**, *11*, 651. [CrossRef] [PubMed]
8. Narayan, R. Biodegradable and biobased plastics: An overview. In *Soil Degradable Bioplastics for a Sustainable Modern Agriculture*; Malinconico, M., Ed.; Springer: Berlin/Heidelberg, Germany, 2017; pp. 23–34.
9. Li, Y.; Chu, Z.; Li, X.; Ding, X.; Guo, M.; Zhao, H.; Yao, J.; Wang, L.; Cai, Q.; Fan, Y. The effect of mechanical loads on the degradation of aliphatic biodegradable polyesters. *Regen. Biomater.* **2017**, *4*, 179–190. [CrossRef] [PubMed]
10. Segura Gonzalez, E.A.; Olmos, D.; Angel Lorente, M.; Velaz, I.; Gonzalez-Benito, J. Preparation and characterization of polymer composite materials based on PLA/TiO$_2$ for antibacterial packaging. *Polymers* **2018**, *10*, 1365. [CrossRef] [PubMed]
11. Li, Y.; Liao, C.; Tjong, S.C. Synthetic biodegradable aliphatic polyester nanocomposites reinforced with nanohydroxyapatite and/or graphene oxide for bone tissue engineering applications. *Nanomaterials* **2019**, *9*, 590. [CrossRef]
12. Boronat, T.; Fombuena, V.; Garcia-Sanoguera, D.; Sanchez-Nacher, L.; Balart, R. Development of a biocomposite based on green polyethylene biopolymer and eggshell. *Mater. Des.* **2015**, *68*, 177–185. [CrossRef]
13. Filgueira, D.; Holmen, S.; Melbo, J.K.; Moldes, D.; Echtermeyer, A.T.; Chinga-Carrasco, G. 3D printable filaments made of biobased polyethylene biocomposites. *Polymers* **2018**, *10*, 314. [CrossRef] [PubMed]
14. Garcia-Garcia, D.; Carbonell-Verdu, A.; Jorda-Vilaplana, A.; Balart, R.; Garcia-Sanoguera, D. Development and characterization of green composites from bio-based polyethylene and peanut shell. *J. Appl. Polym. Sci.* **2016**, *133*, 43940. [CrossRef]
15. Samper-Madrigal, M.D.; Fenollar, O.; Dominici, F.; Balart, R.; Kenny, J.M. The effect of sepiolite on the compatibilization of polyethylene-thermoplastic starch blends for environmentally friendly films. *J. Mater. Sci.* **2015**, *50*, 863–872. [CrossRef]
16. Yu, X.; Wang, X.; Zhang, Z.; Peng, S.; Chen, H.; Zhao, X. High-performance fully bio-based poly(lactic acid)/polyamide11 (PLA/PA11) blends by reactive blending with multi-functionalized epoxy. *Polym. Test.* **2019**, *78*, 105980. [CrossRef]
17. Gandini, A.; Lacerda, T.M.; Carvalho, A.J.F.; Trovatti, E. Progress of polymers from renewable resources: Furans, vegetable oils, and polysaccharides. *Chem. Rev.* **2016**, *116*, 1637–1669. [CrossRef] [PubMed]

18. Abedini, F.; Ebrahimi, M.; Roozbehani, A.H.; Domb, A.J.; Hosseinkhani, H. Overview on natural hydrophilic polysaccharide polymers in drug delivery. *Polym. Adv. Technol.* **2018**, *29*, 2564–2573. [CrossRef]
19. Rajoka, M.S.R.; Zhao, L.; Mehwish, H.M.; Wu, Y.; Mahmood, S. Chitosan and its derivatives: Synthesis, biotechnological applications, and future challenges. *Appl. Microbiol. Biotechnol.* **2019**, *103*, 1557–1571. [CrossRef]
20. Ferrero, B.; Boronat, T.; Moriana, R.; Fenollar, O.; Balart, R. Green composites based on wheat gluten matrix and posidonia oceanica waste fibers as reinforcements. *Polym. Compos.* **2013**, *34*, 1663–1669. [CrossRef]
21. Ferrero, B.; Fombuena, V.; Fenollar, O.; Boronat, T.; Balart, R. Development of natural fiber-reinforced plastics (NFRP) based on biobased polyethylene and waste fibers from Posidonia oceanica seaweed. *Polym. Compos.* **2015**, *36*, 1378–1385. [CrossRef]
22. DeFrates, K.; Markiewicz, T.; Gallo, P.; Rack, A.; Weyhmiller, A.; Jarmusik, B.; Hu, X. Protein polymer-based nanoparticles: Fabrication and medical applications. *Int. J. Mol. Sci.* **2018**, *19*, 1717. [CrossRef] [PubMed]
23. Rai, K.; Sun, Y.; Shaliutina-Kolesova, A.; Nian, R.; Xian, M. Proteins: Natural polymers for tissue engineering. *J. Biomater. Tissue Eng.* **2018**, *8*, 295–308. [CrossRef]
24. Torres-Giner, S.; Montanes, N.; Boronat, T.; Quiles-Carrillo, L.; Balart, R. Melt grafting of sepiolite nanoclay onto poly(3-hydroxybutyrate-co-4-hydroxybutyrate) by reactive extrusion with multi-functional epoxy-based styrene-acrylic oligomer. *Eur. Polym. J.* **2016**, *84*, 693–707. [CrossRef]
25. Haddadi, M.H.; Asadolahi, R.; Negahdari, B. The bioextraction of bioplastics with focus on polyhydroxybutyrate: A review. *Int. J. Environ. Sci. Technol.* **2019**, *16*, 3935–3948. [CrossRef]
26. Zubir, N.H.M.; Sam, S.T.; Zulkepli, N.N.; Omar, M.F. The effect of rice straw particulate loading and polyethylene glycol as plasticizer on the properties of polylactic acid/polyhydroxybutyrate-valerate blends. *Polym. Bull.* **2018**, *75*, 61–76. [CrossRef]
27. Garcia-Garcia, D.; Garcia-Sanoguera, D.; Fombuena, V.; Lopez-Martine, J.; Balart, R. Improvement of mechanical and thermal properties of poly(3-hydroxybutyrate) (PHB) blends with surface-modified halloysite nanotubes (HNT). *Appl. Clay Sci.* **2018**, *162*, 487–498. [CrossRef]
28. Pramanik, N.; Bhattacharya, S.; Rath, T.; De, J.; Adhikary, A.; Basu, R.K.; Kundu, P.P. Polyhydroxybutyrate-co-hydroxyvalerate copolymer modified graphite oxide based 3D scaffold for tissue engineering application. *Mater. Sci. Eng. C Mater. Biol. Appl.* **2019**, *94*, 534–546. [CrossRef] [PubMed]
29. Quiles-Carrillo, L.; Montanes, N.; Jorda-Vilaplana, A.; Balart, R.; Torres-Giner, S. A comparative study on the effect of different reactive compatibilizers on injection-molded pieces of bio-based high-density polyethylene/polylactide blends. *J. Appl. Polym. Sci.* **2019**, *136*, 47396. [CrossRef]
30. Liu, Y.; Wei, H.; Wang, Z.; Li, Q.; Tian, N. Simultaneous enhancement of strength and toughness of PLA induced by miscibility variation with PVA. *Polymers* **2018**, *10*, 1178. [CrossRef]
31. Behera, K.; Sivanjineyulu, V.; Chang, Y.-H.; Chiu, F.-C. Thermal properties, phase morphology and stability of biodegradable PLA/PBSL/HAp composites. *Polym. Degrad. Stab.* **2018**, *154*, 248–260. [CrossRef]
32. Notta-Cuvier, D.; Odent, J.; Delille, R.; Murariu, M.; Lauro, F.; Raquez, J.M.; Bennani, B.; Dubois, P. Tailoring polylactide (PLA) properties for automotive applications: Effect of addition of designed additives on main mechanical properties. *Polym. Test.* **2014**, *36*, 1–9. [CrossRef]
33. Zhang, L.; Lv, S.; Sun, C.; Wan, L.; Tan, H.; Zhang, Y. Effect of MAH-g-PLA on the properties of wood fiber/polylactic acid composites. *Polymers* **2017**, *9*, 591. [CrossRef] [PubMed]
34. Jiang, Y.; Yan, C.; Wang, K.; Shi, D.; Liu, Z.; Yang, M. Super-toughed PLA blown film with enhanced gas barrier property available for packaging and agricultural applications. *Materials* **2019**, *12*, 1663. [CrossRef] [PubMed]
35. Radusin, T.; Tomsik, A.; Saric, L.; Ristic, I.; Baschetti, M.G.; Minelli, M.; Novakovic, A. Hybrid Pla/wild garlic antimicrobial composite films for food packaging application. *Polym. Compos.* **2019**, *40*, 893–900. [CrossRef]
36. Lopusiewicz, L.; Jedra, F.; Mizielinska, M. New poly(lactic acid) active packaging composite films incorporated with fungal melanin. *Polymers (Basel)* **2018**, *10*, 386. [CrossRef] [PubMed]
37. Behera, K.; Chang, Y.-H.; Chiu, F.-C.; Yang, J.-C. Characterization of poly(lactic acid)s with reduced molecular weight fabricated through an autoclave process. *Polym. Test.* **2017**, *60*, 132–139. [CrossRef]
38. Mistro Matos, B.D.; Rocha, V.; da Silva, E.J.; Moro, F.H.; Bottene, A.C.; Ribeiro, C.A.; Dias, D.d.S.; Antonio, S.G.; do Amaral, A.C.; Cruz, S.A.; et al. Evaluation of commercially available polylactic acid (PLA) filaments for 3D printing applications. *J. Therm. Anal. Calorim.* **2019**, *137*, 555–562. [CrossRef]

39. Alturkestany, M.T.; Panchal, V.; Thompson, M.R. Improved part strength for the fused deposition 3D printing technique by chemical modification of polylactic acid. *Polym. Eng. Sci.* **2019**, *59*, E59–E64. [CrossRef]
40. Fairag, R.; Rosenzweig, D.H.; Ramirez-Garcialuna, J.L.; Weber, M.H.; Haglund, L. Three-dimensional printed polylactic acid scaffolds promote bone-like matrix deposition in vitro. *ACS Appl. Mater. Interfaces* **2019**, *11*, 15306–15315. [CrossRef]
41. Sanatgar, R.H.; Cayla, A.; Campagne, C.; Nierstrasz, V. Morphological and electrical characterization of conductive polylactic acid based nanocomposite before and after FDM 3D printing. *J. Appl. Polym. Sci.* **2019**, *136*, 47040. [CrossRef]
42. Song, B.; Li, W.; Chen, Z.; Fu, G.; Li, C.; Liu, W.; Li, Y.; Qin, L.; Ding, Y. Biomechanical comparison of pure magnesium interference screw and polylactic acid polymer interference screw in anterior cruciate ligament reconstruction—A cadaveric experimental study. *J. Orthop. Transl.* **2017**, *8*, 32–39. [CrossRef] [PubMed]
43. Leksakul, K.; Phuendee, M. Development of hydroxyapatite-polylactic acid composite bone fixation plate. *Sci. Eng. Compos. Mater.* **2018**, *25*, 903–914. [CrossRef]
44. Zhan, X.; Guo, X.; Liu, R.; Hu, W.; Zhang, L.; Xiang, N. Intervention using a novel biodegradable hollow stent containing polylactic acid-polyprolactone-polyethylene glycol complexes against lacrimal duct obstruction disease. *PLoS ONE* **2017**, *12*, e0178679. [CrossRef] [PubMed]
45. Chen, Y.; Murphy, A.; Scholz, D.; Geever, L.M.; Lyons, J.G.; Devine, D.M. Surface-modified halloysite nanotubes reinforced poly(lactic acid) for use in biodegradable coronary stents. *J. Appl. Polym. Sci.* **2018**, *135*, 46521. [CrossRef]
46. Dillon, B.; Doran, P.; Fuenmayor, E.; Healy, A.V.; Gately, N.M.; Major, I.; Lyons, J.G. The influence of low shear microbore extrusion on the properties of high molecular weight poly(l-lactic acid) for medical tubing applications. *Polymers* **2019**, *11*, 710. [CrossRef] [PubMed]
47. Haroosh, H.J.; Dong, Y.; Lau, K.-T. Tetracycline hydrochloride (TCH)-loaded drug carrier based on PLA:PCL nanofibre mats: Experimental characterisation and release kinetics modelling. *J. Mater. Sci.* **2014**, *49*, 6270–6281. [CrossRef]
48. Park, J.W.; Shin, J.H.; Shim, G.S.; Sim, K.B.; Jang, S.W.; Kim, H.J. Mechanical strength enhancement of polylactic acid hybrid composites. *Polymers* **2019**, *11*, 349. [CrossRef]
49. Torres-Giner, S.; Torres, A.; Ferrandiz, M.; Fombuena, V.; Balart, R. Antimicrobial activity of metal cation-exchanged zeolites and their evaluation on injection-molded pieces of bio-based high-density polyethylene. *J. Food Saf.* **2017**, *37*, e12348. [CrossRef]
50. Jamroz, E.; Kulawik, P.; Kopel, P. The effect of nanofillers on the functional properties of biopolymer-based films: A review. *Polymers* **2019**, *11*, 675. [CrossRef]
51. Huang, T.; Qian, Y.; Wei, J.; Zhou, C. Polymeric antimicrobial food packaging and its applications. *Polymers* **2019**, *11*, 560. [CrossRef]
52. Sharmeen, S.; Rahman, A.F.M.M.; Lubna, M.M.; Salem, K.S.; Islam, R.; Khan, M.A. Polyethylene glycol functionalized carbon nanotubes/gelatin-chitosan nanocomposite: An approach for significant drug release. *Bioact. Mater.* **2018**, *3*, 236–244. [CrossRef] [PubMed]
53. Stango, A.X.; Vijayalakshmi, U. Electrochemically grown functionalized-Multi-walled carbon nanotubes/hydroxyapatite hybrids on surgical grade 316L SS with enhanced corrosion resistance and bioactivity. *Colloids Surf. B Biointerfaces* **2018**, *171*, 186–196.
54. Van den Broeck, L.; Piluso, S.; Soultan, A.H.; De Volder, M.; Patterson, J. Cytocompatible carbon nanotube reinforced polyethylene glycol composite hydrogels for tissue engineering. *Mater. Sci. Eng. C Mater. Biol. Appl.* **2019**, *98*, 1133–1144. [CrossRef] [PubMed]
55. Osfoori, A.; Selahi, E. Performance analysis of bone scaffolds with carbon nanotubes, barium titanate particles, hydroxyapatite and polycaprolactone. *Biomater. Biomech. Bioeng.* **2019**, *4*, 33–44.
56. Zhang, X.; Zhang, D.; Peng, Q.; Lin, J.; Wen, C. Biocompatibility of nanoscale hydroxyapatite coating on TiO_2 nanotubes. *Materials* **2019**, *12*, 1979. [CrossRef] [PubMed]
57. Beke, S.; Barenghi, R.; Farkas, B.; Romano, I.; Koroesi, L.; Scaglione, S.; Brandi, F. Improved cell activity on biodegradable photopolymer scaffolds using titanate nanotube coatings. *Mater. Sci. Eng. C Mater. Biol. Appl.* **2014**, *44*, 38–43. [CrossRef] [PubMed]
58. Chandanshive, B.B.; Rai, P.; Rossi, A.L.; Ersen, O.; Khushalani, D. Synthesis of hydroxyapatite nanotubes for biomedical applications. *Mater. Sci. Eng. C Mater. Biol. Appl.* **2013**, *33*, 2981–2986. [CrossRef]

59. Zhang, Y.; Nayak, T.R.; Hong, H.; Cai, W. Biomedical applications of zinc oxide nanomaterials. *Curr. Mol. Med.* **2013**, *13*, 1633–1645. [CrossRef]
60. Garcia-Garcia, D.; Ferri, J.M.; Ripoll, L.; Hidalgo, M.; Lopez-Martinez, J.; Balart, R. Characterization of selectively etched halloysite nanotubes by acid treatment. *Appl. Surf. Sci.* **2017**, *422*, 616–625. [CrossRef]
61. Venkatesh, C.; Clear, O.; Major, I.; Lyons, J.G.; Devine, D.M. Faster release of lumen-loaded drugs than matrix-loaded equivalent in polylactic acid/halloysite nanotubes. *Materials* **2019**, *12*, 1830. [CrossRef]
62. Pluta, M.; Bojda, J.; Piorkowska, E.; Murariu, M.; Bonnaud, L.; Dubois, P. The effect of halloysite nanotubes and N,N'-ethylenebis (stearamide) on morphology and properties of polylactide nanocomposites with crystalline matrix. *Polym. Test.* **2017**, *64*, 83–91. [CrossRef]
63. Yin, X.; Wang, L.; Li, S.; He, G.; Yang, Z. Effects of surface modification of halloysite nanotubes on the morphology and the thermal and rheological properties of polypropylene/halloysite composites. *J. Polym. Eng.* **2018**, *38*, 119–127. [CrossRef]
64. Padhi, S.; Achary, P.G.R.; Nayak, N.C. Mechanical and morphological properties of modified halloysite nanotube filled ethylene-vinyl acetate copolymer nanocomposites. *J. Polym. Eng.* **2018**, *38*, 271–279. [CrossRef]
65. Gorrasi, G.; Bugatti, V.; Ussia, M.; Mendichi, R.; Zampino, D.; Puglisi, C.; Carroccio, S.C. Halloysite nanotubes and thymol as photo protectors of biobased polyamide 11. *Polym. Degrad. Stab.* **2018**, *152*, 43–51. [CrossRef]
66. Massaro, M.; Cavallaro, G.; Colletti, C.G.; D'Azzo, G.; Guernelli, S.; Lazzara, G.; Pieraccini, S.; Riela, S. Halloysite nanotubes for efficient loading, stabilization and controlled release of insulin. *J. Colloid Interface Sci.* **2018**, *524*, 156–164. [CrossRef] [PubMed]
67. Sikora, J.W.; Gajdos, I.; Puszka, A. Polyethylene-matrix composites with halloysite nanotubes with enhanced physical/thermal properties. *Polymers* **2019**, *11*, 787. [CrossRef] [PubMed]
68. Therias, S.; Murariu, M.; Dubois, P. Bionanocomposites based on PLA and halloysite nanotubes: From key properties to photooxidative degradation. *Polym. Degrad. Stab.* **2017**, *145*, 60–69. [CrossRef]
69. Saeidlou, S.; Huneault, M.A.; Li, H.; Park, C.B. Poly(lactic acid) crystallization. *Prog. Polym. Sci.* **2012**, *37*, 1657–1677. [CrossRef]
70. Ke, T.Y.; Sun, X.Z. Effects of moisture content and heat treatment on the physical properties of starch and poly(lactic acid) blends. *J. Appl. Polym. Sci.* **2001**, *81*, 3069–3082. [CrossRef]
71. Fischer, E.W.; Sterzel, H.J.; Wegner, G. Investigation of structure of solution grown crystals of lactide copolymers by means of chemical-reactions. *Kolloid Z. Polym.* **1973**, *251*, 980–990. [CrossRef]
72. Li, Y.; Venkateshan, K.; Sun, X.S. Mechanical and thermal properties, morphology and relaxation characteristics of poly(lactic acid) and soy flour/wood flour blends. *Polym. Int.* **2010**, *59*, 1099–1109. [CrossRef]
73. Russo, P.; Cammarano, S.; Bilotti, E.; Peijs, T.; Cerruti, P.; Acierno, D. Physical properties of poly lactic acid/clay nanocomposite films: Effect of filler content and annealing treatment. *J. Appl. Polym. Sci.* **2014**, *131*, 39798. [CrossRef]
74. Prashantha, K.; Lecouvet, B.; Sclavons, M.; Lacrampe, M.F.; Krawczak, P. Poly(lactic acid)/halloysite nanotubes nanocomposites: Structure, thermal, and mechanical properties as a function of halloysite treatment. *J. Appl. Polym. Sci.* **2013**, *128*, 1895–1903. [CrossRef]
75. De Silva, R.T.; Soheilmoghaddam, M.; Goh, K.L.; Wahit, M.U.; Bee, S.A.H.; Chai, S.-P.; Pasbakhsh, P. Influence of the processing methods on the properties of poly(lactic acid)/halloysite nanocomposites. *Polym. Compos.* **2016**, *37*, 861–869. [CrossRef]
76. De Silva, R.T.; Pasbakhsh, P.; Goh, K.L.; Chai, S.P.; Chen, J. Synthesis and characterisation of poly (lactic acid)/halloysite bionanocomposite films. *J. Compos. Mater.* **2014**, *48*, 3705–3717. [CrossRef]
77. Pracella, M.; Haque, M.M.-U.; Puglia, D. Morphology and properties tuning of PLA/cellulose nanocrystals bio-nanocomposites by means of reactive functionalization and blending with PVAc. *Polymer* **2014**, *55*, 3720–3728. [CrossRef]
78. Kontou, E.; Niaounakis, M.; Georgiopoulos, P. Comparative study of PLA nanocomposites reinforced with clay and silica nanofillers and their mixtures. *J. Appl. Polym. Sci.* **2011**, *122*, 1519–1529. [CrossRef]
79. Chen, Y.; Geever, L.M.; Killion, J.A.; Lyons, J.G.; Higginbotham, C.L.; Devine, D.M. Halloysite nanotube reinforced polylactic acid composite. *Polym. Compos.* **2017**, *38*, 2166–2173. [CrossRef]

80. Guo, J.; Qiao, J.; Zhang, X. Effect of an alkalized-modified halloysite on PLA crystallization, morphology, mechanical, and thermal properties of PLA/halloysite nanocomposites. *J. Appl. Polym. Sci.* **2016**, *133*, 44272. [CrossRef]
81. Liu, M.; Zhang, Y.; Zhou, C. Nanocomposites of halloysite and polylactide. *Appl. Clay Sci.* **2013**, *75–76*, 52–59. [CrossRef]
82. Tham, W.L.; Poh, B.T.; Ishak, Z.A.M.; Chow, W.S. Thermal behaviors and mechanical properties of halloysite nanotube-reinforced poly(lactic acid) nanocomposites. *J. Therm. Anal. Calorim.* **2014**, *118*, 1639–1647. [CrossRef]
83. Murariu, M.; Doumbia, A.; Bonnaud, L.; Dechief, A.-L.; Paint, Y.; Ferreira, M.; Campagne, C.; Devaux, E.; Dubois, P. High-performance Polylactide/ZnO nanocomposites designed for films and fibers with special end-use properties. *Biomacromolecules* **2011**, *12*, 1762–1771. [CrossRef] [PubMed]
84. Murariu, M.; Dechief, A.-L.; Paint, Y.; Peeterbroeck, S.; Bonnaud, L.; Dubois, P. Polylactide (PLA)-halloysite nanocomposites: Production, morphology and key-properties. *J. Polym. Environ.* **2012**, *20*, 932–943. [CrossRef]
85. Zhu, T.; Qian, C.; Zheng, W.; Bei, R.; Liu, S.; Chi, Z.; Chen, X.; Zhang, Y.; Xu, J. Modified halloysite nanotube filled polyimide composites for film capacitors: High dielectric constant, low dielectric loss and excellent heat resistance. *RSC Adv.* **2018**, *8*, 10522–10531. [CrossRef]
86. Kumar, R.; Yakubu, M.K.; Anandjiwala, R.D. Biodegradation of flax fiber reinforced poly lactic acid. *Express Polym. Lett.* **2010**, *4*, 423–430. [CrossRef]
87. Mathew, A.P.; Oksman, K.; Sain, M. Mechanical properties of biodegradable composites from poly lactic acid (PLA) and microcrystalline cellulose (MCC). *J. Appl. Polym. Sci.* **2005**, *97*, 2014–2025. [CrossRef]
88. Aguero, A.; Quiles-Carrillo, L.; Jorda-Vilaplana, A.; Fenollar, O.; Montanes, N. Effect of different compatibilizers on environmentally friendly composites from poly(lactic acid) and diatomaceous earth. *Polym. Int.* **2019**, *68*, 893–903. [CrossRef]
89. Paul, M.A.; Delcourt, C.; Alexandre, M.; Degée, P.; Monteverde, F.; Dubois, P. Polylactide/montmorillonite nanocomposites: Study of the hydrolytic degradation. *Polym. Degrad. Stab.* **2005**, *87*, 535–542. [CrossRef]

 © 2019 by the authors. Licensee MDPI, Basel, Switzerland. This article is an open access article distributed under the terms and conditions of the Creative Commons Attribution (CC BY) license (http://creativecommons.org/licenses/by/4.0/).

Article

Characterization on Polyester Fibrous Panels and Their Homogeneity Assessment

Tao Yang [1,*], Ferina Saati [2], Jean-Philippe Groby [3], Xiaoman Xiong [4], Michal Petrů [1], Rajesh Mishra [4], Jiří Militký [4] and Steffen Marburg [2]

1. Institute for Nanomaterials, Advanced Technologies and Innovation, Technical University of Liberec, 461 17 Liberec, Czech Republic; michal.petru@tul.cz
2. Chair of Vibroacoustics of Vehicles and Machines, Department of Mechanical Engineering, Technical University of Munich, Boltzmannstrasse 15, 85748 Garching, Germany; ferina.saati@tum.de (F.S.); steffen.marburg@tum.de (S.M.)
3. Laboratoire d'Acoustique de l'Université du Mans, LAUM-UMR CNRS 6613, Le Mans Université, Avenue Olivier Messiaen, 72085 Le Mans CEDEX 9, France; Jean-Philippe.Groby@univ-lemans.fr
4. Department of Material Engineering, Faculty of Textile Engineering, Technical University of Liberec, 46117 Liberec, Czech Republic; xiaoman.xiong@tul.cz (X.X.); rajesh.mishra@tul.cz (R.M.); Jiri.Militky@tul.cz (J.M.)
* Correspondence: tao.yang@tul.cz

Received: 1 September 2020; Accepted: 14 September 2020; Published: 15 September 2020

Abstract: Nowadays, fibrous polyester materials are becoming one of the most important alternatives for controlling reverberation time by absorbing unwanted sound energy in the automobile and construction fields. Thus, it is worthy and meaningful to characterize their acoustic behavior. To do so, non-acoustic parameters, such as tortuosity, viscous and thermal characteristic lengths and thermal permeability, must be determined. Representative panels of polyester fibrous material manufactured by perpendicular laying technology are thus tested via the Bayesian reconstruction procedure. The estimated porosity and airflow resistivity are found in good agreement with those tested via direct measurements. In addition, the homogeneity of polyester fibrous panels was characterized by investigating the mean relative differences of inferred non-acoustic parameters from the direct and reverse orientation measurements. Some parameters, such as tortuosity, porosity and airflow resistivity, exhibit very low relative differences. It is found that most of the panels can be assumed homogeneous along with the panel thickness, the slight inhomogeneity mostly affecting the thermal characteristic length.

Keywords: Bayesian reconstruction; homogeneity; porous materials; fibrous polyester materials

1. Introduction

It has been demonstrated that polyester fibrous material is a good alternative to conventional sound absorbing material [1]. In textile industry, polyester fiber assemblies are in the form of woven, knitted and nonwoven structures. The woven structure fibrous material is made by using two or more sets of yarn interlaced to each other. Woven structure is generally more durable in comparison with knitted and nonwoven structures. However, woven structure is not widely used in noise treatment at low frequency range due to its relatively small thickness. The sound absorption of polyester woven structure determined by impedance tube and reverberant field method has been reported [2]. The single layer woven structure with a small thickness (i.e., 2.16–2.41 mm) exhibits poor sound absorption at low frequency range. Knitted structures are made of interlocking loops by using one or more yarns. Spacer fabrics, a special type of knitted structure, attracted great attention for sound absorption because of their thick structure possibility and designable appearances [3,4].

Unlike woven and knitted structures, nonwoven structure fibrous material has more advantages as a sound absorbing material, such as high porosity, economical price, light weight, recyclability, good elasticity and a large thickness range. Thus, nonwoven structure is more widely used for different noise reduction requirements. Nonwoven structure is produced in three stages: web formation, web bonding and finishing. Drylaid (i.e., carded, airlaid), wet-laid, spunmelt are the main technologies for web formation. Web bonding can be achieved through mechanical, thermal and chemical methods. The finishing aims to improve the outward appearance and the quality of the fibrous structure. Finishing methods are various according to different required specific properties. Nonwovens have an important role in sound absorption within the automotive, construction and a variety of industrial uses [5]. Nonwoven structure and its combination with other materials (such as woven structure, polyurethane foam, polypropylene foam, etc.) are used in seating area, headliners, side panels, carpets, trunks, bonnet liners for interior vehicle noise control [6]. Nonwoven structure material coupled with a hard wall can significantly reduce noise transmission and reduce the reverberation by improving the sound absorption [7,8].

The polyester nonwoven structure used in noise reduction is normally in the form of panel. The acoustic properties of polyester nonwoven panels have been well studied [9–11]. Kino et al. investigated the effect of various cross-sectional shapes polyester nonwoven structure on the acoustic and non-acoustic properties [10]. They concluded that cross-sectional shape has a slight effect on sound absorption, while there is a significant effect on the airflow resistivity, the thermal and viscous characteristic lengths. The accuracy of several prediction models for polyester materials was investigated by Garai and Pompoli, they recommended a new model to predict the acoustic characteristics of polyester fibrous materials [12].

Investigation of the homogeneity of an acoustic absorber can facilitate optimization of acoustic properties in practical applications. For instance, an absorber which is inhomogeneous in thickness direction will usually exhibit different acoustic performance at direct and reverse orientations. However, very few research studied characterization of the homogeneity of porous materials. In one paper, researchers applied X-ray computerized tomography (CT) to characterize the homogeneity of prebaked anode by analyzing the distribution of coke, pitch and porosity throughout the anodes as well as the variations in binder matrix thickness [13].

High-loft nonwoven panel is known to have useful acoustic properties [14]. Although some studies related to the acoustic properties of this material can be found in the literature [14,15], there is a lack of research on recovered non-acoustic parameters via inverse method for this type of material. Moreover, there are only a few publications focusing on characterization of the homogeneity of nonwoven panels. This paper presents an investigation of estimating non-acoustic parameters of polyester high-loft nonwoven panel via Bayesian reconstruction procedure. In addition, the homogeneity of polyester panels has been numerically assessed by comparing the results from panels' direct and reverse orientations.

2. Experiment

2.1. Materials

The polyester fibrous panels (Technical University of Liberec, Liberec, Czech Republic) are composed of 45 wt.% staple polyester, 30 wt.% hollow polyester and 25 wt.% bicomponent polyester. Low-melting polyester fiber consists of the sheath part of bicomponent fiber which is used to thermally bond the fiber structure. The cross-sectional images of three types of polyester fibers are shown in Figure 1. A microscope was used to measure fiber diameter, and fifty fibers were measured for each type of fiber to obtain an accurate average value. The mean diameters of the staple, hollow and bicomponent fibers are 13.19, 24.45 and 17.94 µm, respectively.

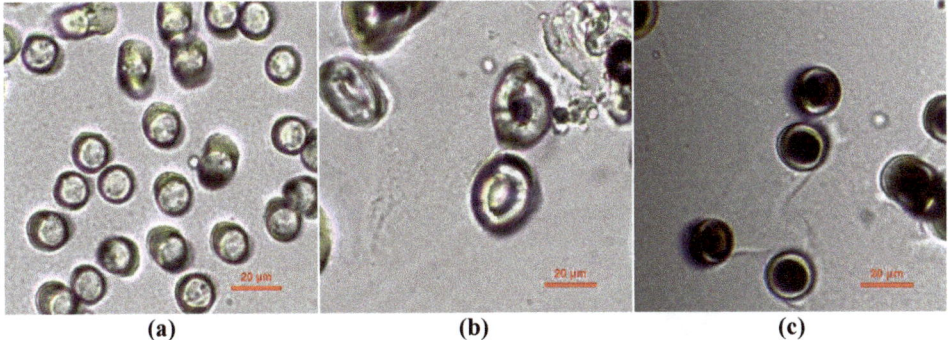

Figure 1. Cross-sectional images of fibers. (**a**) Staple fiber; (**b**) hollow fiber; (**c**) bicomponent fiber [9].

Polyester fibrous panel samples were produced by perpendicular laying technology which consists of carded and thermal bonding procedures (see Figure 2) [16,17]. The carded web, consisting of a proportion of bicomponent polyester fibers in the blend, is fed onto the conveyor belt. The reciprocating forming comb and pressure bar are two main working elements used to create vertical folds. The forming comb strokes the lower part of the carded web and pushes the carded web to form a vertical fold. The reciprocating pressure bar moves the folded web along with the wire guide and the conveyor belt to the batt layer. With the movement of the pressure bar, the needles placed on the pressure bar penetrate the folded web and strengthens the folds, which improves the vertical orientation of the fibers in the nonwoven structure. The web is subsequently stabilized by melting the bonding fibers present in the fiber blend when it passes through the through-air bonding chamber. Thereafter, the nonwoven fabric is cooled. Panel thickness is controllable by setting the distance between the grid and conveyor as well as the dimension of the pressure bar. Panel density is adjusted via the velocity of the conveyor belt. By selecting proper fiber blend and adjusting the lapping device, various end products providing high absorption and insulation performance to meet a variety of applications could be achieved. An example of so manufactured polyester fibrous panel is illustrated in Figure 3.

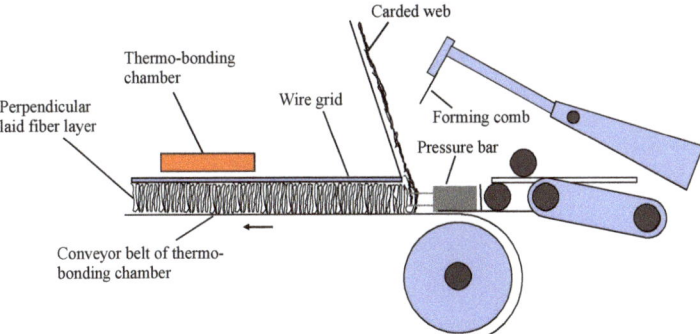

Figure 2. Sketch of perpendicular laying technology [17].

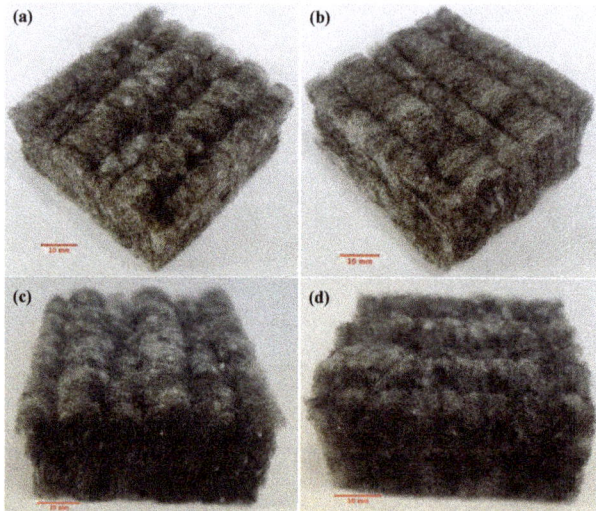

Figure 3. Example of manufactured polyester fibrous panel. Images are taken from different directions: (**a**,**b**) top view, and (**c**,**d**) side view.

2.2. Direct Characterization

Some non-acoustic parameters, such as thickness, porosity and airflow resistivity, can be easily determined according to ASTM D1777-96, ASTM C830-00 and ISO 9053-1, respectively [18–20]. The micro-CT (micro computed tomography), SEM (scanning electron microscope) and length weighted methods can be applied to determine fiber mean diameter [21,22]. The tomographic reconstruction method used to measure tortuosity of porous materials has been reported in 2009 [23].

Fiber diameter, thickness, porosity and airflow resistivity were directly measured in this study. The characteristics of the polyester fibrous panel specimens are listed in Table 1. An Alambeta device (SENSORA, Liberec, Czech Republic) was used to determine the panel thicknesses. Sample porosities were determined according to ASTM C830-00 [18]. In this standard, the porosity was determined as $\phi = 1 - \rho/\rho_f$, where ρ is the fabric bulk density and ρ_f is the fiber density which is 1141.82 kg/m^3. The density of three types of fibers was measured by liquid pycnometry technique [24]. Closed pores have less or no effect on airflow resistivity and sound absorption compared to open pores [25]. As a consequence, the voids in hollow fibers were not included in the analysis.

Table 1. Characteristics of polyester materials.

Samples	Porosity (%)	Bulk Density (kg/m^3)	Thickness (mm)	Airflow Resistivity (Pa·s/m^2)
Sample 1	98.52	16.93	27.48	4108 ± 199
Sample 2	98.29	19.49	23.87	5357 ± 217
Sample 3	98.03	22.49	20.69	7029 ± 356
Sample 4	97.94	23.54	20.32	7498 ± 333
Sample 5	97.86	24.45	20.76	7319 ± 243
Sample 6	97.85	24.54	19.49	10,978 ± 329
Sample 7	97.66	26.71	19.00	7530 ± 408
Sample 8	97.59	27.54	18.43	9829 ± 376
Sample 9	97.58	27.61	16.85	10,181 ± 259
Sample 10	97.29	30.94	15.46	13,397 ± 329
Sample 11	96.94	34.95	13.31	12,868 ± 199
Sample 12	96.89	35.56	14.27	14,989 ± 285
Sample 13	96.86	35.87	14.15	15,414 ± 167
Sample 14	96.59	38.98	12.27	19,751 ± 442
Sample 15	96.09	44.60	10.43	20,474 ± 687
Sample 16	96.01	45.56	11.14	19,733 ± 688

The 100 mm diameter circular specimens were punched by an Elektronische Stanzmaschine Type 208 machine for the standard airflow resistivity test. The airflow resistivity of samples was measured on AFD300 AcoustiFlow device (Gesellschaft für Akustikforschung Dresden mbH, Dresden, Germany) according to ISO 9053:1991 [20]. Ten samples were measured for each polyester nonwoven panel. Normally, the decrease of porosity results in the increase of airflow resistivity for the samples made by the same fiber content and manufacturing technology. While the airflow resistivity exhibits a drop in Samples 7–9, the phenomenon can be attributed to the slightly different manufacturing technology for these three samples compared to other samples.

2.3. Acoustic Characterization

The acoustic properties of the materials can be directly evaluated via reverberant chamber measurements, steady-state measurements and impedance tube measurements. Moreover, acoustic methods can be used to characterize the morphologic characterizations of reticulated foams, fibrous materials, granular materials and sandy soils [26].

2.3.1. Impedance Tube Measurement

A four-microphone impedance tube was applied to carry out the measurements to recover the reflection R and transmission T coefficients. Then, the dynamic density $\widetilde{\rho}_{eq}$ and dynamic bulk modulus $\widetilde{K}_{eq}$ can be easily recovered [27–29].

Figure 4 illustrates the 30 mm impedance tube used for this study. It consists of two 1/4 in. G.R.A.S. microphones flush mounted on both sides of the test sample and the tube ends with an anechoic termination. The distances between microphone positions and the sample front surface are 50, 30, 150 and 170 mm, respectively. The excitation signal was a logarithmic swept sine, over the frequency range of 800–5500 Hz. To achieve accurate results, the microphones were calibrated, and phase matched with each other [27].

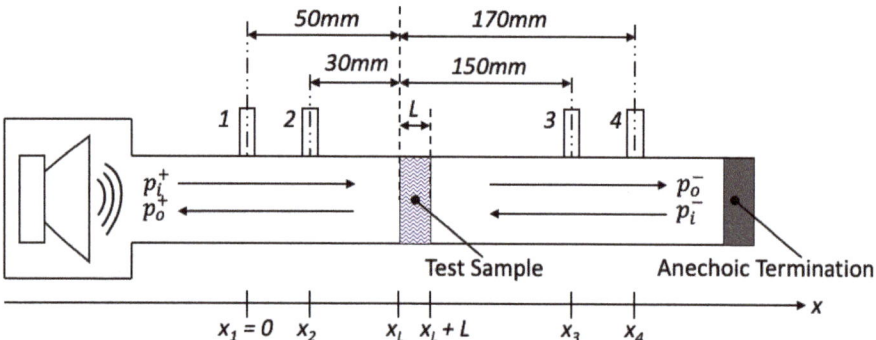

Figure 4. Four-microphone impedance tube configurations.

2.3.2. The Johnson-Champoux-Allard-Lafarge Model

In the Johnson-Champoux-Allard-Lafarge (JCAL) model, the equivalent dynamic density is associated with the viscous losses and the equivalent dynamic bulk modulus with the thermal losses [27]. The dynamic density of porous media was first proposed by Johnson et al. in 1987 [30]. Champoux, Allard and Lafarge et al. then modified the dynamic bulk modulus of porous media [31,32].

If saturating fluid is air, the JCAL model assumes the porous media are rigid and motionless at the frequency over the phase decoupling frequency. In the model, the equivalent dynamic density is described as:

$$\widetilde{\rho}_{eq} = \frac{\rho_0}{\phi}\widetilde{\alpha}(\omega) \tag{1}$$

where $\widetilde{\alpha}(\omega)$ is the dynamic tortuosity, given by:

$$\widetilde{\alpha}(\omega) = \alpha_\infty + \frac{jv}{\omega}\frac{\phi}{k_0}\sqrt{1 - \frac{j\omega}{v}\left(\frac{2\alpha_\infty k_0}{\phi\Lambda}\right)^2} \tag{2}$$

where ρ_0 is the density of the saturating fluid, ϕ is the open porosity, α_∞ is the dynamic tortuosity, j is the complex number, $v = \eta/\rho_0$ is the kinematic viscosity, where η is the dynamic viscosity, $\omega = 2\pi f$ is the angular frequency, $k_0 = \eta/\sigma$ is the static viscous permeability, where σ is the airflow resistivity, and Λ is the viscous characteristic length [30].

The dynamic bulk modulus is described as:

$$\widetilde{K}_{eq} = \frac{\gamma P_0}{\phi}\left(\gamma - \frac{\gamma - 1}{\widetilde{\alpha}'(\omega)}\right)^{-1} \tag{3}$$

where $\widetilde{\alpha}'(\omega)$ is the thermal tortuosity, given by

$$\widetilde{\alpha}'(\omega) = 1 + \frac{jv'}{\omega}\frac{\phi}{k_0'}\sqrt{1 - \frac{j\omega}{v'}\left(\frac{2k_0'}{\phi\Lambda'}\right)^2} \tag{4}$$

where $v' = \frac{v}{Pr}$, where Pr is the Prandtl number, k_0' is the static thermal permeability, Λ' is the thermal characteristic length [31,32].

2.3.3. The Inversion Procedure

Inverse characterization methods are becoming more popular since they are able to simultaneously reckon several parameters [27]. The comparison between inferred porosity, airflow resistivity and characteristic lengths of porous materials using a commercial inversion software (i.e., FOAM-X) and those obtained from ultrasonic measurements, Bies-Allard and Kino-Allard models was studied in

2012 [33]. The deterministic inverse methods normally fit the measurements and predictions through models, then find the parameter that best describes the material. The Bayesian approach was applied to carry out the inversion procedure in this paper. Bayesian approach can inversely determine some physical parameters of porous materials from the impedance tube measurement [34]. The unknown parameters are assumed as random variables distributed according to a probability distribution in the Bayesian approach. This distribution is based on the existing knowledge or experience about the parameter values. However, for many of the non-acoustic parameters, it is possible to specify lower and upper bounds. Therefore, the prior parameter distribution can be defined. Posterior to collecting data, the parameter distribution is given by Bayes' theorem. The posterior parameter distribution depends both on the prior distribution and on the measurements. Thus, this distribution contains all the available information about the parameters. By using the optimization method, good estimation of parameters can be obtained. One optimization technique called Markov chain Monte Carlo (MCMC) has been well applied to solve the problem of non-acoustic parameters estimation. Not only the probability density function (pdf) of each parameter, but also the joint probability density functions of the parameters can be retrieved according to the Bayesian approach in conjunction with MCMC [35].

2.4. Homogeneity Assessment

A homogeneous material has the same properties everywhere, i.e., uniform without irregularities. Some macroscopic pore structure parameters (e.g., permeability or porosity) can be used to verify the homogeneity of porous media [36]. However, homogeneity strongly depends on the selected sample size. At the initial increase of sample size, those parameters vary with the sample size and exhibit random fluctuations. As the sample size increases, the amplitude of the fluctuations gradually diminishes. The values of the pore structure parameters remain stable once a certain sample size is reached. Dullien [36] stated that when the structure parameters of a porous material maintain close values to the increasing sample size, the media is said to be macroscopically, or statistically, homogeneous.

As stated above, researchers characterized the homogeneity of prebaked anodes by analyzing the distribution of pores and different types of particles based on X-ray computerized tomography and image analysis [13]. However, the first step, i.e., computerized tomography on porous material, is a time-consuming work. For instance, one cubic sample with 3 mm × 4 mm × 10 mm cube lengths takes 6 to 8 h to accomplish the tomography. By contrast, the acoustic method is much more efficient. The inferred non-acoustic parameters (e.g., porosity, tortuosity, airflow resistivity) can represent the geometric structure in the polyester panels. Moreover, homogeneity in through-plane orientation is more important than in-plane orientation since sound waves mainly propagate from surface to inner structure. Thus, the homogeneity at thickness direction of polyester panels will be analyzed by comparing the inferred non-acoustic parameters.

3. Results

In order to figure out the homogeneity of polyester panels, the measurements from both direct and reverse orientations of samples have been carried out. The surface texture of front and back sides is shown in Figure 5.

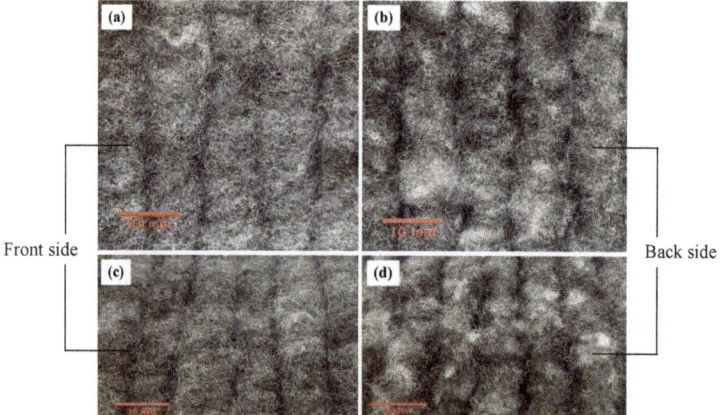

Figure 5. The surface texture of polyester fibrous panel sample: (**a**,**c**) front side, and (**b**,**d**) back side.

In order to prepare specimens with proper radius, specimens are carefully cut by scissors. A transparent plastic tube having the same radius as the impedance tube was adapted to ensure the specimens fitting exactly into the tube's cross-section. The samples are cautiously mounted in the connecting tube in the middle of four microphones before testing. The surface of each specimen is at the same level to the edge of the connecting tube. When changing the measurement direction, samples are rotated to another direction to avoid displacement. The front surface orientated to the speaker was referred to as direct orientation. The recovered parameters are listed in Table 2. Since airflow resistivity was used more often compared to static viscous permeability and these two parameters can be easily converted through the formula $k_0 = \eta/\sigma$, only the airflow resistivity is presented in Table 2. In addition, the values of the recovered porosity (ϕ), tortuosity (α_∞), viscous characteristic length (Λ), thermal characteristic length (Λ') and static thermal permeability (k_0') are presented. The standard deviations for each value are reported in brackets. The tortuosity of common fibrous absorbent ranges from 1 to 1.06 [37]. As porosity approaches the value of 1, tortuosity reduces to the minimal value of 1 [38]. Some existing empirical formula between porosity and tortuosity, $\alpha_\infty = 1/\sqrt{\phi}$ and $\alpha_\infty = 1 + (1 - \phi)/2\phi$, can also explain this correlation [39,40]. The reconstructed tortuosity for all of the samples is 1 with low standard deviation as shown in Table 2. Although the recovered porosities seemingly display big differences compared to the directly measured values, the maximum relative error of inferred porosity is less than 4%. It was considered that the results with an error less than 10% are accurate enough for inverse analysis, as the value of porosity for a porous material can vary due to several uncertainties during measurements. Thus, it can be concluded that the inferred porosity is reasonable.

Table 2. Inferred non-acoustic parameters of polyester panels.

Samples	Orientations	ϕ (%)	α_∞	Λ (μm)	Λ' (μm)	σ (Pa·s/m$_2$)	k'_0 (10^{-9} m$_2$)
Sample 1	Direct	99.8 (0.19)	1(0.0017)	184 (3.2)	354(30.8)	5225	6.98 (0.53)
	Reverse	99.8 (0.18)	1(0.0017)	187 (4.0)	414(32.0)	5202	6.95 (0.50)
Sample 2	Direct	99.0 (0.86)	1 (0.0026)	146 (5.1)	333(71.0)	6527	4.56 (0.36)
	Reverse	99.8 (0.19)	1 (0.0018)	153 (3.3)	239 (16.9)	6638	4.81 (0.28)
Sample 3	Direct	99.6 (0.30)	1 (0.0027)	123 (2.3)	241 (16.7)	8052	4.02 (0.24)
	Reverse	99.6 (0.35)	1(0.0029)	127 (2.6)	231 (15.6)	8129	3.83 (0.21)
Sample 4	Direct	99.7 (0.27)	1 (0.0027)	110 (2.0)	649 (230.0)	9147	3.51 (0.15)
	Reverse	99.7 (0.18)	1 (0.0016)	131 (2.2)	146 (7.7)	9538	2.98 (0.18)
Sample 5	Direct	99.8 (0.18)	1 (0.0016)	118 (1.8)	211 (13.7)	9650	4.27 (0.30)
	Reverse	99.8 (0.17)	1(0.0012)	114 (1.5)	227 (12.2)	9641	4.64 (0.33)
Sample 6	Direct	98.9 (0.54)	1 (0.0035)	91 (1.8)	1025 (325.0)	10,754	4.07 (0.24)
	Reverse	100 (0.03)	1 (0.0012)	124 (2.0)	125 (2.2)	11,576	3.41 (0.30)
Sample 7	Direct	99.8 (0.13)	1 (0.0011)	105 (1.6)	184 (13.5)	11,313	3.68 (0.18)
	Reverse	99.8 (0.14)	1 (0.0014)	103 (1.4)	206 (12.9)	11,213	3.77 (0.23)
Sample 8	Direct	99.8 (0.18)	1 (0.0012)	98.7 (1.5)	219 (17.4)	12,283	3.97 (0.19)
	Reverse	99.8 (0.15)	1 (0.0013)	101 (1.3)	204 (16.3)	12,259	4.07 (0.35)
Sample 9	Direct	97.0 (0.95)	1 (0.0024)	103 (3.4)	276 (44.4)	11,434	4.10 (0.31)
	Reverse	99.8 (0.25)	1 (0.0022)	107 (1.8)	214 (15.3)	11,491	4.00 (0.21)
Sample 10	Direct	99.7 (0.24)	1 (0.0020)	76 (1.0)	259 (28.9)	14,528	3.13 (0.17)
	Reverse	99.8 (0.12)	1 (0.0011)	94 (1.4)	104 (4.0)	14,935	2.19 (0.14)
Sample 11	Direct	99.7 (0.23)	1 (0.0018)	78 (0.9)	330 (38.3)	13,613	3.49 (0.18)
	Reverse	99.8 (0.18)	1 (0.0012)	83 (1.0)	180 (8.9)	13,763	3.18 (0.15)
Sample 12	Direct	99.8 (0.13)	1 (0.0013)	76 (0.9)	383 (34)	14,535	4.03 (0.19)
	Reverse	99.8 (0.14)	1 (0.0018)	87 (1.3)	159 (9.6)	15,004	3.58 (0.20)
Sample 13	Direct	99.8 (0.24)	1 (0.0011)	70 (0.9)	398 (45)	16,167	3.51 (0.14)
	Reverse	99.8 (0.15)	1 (0.0020)	79 (1.0)	134 (7.6)	16,567	2.95 (1.77)
Sample 14	Direct	99.8 (0.16)	1 (0.0017)	68 (1.0)	246 (19.4)	17,734	3.24 (0.22)
	Reverse	99.9 (0.08)	1 (0.0014)	85 (1.4)	86 (1.6)	18,205	1.89 (0.95)
Sample 15	Direct	98.3 (0.76)	1 (0.0041)	53 (1.1)	206 (21.9)	23,241	2.67 (0.18)
	Reverse	99.8 (0.16)	1 (0.0016)	56 (0.81)	130 (6.9)	23,432	2.60 (0.14)
Sample 16	Direct	99.8 (0.14)	1 (0.0020)	66 (0.80)	143 (8.0)	20,712	2.43 (0.12)
	Reverse	99.8 (0.14)	1(0.0017)	72 (1.2)	86 (4.1)	20,619	1.81 (0.09)

Figure 6 presents the comparison of some recovered non-acoustic parameters between two orientations. It can be easily seen that, with an increase of density the viscous characteristic length decreases, while the airflow resistivity increases. However, no clear trend can be found between thermal characteristic length, thermal permeability and density. Moreover, the difference on viscous characteristic length, airflow resistivity and thermal permeability are relatively small. It is obviously found that the inferred thermal characteristic length yields a significant difference on the two sides, especially for Samples 4 and 6 with densities of 23.54 kg/m^3 and 24.54 kg/m^3. In addition, the standard deviation is extremely high. This phenomenon can be attributed to the following reasons: interface difference on the two sides, complication on determination of thermal characteristic length and small frequency range or large measurement uncertainty [27]. The front side of samples is more even and continuous, while the back side is rough and uneven (see Figure 5). In addition, slight inhomogeneity can significantly affect thermal characteristic length. Estimation of the thermal characteristic length based on impedance measurement is very sensitive to boundary conditions. If the sample radius does not properly fit the impedance tube's radius, the sample can be compressed, or air leakage existed between the sample and the tube. Consequently, erroneous values will occur because of the modified sample microstructure or the influence of air leakage on the inversion procedure.

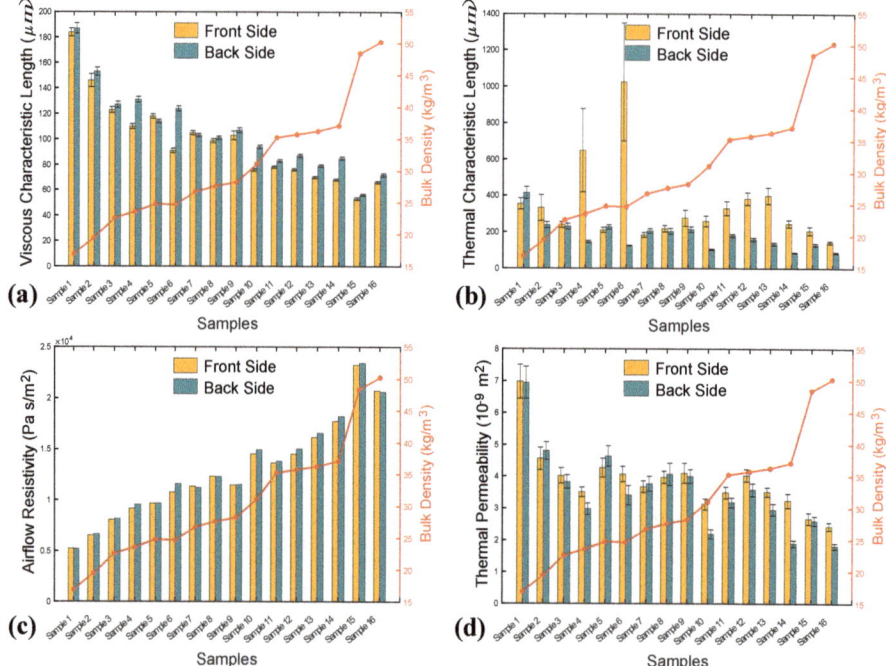

Figure 6. Inferred non-acoustic parameters from direct and reverse orientations: (**a**) viscous characteristic length Λ, (**b**) thermal characteristic length Λ', (**c**) airflow resistivity σ, and (**d**) thermal permeability k'_0.

Since the values of inferred thermal characteristic length on the direct orientation are not reliable, the inferred airflow resistivity on the reverse direction was chosen to compare with the measured value in Figure 7. The correlation between the directly measured and recovered values is presented. It can be seen that the regression line has slope value close to 1 and the coefficient of determination is over 0.94. It can be further demonstrated that the recovering method can accurately estimate airflow resistivity of polyester fibrous panel. The relative difference was defined as $\delta = |\sigma_{meas} - \sigma_{infer}|/\sigma_{meas}$, where σ_{meas} is the measured airflow resistivity and σ_{infer} is the inferred value. The relative difference exhibits the biggest value for Sample 7 with a value of 0.489. This phenomenon can be explained by the small change that exists in manufacturing technology, as was stated above. The most accurate inversion of airflow resistivity occurs on Sample 12 with the smallest δ which is 0.001. As seen in Table 1, from Sample 1 to Sample 16 the density of the panels increases from 16.93 to 45.56 kg/m^3. It can be found that δ is relatively lower when the samples are denser. For instance, Samples 9–16 have lower relative difference (e.g., <0.2). Meanwhile, among the low density panels (Samples 1–8) only two samples reach this relative difference level. It can be concluded that the inversion method of airflow resistivity is more accurate for denser polyester panel materials.

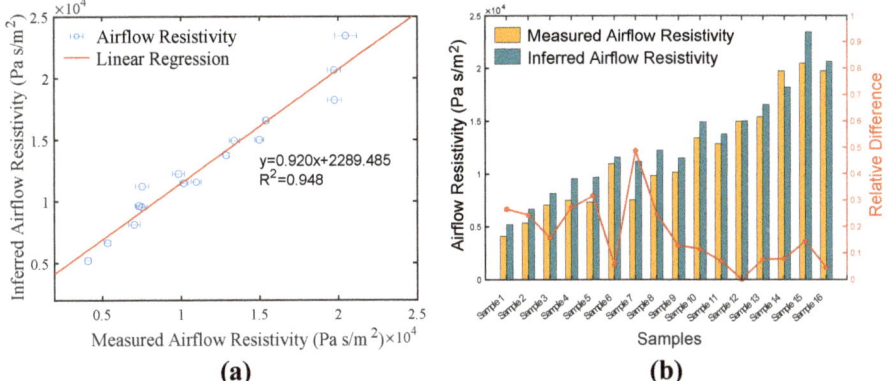

Figure 7. Comparison between measured and inferred airflow resistivity: (**a**) correlation between the measured and inferred values, and (**b**) their relative difference.

To compare the inferred non-acoustic parameters from direct and reverse orientations, the mean relative difference was calculated according to the following equation:

$$\delta = \frac{1}{N}\sum_{n=1}^{N}\delta_n = \frac{1}{N}\sum_{n=1}^{N}\frac{|x_{f,n} - x_{b,n}|}{x_{f,n}} \quad (5)$$

where δ is the mean relative difference, x_f and x_b are, respectively the inferred non-acoustic parameters from direct and reverse orientations, and N is the total number of studied material specimens ($N = 16$).

The mean relative differences of inferred porosity (ϕ), tortuosity (α_∞), viscous characteristic length (Λ), thermal characteristic length (Λ'), airflow resistivity (σ) and thermal permeability (k'_0) are presented in Figure 8. The values of mean relative differences are demonstrated on the top of each bar. It was assumed that the material is homogeneous when the physical parameters having variances less than 0.05 [41]. The mean relative differences of tortuosity, porosity and airflow resistivity are much smaller than the critical value. The differences of viscous characteristic length and thermal permeability are 0.109 and 0.121, respectively. However, the inferred thermal characteristic length exhibits the highest mean relative difference with the value of 0.397. Due to the sensitivity of estimating thermal characteristic length, its inferred values are not recommended for assessing materials homogeneity. Furthermore, tortuosity, porosity and airflow resistivity can well represent the pore size, fiber size and their distributions. Thus, the polyester fibrous sample panel is nearly homogeneous or slightly inhomogeneous at thickness direction. Moreover, the acoustic method can be an alternative approach to characterize the homogeneity of porous material. It can be used not only in the thickness direction, but also in arbitrary directions. Nevertheless, the size of the sample and the suitability of the JCAL model for the porous material of interest should be considered.

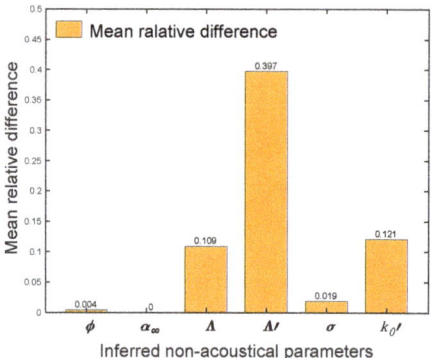

Figure 8. Mean relative differences of inferred non-acoustic parameters.

4. Conclusions

This work applied a Bayesian approach on polyester fibrous materials to inversely estimate some non-acoustic parameters. Meanwhile, the homogeneity of polyester nonwoven panels was assessed by comparing the inferred non-acoustic parameters from direct and reverse orientations. The Johnson-Champoux-Allard-Lafarge model and Markov chain Monte Carlo optimization technique were chosen to implement the Bayesian approach. Polyester samples with density ranging from 16.93–45.56 kg/m^3 were selected in this study. Measurements of reflection coefficient and transmission coefficient were carried out in a four-microphone impedance tube. The mean relative differences of inferred parameters between two orientations were used to analyze the homogeneity. The results indicated that the estimated porosity and tortuosity are reasonable. Moreover, other assessed parameters have generally lower contrast and standard deviation by looking at the qualities from two directions, while a sizable difference of thermal characteristic length was found. In addition, the increase in density decreases estimates of viscous characteristic length and increases estimates of airflow resistivity. While density does not show clear connections with thermal characteristic length and thermal permeability. Measured airflow resistivity and inferred values are very close. The inverse method can accurately estimate the non-acoustic parameters of denser polyester fibrous panel material (i.e., density > 28 kg/m^3). Slight inhomogeneity could significantly affect the determination of thermal characteristic length. The mean relative differences of inferred tortuosity, porosity and airflow resistivity are, respectively 0, 0.004 and 0.019 which means the polyester material is nearly homogeneous. The applied acoustic method is an efficient way to characterize homogeneity of porous materials by comparing with computerized tomography technology. Hence, the proposed Bayesian inversion based on impedance tube measurement is valuable for the study on characterization of the homogeneity of porous material.

Author Contributions: Conceptualization, Conceptualization, T.Y. and J.-P.G.; Methodology, T.Y. and J.-P.G.; Software, J.-P.G. and T.Y.; Validation, T.Y., F.S., J.-P.G. and S.M.; Formal analysis, T.Y., J.-P.G. and F.S.; Investigation, T.Y. and J.-P.G.; Resources, T.Y., M.P. and J.M.; Data curation, T.Y., J.-P.G. and X.X.; Writing—original draft preparation, T.Y.; Writing—review and editing, T.Y. and X.X.; Visualization, T.Y. and J.M.; Supervision, J.-P.G., S.M., M.P., R.M. and J.M.; Project administration, M.P.; Funding acquisition, M.P. and T.Y. All authors have read and agreed to the published version of the manuscript.

Funding: This article is based upon work from COST Action DENORMS CA15125, supported by COST (European Cooperation in Science and Technology) and the European Union (European Structural and Investment Funds-Operational Program Research, Development and Education) in the frames of the project "Modular platform for autonomous chassis of specialized electric vehicles for freight and equipment transportation", Reg. No. CZ.02.1.01/0.0/0.0/16_025/0007293.

Acknowledgments: The authors would like to gratefully acknowledge the support from the Laboratoire d'Acoustique de L'Université du Mans (LAUM, UMR CNRS 6613). The authors would like to thank

Kirill V Horoshenkov, Alistair I. Hurrell and Mohan Jiao for their help throughout part of the airflow resistivity measurements.

Conflicts of Interest: The authors declare no conflict of interest.

References

1. Narang, P.P. Material parameter selection in polyester fibre insulation for sound transmission and absorption. *Appl. Acoust.* **1995**, *45*, 335–358. [CrossRef]
2. Kang, Y.A.; Lee, E.N.; Lee, K.C.; Choi, S.M.; Shin, E.J. Acoustic properties of sound-absorbing polyester fabrics woven with thick staple and thin draw textured yarn for use in interior decoration. *J. Text. Inst.* **2019**, *110*, 202–210. [CrossRef]
3. Liu, Y.; Hu, H. Sound Absorption Behavior of Knitted Spacer Fabrics. *Text. Res. J.* **2010**, *80*, 1949–1957.
4. Arumugam, V.; Mishra, R.; Militky, J.; Novak, J. Thermo-acoustic behaviour of 3D knitted spacer fabrics. *Fibers Polym.* **2015**, *16*, 2467–2476. [CrossRef]
5. Çelikel, D.C.; Babaarslan, O. Acoustic Insulation Behavior of Composite Nonwoven. In *Engineered Fabrics*; IntechOpen: London, UK, 2019.
6. Arenas, J.P. Applications of Acoustic Textiles in Automotive/Transportation. In *Acoustic Textiles*; Springer: Singapore, 2016; pp. 143–163.
7. Setunge, S.; Gamage, N. Application of acoustic materials in civil Engineering. In *Acoustic Textiles*; Springer: Singapore, 2016; pp. 165–183.
8. Patnaik, A.; Mvubu, M.; Muniyasamy, S.; Botha, A.; Anandjiwala, R.D. Thermal and sound insulation materials from waste wool and recycled polyester fibers and their biodegradation studies. *Energy Build.* **2015**, *92*, 161–169. [CrossRef]
9. Yang, T.; Saati, F.; Horoshenkov, K.V.; Xiong, X.; Yang, K.; Mishra, R.; Marburg, S.; Militký, J. Study on the sound absorption behavior of multi-component polyester nonwovens: Experimental and numerical methods. *Text. Res. J.* **2019**, *89*, 3342–3361. [CrossRef]
10. Kino, N.; Ueno, T. Evaluation of acoustical and non-acoustical properties of sound absorbing materials made of polyester fibres of various cross-sectional shapes. *Appl. Acoust.* **2008**, *69*, 575–582. [CrossRef]
11. Shoshani, Y.; Yakubov, Y. Numerical assessment of maximal absorption coefficients for nonwoven fiberwebs. *Appl. Acoust.* **2000**, *59*, 77–87. [CrossRef]
12. Garai, M.; Pompoli, F. A simple empirical model of polyester fibre materials for acoustical applications. *Appl. Acoust.* **2005**, *66*, 1383–1398. [CrossRef]
13. Azari, K.; Majidi, B.; Alamdari, H.; Ziegler, D.; Fafard, M. Characterization of Homogeneity of Green Anodes Through X-Ray Tomography and Image Analysis. In *Light Metals*; Grandfield, J., Ed.; Springer: Cham, Switzerland, 2014.
14. Suvari, F.; Ulcay, Y.; Pourdeyhimi, B. Sound absorption analysis of thermally bonded high-loft nonwovens. *Text. Res. J.* **2015**, *86*, 837–847. [CrossRef]
15. Tascan, M.; Vaughn, E.A.; Stevens, K.A.; Brown, P.J. Effects of total surface area and fabric density on the acoustical behavior of traditional. *J. Text. Inst.* **2011**, *102*, 746–751. [CrossRef]
16. Jirsák, O.; Wadsworth, L.C. *Nonwoven Textiles*; Carolina Academic Press: Durham, NC, USA, 1997.
17. Xiong, X.; Mohanapriya, V.; Darina, J.; Tao, Y.; Rajesh, M.; Jiří, M.; Michal, P. An Experimental Evaluation of Convective Heat Transfer in Multi-Layered Fibrous Materials Composed by Different Middle Layer Structures. *J. Ind. Text.* **2019**. [CrossRef]
18. ASTM D1777-96. *Standard method for Thickness of Textile Materials*; American Society for Testing and Materials: West Conshohocken, PA, USA, 2019.
19. ASTM C830-00. *Standard Test Methods for Apparent Porosity, Liquid Absorption, Apparent Specific Gravity, and Bulk Density of Refractory Shapes by Vacuum Pressure*; American Society for Testing and Materials: West Conshohocken, PA, USA, 2010.
20. ISO 9053-1. *Acoustics-Materials for Acoustical Applications-Determination of Airflow Resistance*; International Organization for Standardization: Geneva, Switzerland, 2018.
21. Xue, Y.; Bolton, J.S.; Gerdes, R.; Lee, S.; Herdtle, T. Prediction of airflow resistivity of fibrous acoustical media having two fiber components and a distribution of fiber radii. *Appl. Acoust.* **2018**, *134*, 145–153. [CrossRef]

22. Yang, T.; Mishra, R.; Horoshenkov, K.V.; Hurrell, A.; Saati, F.; Xiong, X. A study of some airflow resistivity models for multi-component polyester fiber assembly. *Appl. Acoust.* **2018**, *139*, 75–81. [CrossRef]
23. Gommes, C.J.; Bons, A.-J.; Blacher, S.; Dunsmuir, J.H.; Tsou, A.H. Practical Methods for Measuring the Tortuosity of Porous Materials from Binary or Gray-Tone Tomographic Reconstructions. *AIChE J.* **2014**, *55*, 2000–2012. [CrossRef]
24. Maciel, D.O.R.; Ribeiro, C.G.D.; Ferreira, J.; Margem, F.M.; Monteiro, S.N. Characterization of Minerals, Metals, and Materials 2017. In *The Minerals, Metals & Materials Series*; Springer International Publishing: Cham, Switzerland, 2017.
25. McRae, J.D.; Naguib, H.E.; Atalla, N. Mechanical and acoustic performance of compression molded open cell polypropylene foams. *Smart Mater. Adapt. Struct. Intell. Syst.* **2008**, *43314*, 76–84.
26. Horoshenkov, K.V. A review of acoustical methods for porous material characterization. *Int. J. Acoust. Vib.* **2017**, *22*, 92–103.
27. Niskanen, M.; Groby, J.-P.; Duclos, A.; Dazel, O.; Le Roux, J.C.; Poulain, N.; Huttunen, T.; Lahivaara, T. Deterministic and statistical characterization of rigid frame porous materials from impedance tube measurements. *J. Acoust. Soc. Am.* **2017**, *142*, 2407–2418. [CrossRef]
28. Groby, J.-P.; Ogam, E.; De Ryck, L.; Sebaa, N.; Lauriks, W. Analytical method for the ultrasonic characterization of homogeneous rigid porous materials from transmitted and reflected coefficients. *J. Acoust. Soc. Am.* **2010**, *127*, 764–772. [CrossRef]
29. Smith, D.R.; Schultz, S.; Markoš, P.; Soukoulis, C.M. Determination of effective permittivity and permeability of metamaterials from reflection and transmission coefficients. *Phys. Rev. B* **2002**, *65*, 1–5. [CrossRef]
30. Johnson, D.L.; Koplik, J.; Dashen, R. Theory of dynamic permeability and tortuosity in fluid saturated porous media. *J. Fluid Mech.* **1987**, *176*, 379–402. [CrossRef]
31. Champoux, Y.; Allard, J.F. Dynamic tortuosity and bulk modulus in air-saturated porous media. *J. Appl. Phys.* **1991**, *70*, 1975–1979. [CrossRef]
32. Lafarge, D.; Lemarinier, P.; Allard, J.F.; Tarnow, V. Dynamic compressibility of air in porous structures at audible frequencies. *J. Acoust. Soc. Am.* **1997**, *102*, 1995–2006. [CrossRef]
33. Kino, N. A Comparison of Two Acoustical Methods for Estimating Parameters of Glass Fibre and Melamine Foam Materials. *Appl. Acoust.* **2012**, *73*, 590–603. [CrossRef]
34. Fackler, C.J.; Xiang, N.; Horoshenkov, K.V. Bayesian acoustic analysis of multilayer porous media. *J. Acoust. Soc. Am.* **2018**, *144*, 3582–3592. [CrossRef]
35. Chazot, J.-D.; Zhang, E.; Antoni, J. Acoustical and mechanical characterization of poroelastic materials using a Bayesian approach. *J. Acoust. Soc. Am.* **2012**, *131*, 4584–4595. [CrossRef]
36. Dullien, F.A.L. *Porous Media: Fluid Transport and Pore Structure*; Academic Press: San Diego, CA, USA, 1979; pp. 86–88.
37. Cox, T.J.; D'Antonio, P. *Acoustic Absorbers and Diffusers: Theory, Design and Application*; CRC Press: Boca Raton, FL, USA, 2018; p. 205.
38. Vallabh, R.; Banks-Lee, P.; Seyam, A.-F. New Approach for Determining Tortuosity in Fibrous Porous Media. *J. Eng. Fibers Fabr.* **2010**, *5*, 7–15. [CrossRef]
39. Attenborough, K. Models for the acoustical characteristics of air filled granular materials. *Acta Acust* **1993**, *1*, 213–226.
40. Berryman, J.G. Confirmation of Biot's theory. *Appl. Phys. Lett.* **1980**, *37*, 382–384. [CrossRef]
41. Pauwels, J.; Lamberty, A.; Schimmel, H. Homogeneity testing of reference materials. *Accredit. Qual. Assur.* **1998**, *3*, 51–55. [CrossRef]

© 2020 by the authors. Licensee MDPI, Basel, Switzerland. This article is an open access article distributed under the terms and conditions of the Creative Commons Attribution (CC BY) license (http://creativecommons.org/licenses/by/4.0/).

Article

Evaluation of Different Compatibilization Strategies to Improve the Performance of Injection-Molded Green Composite Pieces Made of Polylactide Reinforced with Short Flaxseed Fibers

Ángel Agüero [1], David Garcia-Sanoguera [1], Diego Lascano [1,2,*], Sandra Rojas-Lema [1,2], Juan Ivorra-Martinez [1], Octavio Fenollar [1] and Sergio Torres-Giner [3,*]

1. Technological Institute of Materials (ITM), Universitat Politècnica de València (UPV), Plaza Ferrándiz y Carbonell 1, 03801 Alcoy, Spain; anagrod@epsa.upv.es (Á.A.); dagarsa@dimm.upv.es (D.G.-S.); sanrole@epsa.upv.es (S.R.-L.); juanivorra@outlook.com (J.I.-M.); ocfegi@epsa.upv.es (O.F.)
2. Escuela Politécnica Nacional, 17-01-2759 Quito, Ecuador
3. Novel Materials and Nanotechnology Group, Institute of Agrochemistry and Food Technology (IATA), Spanish National Research Council (CSIC), Calle Catedrático Agustín Escardino Benlloch 7, 46980 Paterna, Spain
* Correspondence: dielas@epsa.upv.es (D.L.); storresginer@iata.csic.es (S.T.-G.); Tel.: +34-966-528-433 (D.L.); +34-963-900-022 (S.T.-G.)

Received: 12 March 2020; Accepted: 31 March 2020; Published: 4 April 2020

Abstract: Green composites made of polylactide (PLA) and short flaxseed fibers (FFs) at 20 wt % were successfully compounded by twin-screw extrusion (TSE) and subsequently shaped into pieces by injection molding. The linen waste derived FFs were subjected to an alkalization pretreatment to remove impurities, improve the fiber surface quality, and make the fibers more hydrophobic. The alkali-pretreated FFs successfully reinforced PLA, leading to green composite pieces with higher mechanical strength. However, the pieces also showed lower ductility and toughness and the lignocellulosic fibers easily detached during fracture due to the absence or low interfacial adhesion with the biopolyester matrix. Therefore, four different compatibilization strategies were carried out to enhance the fiber–matrix interfacial adhesion. These routes consisted on the silanization of the alkalized FFs with a glycidyl silane, namely (3-glycidyloxypropyl) trimethoxysilane (GPTMS), and the reactive extrusion (REX) with three compatibilizers, namely a multi-functional epoxy-based styrene-acrylic oligomer (ESAO), a random copolymer of poly(styrene-*co*-glycidyl methacrylate) (PS-*co*-GMA), and maleinized linseed oil (MLO). The results showed that all the here-tested compatibilizers improved mechanical strength, ductility, and toughness as well as the thermal stability and thermomechanical properties of the green composite pieces. The highest interfacial adhesion was observed in the green composite pieces containing the silanized fibers. Interestingly, PS-*co*-GMA and, more intensely, ESAO yielded the pieces with the highest mechanical performance due to the higher reactivity of these additives with both composite components and their chain-extension action, whereas MLO led to the most ductile pieces due to its secondary role as plasticizer for PLA.

Keywords: PLA; flax; green composites; fiber pretreatment; reactive extrusion

1. Introduction

In the field of polymers technology, one of the most sought and researched objectives nowadays is the replacement of petroleum derived materials with natural and renewable ones. Moreover, international environmental legislation and modern society increasingly require the manufacture of materials and end products that follow a circular model by design [1,2]. Polymer composites are usually a recurrent solution in materials engineering due to its low weight, easy manufacturing, and high mechanical strength [3,4]. The current trend in polymer composites is focused on the use of biopolymers and natural fillers to develop the so-called "green composites" [5]. Over the last years, a wide range of biodegradable polymers derived from both natural and petrochemical sources have been studied, which undergo hydrolytic or enzymatic decomposition by interaction with microorganisms in compost conditions [6,7]. Among them, polylactide (PLA) has been widely employed as the green composite matrix and it is currently considered the front-runner in the bioplastics market with an annual production of approximately 140,000 tons [8].

Fillers have been widely used in polymer composites [9,10]. Natural fillers can be derived from minerals, animals, or plants. In particular, plants derived fillers, both particles and fibers, can find a wide range of applications in green composites due to their abundance and low cost [11,12]. They can also increase the biodesintegration rate of biodegradable polymers in soil due to an increase in active surface area available for the action by degrading microorganisms [13]. Moreover, they can be obtained from agricultural and industrial wastes or food processing by-products, allowing the valorization of discarded materials [14]. Consequently, several green composites have been recently developed using a wide variety of fillers derived from plants and trees such as pineapple leaf fibers [15], babassu nut shell flour [16], jute fabric [17], banana stem fiber [18], orange peel [19], coconut fibers [20], rice husk [21], almond shell [22] walnut shell [23], etc. It is evident that each type of plant has a specific chemical composition and structure. However, it is also worth noting that every elementary vegetable fiber consists on a single plant cell that has a common microstructure. This structure is based on oriented highly crystalline cellulose microfibrils embedded in a matrix formed by lignin and amorphous hemicellulose, which is organized around an empty space denominated "lumen" [24,25]. These two differentiated cell parts, that is, lignin and hemicellulose (matrix) and cellulose microfibrils (reinforcement), result in different behaviors. While the amorphous matrix (primary wall) keeps the fiber integrity, the oriented microfibrils (secondary wall) receive and transfer the mechanical stresses [26]. In fact, the orientation of the microfibrils highly determines the physical and mechanical properties of the fiber [27]. The microfibrils angle is specific of each type of plant and there may be slight differences among fibers belonging to the same plant as well as the cellulose, lignin, and hemicellulose contents [28]. Hence, due to these microstructural variability, vegetable fibers associate some disadvantages and they rarely fulfill the potential expected in the polymer composites or, even, their behavior as reinforcement shows uncertainty under short- and long-term loadings [29]. Furthermore, they habitually present low thermal stability, showing degradation when subjected to temperatures above 200–250 °C and limiting the processing of the composites by melting routes such as extrusion or injection molding [30].

Linen (*Linum usitatissimum*) is traditionally one of the most usable and profitable plants for industrial applications. This family of plants has been cultivated and used for millennia, there existing around 230 species perfectly adapted to both cold and warm weather throughout all the continents [31]. It shows a worldwide production of approximately 3 million tons in 2017, being Canada, China, India, USA, and Ukraine the main producers [32]. Linen varieties are usually classified as flax, for bast fiber production in the textile industry, and linseed, for seed oil extraction in the chemical and food industry. Flax fiber (FF) stands out due to its high cellulose percentage (~78 wt%) [33] and the disorientation of its cellulose microfibrils, which show an angle of 10° with the longitudinal axis of the cell [34]. These properties translate into a high mechanical performance compared to other vegetables fibers and makes FF an interesting natural sourced option for fabrics. Indeed, FF is one of the oldest plant fibers used in household textiles or garments [35]. Furthermore, it can also be applied as a suitable reinforcement to replace glass fiber (GF) in polymer composites [36]. Nowadays, nonwovens based on FFs are finding new applications in panel boards, insulation or asbestos, and novel multilayer packaging structures due to the importance of the eco-design [37].

However, it is widely known that polymer composites based on lignocellulosic particles or fibers habitually present low mechanical and thermomechanical performances due to the absence or poor interfacial adhesion between the fillers and the matrix. This issue has been ascribed to the difference in hydrophobicity between the polymers and the plant derived fillers [38–40]. Therefore, to achieve adequate performance, it is essential to carry out a filler surface pretreatment [41] or use a coupling agent [42] to ensure an enhanced adhesion with the polymer matrix. Furthermore, the modification of the natural fillers can also influence some characteristics such as their hydrophobicity and porosity [43]. These methods are generally based on cleaning and/or activating the filler surface and they include both the use of chemicals and physical methods (e.g., ultraviolet (UV) radiation) [44]. Among the different fiber surface pretreatments based on chemical substances, alkalization, also called "mercerization", is the most standardized one due to it reduces hydrophilicity and additionally increases the effectiveness of further chemical treatments [45,46]. In this regard, silanes and chemically modified silanes are some of the most used coupling agents employed in polymer-reinforced composites [47]. Additionally, the use of reactive compatibilizers represents a straightforward strategy to enhance the particle–matrix adhesion. These additives are generally based on polymers with low molecular weight (M_w), oligomers, or even, more recently, functionalized vegetable oils with several reactive groups that are prone to form chemical bonds during melt processing [48,49]. In this sense, some commercially available copolymers and oligomers based on epoxy, maleic, or glycidyl methacrylate groups have successfully compatibilized biopolyesters and their green composites with lignocellulosic particles [22,50–52].

The present study evaluated and compared the effect of surface treatments and different reactive compatibilizers on the performance of green composites based on PLA and short FFs prepared by twin-screw extrusion (TSE) and injection molding. To this end, the lignocellulosic fillers were initially subjected to an alkali pretreatment and drying. Then, four different compatibilization routes were explored, namely silanization and reactive extrusion (REX) with a random copolymer based on glycidyl methacrylate (GMA), an oligomer containing multiple epoxy groups, and linseed oil multi-functionalized with maleic anhydride (MA). The mechanical, morphological, thermal, and thermomechanical properties of the injection-molded green composite pieces were analyzed and related to the different compatibilization strategies explored.

2. Materials and Methods

2.1. Materials

PLA Ingeo™ Biopolymer 6201D was obtained from NatureWorks LLC (Minnetonka, MN, USA). It was supplied in pellets, having a melt flow rate (MFR) of 15–30 g/10 min measured at 210 °C and 2.16 kg and a true density of 1.24 g/cm^3. This resin is suitable for injection molding and it has been previously screened for preparing green composites due to its high fluidity [19,23,53,54]. Schwarzwälder Textil-Werke Heinrich Kautzmann GmbH (Schenkenzell, Germany) delivered FF as NATURAL FIBRE. The manufacturer supplied the fibers with a length of 6 mm in yarns, having the aspect of a "wool ball", as a processing by-product of the linen industry. A natural scale view of the as-received short FFs is shown in Figure 1.

Figure 1. As-received yarn of short flaxseed fibers (FFs).

Sodium hydroxide (NaOH), also known as caustic soda, with a purity of 99%, was obtained from Cofarcas, S.A. (Burgos, Spain) and it was used for preparing the alkali solution. For the silanization treatment, (3-glycidyloxypropyl) trimethoxysilane (GPTMS), also called GLYMO, with CAS 2530-83-8, was obtained from Sigma-Aldrich S.A. (Madrid, Spain). It shows a M_W of 236.34 g/mol and a density of 1.07 g/cm^3.

Joncryl® 4368C, a multi-functional epoxy-based styrene-acrylic oligomer (ESAO), was supplied in solid flakes by BASF S.A. (Barcelona, Spain). This petrochemical oligomer presents values of M_w of 680 g/mol, T_g of 54 °C, and an epoxy equivalent weight (EEW) of 285 g/mol. The manufacturer recommends a dosage of 0.1–1 wt % for appropriate processability of biodegradable polyesters and complying with food contact regulations. Xibond™ 920, a poly(styrene-co-glycidyl methacrylate) random copolymer (PS-co-GMA), was obtained from Polyscope (Geleen, The Netherlands). This petrochemical copolymer has a M_w value of 50,000 g/mol and a glass transition temperature (T_g) of 95 °C. The manufacturer recommends a dosage level of 0.1–5 wt %, while not exceeding 330 °C to avoid thermal degradation. Maleinized linseed oil (MLO), CAS number 68309-51-3, was provided by Vandeputte (Mouscron, Belgium) as VEOMER LIN. It presents an acid value of 105–130 mg potassium hydroxide (KOH)/g and a viscosity of approximately 1000 cP at 20 °C.

2.2. Pretreatment of the Flaxseed Fibers

The as-received short FFs were first washed in distilled water to remove residuals and/or dirtiness. Then, the fibers were subjected to an alkali pretreatment in a water bath containing NaOH at 2 wt % at room temperature for 1 h, following the procedure described by Sever et al. [55]. Thereafter, the fibers were further washed with distilled water repeatedly and dried at 60 °C for 24 h in an air circulation oven Carbolite model PN from Carbolite Gero Ltd. (Hope Valley, UK).

Part of the alkali-pretreated FFs was also subjected to silanization. To this end, the alkalized fibers were dipped in distilled water bath containing 1 wt % of GPTMS, referred to the total amount of FF, and then magnetically stirred for 2 h at room temperature. After this, the silanized FFs were removed from the bath and dried in a stove at 50 °C for 2 days. The concentration of the glycidyl-silane reactive coupling agent was selected based on our previous studies [56–58].

2.3. Reactive Extrusion of the Green Composites

Prior to melt processing, the PLA pellets were dried in a dehumidifier MDEO from Industrial Marsé (Barcelona, Spain) at 60 °C for 24 h to eliminate residual moisture. Then, the PLA pellets and the pretreated FFs were mixed and fed at a constant ratio of 20 wt % into the main hopper of a twin-screw co-rotating extruder from Dupra S.L. (Alicante, Spain). The extruder features two screws of 25 mm diameter (D) with a length-to-diameter ratio (L/D) of 24. The modular barrel is equipped with 4 individual heating zones coupled to a strand die. Figure 2 shows the extruder layout with its different sections and screw configuration. In the feed section, the pellets enter the barrel with the rotating screws. The compression section contains two mixing zones, a conveying section where the pitch reduces in order to shear the molten plastic at the barrel wall followed by a series of kneading blocks (KBs, 8 × 90° R and 5 × 44° L) to enhance filler dispersion. In the metering zone, the channel depth is reduced and the melts moves in a spiraling motion down the screw channels. The melt is finally pumped out of the extruder through an annular die.

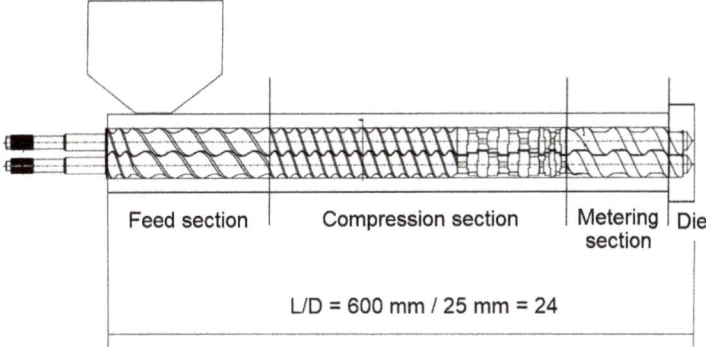

Figure 2. Twin-screw co-rotating extruder and its screw configuration.

The three tested compatibilizers were added as different parts per hundred resin (phr) of composite. Particularly, PS-co-GMA and ESAO were both added at 1 phr, as this content has reported to enhance the interaction between biopolymers without leading undesired effects such as gel formation [50,59]. The selected MLO content was 5 phr since higher oil loadings could exert a negative effect by saturation [60]. A neat PLA sample was also prepared in the same conditions as the control material. In all cases, temperature profile was set at 165 °C at the feed section, 170 °C, 175 °C, and 180 °C at the die. The rotating speed was adjusted to 16 rpm and the residence time was kept at approximately 1 min, being estimated by means of a blue masterbatch [20]. The extruded strands were cooled down in air and granulated in pellets by an air-knife unit. Table 1 summarizes the pretreatments and compositions for all the compounded materials.

Table 1. Code and composition of each sample according to the weight content (wt%) of polylactide (PLA) and flaxseed fiber (FF) in which epoxy-based styrene-acrylic oligomer (ESAO), poly(styrene-co-glycidyl methacrylate) (PS-co-GMA), and maleinized linseed oil (MLO) were added as parts per hundred resin (phr) of composite. Fiber pretreatments consisted of alkalization, in all cases, and silanization with (3-glycidyloxypropyl) trimethoxysilane (GPTMS).

Sample	PLA (wt %)	FF (wt %)	Fiber Pretreatment	Compatibilizer (phr)
PLA	100	0	-	-
PLA/FF	80	20	Alkalization	-
PLA/FF + GPTMS	80	20	Alkalization + Silanization	-
PLA/FF + ESAO	80	20	Alkalization	1
PLA/FF + PS-co-GMA	80	20	Alkalization	1
PLA/FF + MLO	80	20	Alkalization	5

2.4. Injection Molding of the Green Composites

The resultant pellets were dried at 60 °C for 72 h to remove moisture since PLA has a high sensitivity to hydrolysis. Each sample was processed by injection molding in a Meteor 270/75 from Mateu & Solé (Barcelona, Spain). The temperature profile was set, from the feed section to the injection nozzle, as 180–175–170–170 °C. The cavity filling and packing times were 1 s and 10 s, respectively. The clamping force was 75 tons and the samples were cooled in the mold for 10 s. Pieces with a thickness of approximately 4 mm were obtained for characterization.

2.5. Characterization of the Green Composite Pieces

2.5.1. Microscopy

Optical microscopy of the as-received FFs was performed using a stereomicroscope system SZX7 model from Olympus (Tokyo, Japan) with an ocular magnifying glass of 12.5× and equipped with a KL1500-LCD light source. The morphology of the as-received FFs and the fracture surfaces of the injection-molded PLA and PLA/FF pieces obtained from the impact strength test were also observed by field emission scanning electron microscopy (FESEM) in a ZEISS ULTRA 55 FESEM microscope from Oxfrod Instruments (Abingdon, UK). The microscope worked at an acceleration voltage of 2 kV. Prior to analysis, the fracture surfaces were coated with a gold-palladium alloy in a Quorun Technologies Ltd. EMITECH mod. SC7620 sputter coater (East Sussex, UK).

2.5.2. Mechanical Tests

Tensile tests of the PLA and PLA/FF composite pieces were carried out in a universal test machine ELIB 30 from S.A.E. Ibertest (Madrid, Spain) using samples sizing 150 mm × 10 mm × 4 mm and following the guidelines of UNE-EN-ISO 527-1 standard. The selected load cell was 5 kN whereas the cross-head speed was set to 5 mm/min. Impact strength was tested in a 6-J Charpy pendulum from Metrotec S.A. (San Sebastian, Spain) using unnotched pieces with dimensions of 80 mm × 10 mm × 4 mm and as indicated in the ISO 179. Finally, hardness was measured in a Shore D durometer mod. 673-D from J. Bot S.A. (Barcelona, Spain), following the ISO 868 standard. At least six different samples were tested for each mechanical test.

2.5.3. Thermal Tests

Thermal transition of the PLA and PLA/FF composites was studied by differential scanning calorimetry (DSC) in a Mettler-Toledo 821 calorimeter (Schwerzenbach, Switzerland). An average weight of 5–7 mg of each sample was placed in 40-µL aluminum-sealed crucibles and subjected to a heating ramp from 25 °C to 200 °C at 10 °C/min in inert atmosphere of nitrogen with a constant flow of 30 mL/min. The percentage of crystallinity (X_c) was calculated using Equation (1):

$$X_C = \left[\frac{\Delta H_m - \Delta H_{CC}}{\Delta H_m^0 \cdot (1-w)} \right] \times 100 \quad (1)$$

where ΔH_m (J/g) and ΔH_{CC} (J/g) correspond to the melting and cold crystallization enthalpies of PLA, respectively. ΔH_m^0 (J/g) is the theoretical value of a fully crystalline PLA, that is, 93.0 J/g [61] and $1-w$ indicates the weight fraction of PLA in the green composites.

Thermogravimetric analysis (TGA) was carried out to evaluate the thermal stability of the PLA and PLA/FF composite pieces. A TGA/SDTA 851 thermobalance from Mettler-Toledo Inc. (Schwerzenbach, Switzerland) was employed using a single-step thermal program from 30 to 700 °C at a heating rate of 20 °C/min in an air atmosphere. Both thermal analyses were performed in triplicate.

2.5.4. Thermomechanical Tests

The effect of temperature on the mechanical properties of the PLA and PLA/FF composite pieces was studied by dynamic mechanical thermal analysis (DMTA) in an oscillatory rheometer AR-G2 from TA Instruments (New Castle, DE, USA) equipped with an especial clamp system for solid samples that works with a combination of shear-torsion stresses. Pieces with a dimension of 40 mm × 10 mm × 4 mm were tested and the maximum shear deformation (%) was set to 0.1% at a frequency of 1 Hz. The thermal sweep was scheduled from 30 °C to 140 °C at a heating rate of 2 °C/min. The DMTA tests were carried out in triplicate.

3. Results and Discussion

3.1. Morphology of the Flaxseed Fibers

Figure 3 shows the morphology of the as-received yarn of FF and the surfaces of the FFs after the alkali and subsequent silane pretreatments. In the optical microscopy image included in Figure 3a, one can see that the as-received yarn of FF consisted of long individual fibers with a heterogeneous fiber diameter. FESEM analysis of the alkali-pretreated FFs prior and after silanization, respectively, shown in Figure 3b,c, indicated that fiber diameter varied in the 10–25 µm range and it was unaltered during silanization. Images taken at higher magnification, that is, 1000×, which are included in Figure 3d,e, revealed that the fiber surface was rough and fluted. This morphology is in agreement with previous studies, indicating that fibers derived from the flax plant are grouped in conglomerates of elementary fibers bonded by a smooth pectin and lignin phase to form bast technical fibers with a diameter of 50–100 µm [62]. Furthermore, FFs are long plant cells in which perpendicular dislocations, generally termed "kink bands", occur naturally but they can also be formed by physical stresses [63,64]. These kink bands were easily visible in the FESEM micrographs, being distributed randomly along the fiber axis. These dislocations are known to result in a reduction of the fiber tensile properties, and they, furthermore, contribute to a non-linear straining behavior [29,65].

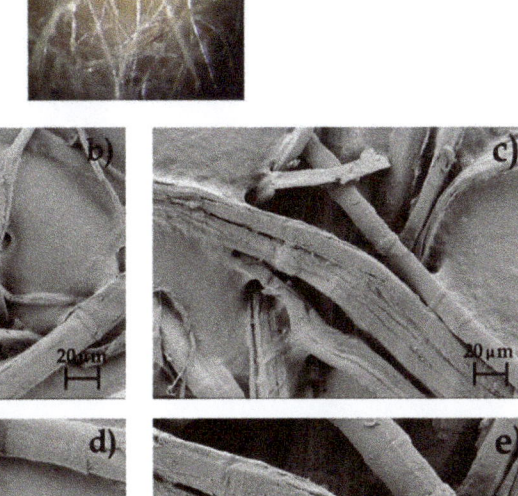

Figure 3. Morphology of the: (**a**) As-received yarn of flaxseed fibers (FFs) taken by optical microscopy at 12.5× with scale maker of 1 mm; (**b**) Alkali-pretreated FFs; (**c**) Alkali-pretreated FFs after silanization taken by field emission scanning electron microscopy (FESEM) at 500× with scale makers of 20 µm; (**d**) Alkali-pretreated FFs; (**e**) Alkali-pretreated FFs after silanization taken by FESEM at 1000× with scale makers of 10 µm.

As commented above, these natural fibers are composed of cellulose chains strongly linked to lignin and hemicellulose. The alkali pretreatment of the fiber was performed to remove the cemented material, that is, lignin and hemicellulose, and modified the cellulose structure [66]. For instance, Valadez-Gonzales et al. [46] showed that henequén fibers with a diluted alkaline solution promoted the partial removal of hemicellulose, waxes, and lignin on the fiber surface. This process led to some changes in their morphology and chemical composition. This particular microfibrillar structure can be observed in the FESEM micrographs shown in Figure 3b,d in which the alkali-pretreated FFs were rugged due to the decomposition of its constituent compounds. In particular, the alkalized FFs showed clean and smoother surfaces without impurities, which prove the effectiveness of the diluted NaOH solution to improve the overall fiber surface quality. After silanization, one can observe in Figure 3c,e that the fibers showed a similar morphology but they developed rougher surfaces due to the formation of nano-sized precipitates that correspond to the spherical silane molecules. This surface has been specifically ascribed to the presence of thin silane layers produced by the formation of a siloxane bonds (Si–O–Si) through reaction between the hydrolyzed alkoxy groups (R–O) in the silane and hydroxyl functional groups (–OH) available on the fibers surface [57]. As a result of this surface modification, the FFs contained suitable glycidyl functional groups to react with the biopolymer matrix of the green composite.

3.2. Mechanical Properties of the Green Composite Pieces

Table 2 gathers the results of the mechanical tests of the injection-molded pieces of the neat PLA and the PLA/FF composites. In relation to the neat PLA, one can observe that the piece showed mechanical properties of a hard but brittle material with values of elastic modulus (E) of 1194.2 MPa, tensile strength at yield (σ_y) of 64.7 MPa, and elongation at break (ε_b) of 8.1%. Its brittleness was also reflected in terms of impact strength, showing a value of 34.5 kJ/m^2. As one expected, the addition of the alkali-pretreated flax fibers yielded an increment in the rigidity and brittleness, resulting in green composite samples with an average E and σ_y values of ~1750 MPa and ~39 MPa, respectively. Moreover, the incorporation of the lignocellulosic fillers induced a reduction in ductility, showing a ε_b value of 3.4% and impact strength of 5.8 kJ/m^2. Additionally, hardness increased from 75.8, for the neat PLA piece, to 79.5, for the PLA/FF composite piece. This ductility impairment can be ascribed to the hydrophilic behavior of the lignocellulosic fibers that may result in a poor dispersion and incompatibility with the hydrophobic PLA matrix, leading to a poor stress transfer between the two components of the green composite [67–69]. Although FF was subjected to an alkaline pretreatment, this modification was mainly aimed to remove impurities and made the fiber surface cleaner and rougher. Similar pretreatments have resulted in a higher aspect ratio of the fiber, which is a positive aspect for fiber reinforcement, but the particle and matrix remained incompatible since no chemical modification was performed [5,11].

Table 2. Mechanical properties of the polylactide (PLA)/flaxseed fiber (FF) pieces compatibilized with (3-glycidyloxypropyl) trimethoxysilane (GPTMS), epoxy-based styrene-acrylic oligomer (ESAO), poly(styrene-co-glycidyl methacrylate) (PS-co-GMA), and maleinized linseed oil (MLO) in term of elastic modulus (E), tensile strength at yield (σ_y), elongation at break (ε_b), impact strength, and Shore D hardness.

Piece	Tensile Test			Impact Strength (kJ/m^2)	Shore D Hardness
	E (MPa)	σ_y (MPa)	ε_b (%)		
PLA	1194.2 ± 27.4	64.7 ± 1.2	8.1 ± 0.5	34.5 ± 2.7	75.8 ± 0.9
PLA/FF	1749.9 ± 32.9	39.1 ± 5.8	3.4 ± 0.2	5.8 ± 0.4	79.5 ± 1.4
PLA/FF + GPTMS	1676.9 ± 45.7	57.9 ± 4.9	5.7 ± 0.9	21.9 ± 2.1	83.1 ± 0.7
PLA/FF + ESAO	1685.1 ± 27.6	65.2 ± 4.1	6.8 ± 1.1	24.1 ± 2.7	83.2 ± 0.7
PLA/FF + PS-co-GMA	1719.7 ± 33.8	61.8 ± 5.2	5.8 ± 0.8	21.3 ± 3.7	78.8 ± 0.8
PLA/FF + MLO	1192.7 ± 41.4	24.8 ± 5.5	10.7 ± 1.1	20.4 ± 3.8	76.2 ± 0.9

One can observe that the green composite pieces prepared with the alkalized FFs subjected to silanization, the here so-called PLA/FF + GPTMS, showed a slight reduction in elasticity but a remarkable increase in strength and ductility when compared to the PLA/FF pieces. In particular, the pieces showed a value of E of 1676.9 MPa, still higher than that of PLA, and σ_y and ε_b values of 57.9 MPa and 5.7%, respectively, representing an increment of 48% and 67% compared with the uncompatibilized PLA/FF piece, although both values were still lower than those of the neat PLA. These results suggest that the glycidyl silane modification successfully promoted the interfacial interaction between FF and PLA. Other research works ascribed the effectiveness of silanization to the formation of Si–C, C–H, and N–H stretching bonds between the hydrolyzed silane and the terminal –OH groups present in both cellulose and PLA [22,70–72]. Therefore, this chemical bonding can potentially play a key role in forming a link between the lignocellulosic fibers and PLA, with consequent enhancement of the mechanical properties. This effect was also reflected in the impact strength, which showed a notably increment to 21.9 kJ/m^2 and an increase in hardness, reaching a value of 83.1.

Furthermore, the incorporation of the two petrochemical reactive additives, that is, PS-co-GMA and ESAO, also produced an improvement in the mechanical properties of the green composite. The addition of ESAO yielded pieces with E and σ_y values of 1685.1 MPa and 65.2 MPa, respectively, while ε_b significantly increased to 6.8%. Alternatively, the green composite pieces processed with PS-co-GMA showed E and σ_y values of nearly 1720 MPa and 62 MPa, respectively, while ε_b was 5.8%. Toughness also improved in both formulations, showing values of 24.1 kJ/m^2, for the pieces processed with ESAO, and 21.3 kJ/m^2, for those with PS-co-GMA. The hardness values were also higher in the case of ESAO, showing values of approximately 83.2 and 78.8, respectively. Therefore, although both reactive additives enhanced the mechanical performance of the green composites, the improvement achieved by the addition of ESAO was slightly higher than that attained with PS-co-GMA. This difference can be ascribed to both the different chemical reactive groups and also the differences in the M_w of these two additives [68,73]. In the case of PS-co-GMA, Anakabe et al. [74] reported that the –OH end groups of PLA can react with the MA groups present in the random copolymer, leading to ester and ether linkages. For ESAO, its multiple epoxy groups can react with the carboxyl (–COOH) and –OH end groups of the PLA chains based on a linear chain-extended, branched or even cross-linked structure depending on the functionality whereas the remaining epoxy groups are also able to react with the –OH groups present on the surface of the cellulosic fillers [73,75]. In this regard, Battegazzore et al. [76,77] reported the ability of ESAO to increase the interfacial adhesion between cotton fabric with PLA/poly(3-hydroxybutyrate-co-3-hydroxyhexanoate) (PHBH) blends as well as to produce multilayer cotton fabric composites of PLA/PHBH with less porosity and improved stress transfer capacity. This dual reactivity of ESAO of coupling agent and melt strength additive can potentially lead to green composites of higher improvement. Furthermore, the lower M_w of oligomer could also favor its higher miscibility and reactivity within the PLA matrix.

In relation to the oil plant derived multi-functionalized oil, the green composites pieces processed with MLO presented significantly lower values of E and σ_y, that is, 1192.7 MPa and 24.8 MPa, respectively. However, interestingly, the ε_b value raised to 10.7%, being the highest value attained for the green composite in the present study. This ductility enhancement can be ascribed to the plasticization of the PLA matrix in which the molecular forces are reduced and the free volume increased [78]. Indeed, MLO has been reported as an effective plasticizer for PLA materials, promoting an additional interfacial adhesion enhancement with both organic and inorganic fillers based on its multi-functionality [22,56,60]. In particular, the MAH groups present in MLO can result in the formation of carboxylic ester bonds by their reaction with the –OH end groups of both the PLA molecules and cellulose, resulting in a cellulose-grafted PLA (cellulose-g-PLA) structure [22]. It is also worthy to note that the green composite pieces processed with MLO showed the lowest value of impact strength, that is, 20.4 kJ/m^2, and also hardness, that is, 76.2, among all the tested compatibilized green composites. This may seem contradictory in relation to values of ductility, but other mechanical properties such as tensile strength also influences in toughness.

3.3. Morphology of the Green Composite Pieces

The FESEM images taken at 1000× of the fracture surfaces after the impact test of the pieces are shown in Figure 4. The micrograph shown in Figure 4a corresponds to the PLA sample, in which a smooth and homogeneous surface with small plastic deformation filaments can be observed. This morphology correlates well the previous mechanical tests that indicated that the PLA piece was brittle.

Figure 4b shows the fracture surface of the PLA/FF piece. One can observe that, even though the fibers were pretreated with NaOH, their interaction with the PLA matrix was low. A considerable gap between the lignocellulosic fibers and the biopolyester matrix can be noticed as well as some holes due to fiber detachment during fracture. This type of morphology has been previously described for green composites, reflecting an increase in rigidity due to fiber reinforcement but also reduction in ductility since the stress is not effectively transferred from the matrix to the fillers [70,79].

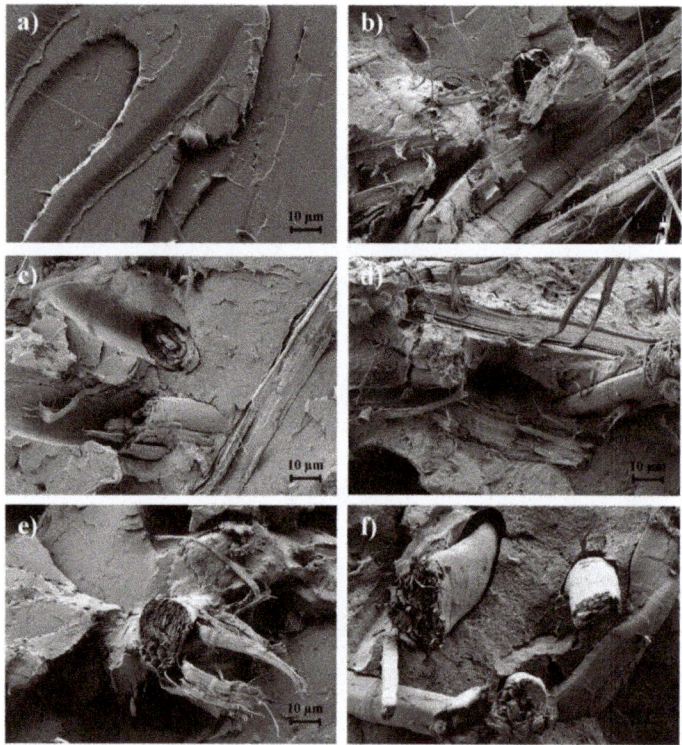

Figure 4. Field emission scanning electron microscopy (FESEM) images of the fracture surfaces of the polylactide (PLA)/flaxseed fiber (FF) pieces of: (**a**) PLA; (**b**) PLA/FF; (**c**) PLA/FF + (3-glycidyloxypropyl) trimethoxysilane (GPTMS); (**d**) PLA/FF + epoxy-based styrene-acrylic oligomer (ESAO); (**e**) PLA/FF + poly(styrene-*co*-glycidyl methacrylate) (PS-*co*-GMA); (**f**) PLA/FF + maleinized linseed oil (MLO). Images were taken at 1000× with scale makers of 10 µm.

As can be seen in Figure 4c, the silanization performed on FFs promoted a high interfacial interaction with the PLA matrix due to the nearly absence of gap. A similar interfacial enhancement has been achieved for kenaf, pineapple, and *Phormium Tenax* (New Zealand flax) fibers subjected to alkaline and silane combined pretreatments [70–72,79]. The FESEM micrograph shows that some fibers remained embedded in the PLA matrix after fracture, which further suggests their strong adhesion to the matrix. This morphology can be correlated with the mechanical improvement described above, in which the green composite presented high strength and toughness. However, the fracture surface was relatively smooth, similar to that of the neat PLA piece, indicating that the fiber-biopolymer interface was not strong enough to modify the fracture energy. It has been reported that covalent bonds between the fillers and matrix created by special reactive surface treatments are very strong (60–80 kJ/mol), but in practice the strength of the interaction is somewhere between this and secondary and weaker van der Waals forces such as dipole–dipole (Keesom), induced dipole (Deby), and dispersion (London) interactions so that the ultimate mechanical strength of the composite is lower and more complex than the one expected [80]. Figure 4d shows that the addition of ESAO produced a morphological change in the fracture surface of the green composite piece. In particular, the incorporation of the reactive oligomer yielded a rough and irregular surface fracture. In addition, the stress cracks were shorter, indicating that the green composite developed a fracture with certain plastic deformation. However, the FESEM micrograph revealed that the interface between the PLA matrix and the lignocellulosic fibers was not promoted to a high extent and the gap around the embedded fibers was higher than that observed in the PLA/FF + GPTMS piece. Moreover, the presence of large cylindrical cavities indicates that fibers detached from the matrix during fracture. Therefore, the mechanical improvement attained in these sample pieces can be more related to the effect of ESAO as chain extender than an interfacial adhesion improvement, since it can contribute to an increase in the mechanical strength and favor a crack arresting mechanism by an increase of the PLA's M_w [81]. In relation to PS-co-GMA, the fracture surface of the green composite piece processed with the random copolymer is displayed in Figure 4e. The morphology of the PLA/FF + PS-co-GMA piece was rough, being relatively similar to that of PLA/FF + ESAO; however, the crack edges appeared more pronounced. This small difference in the green composite morphology correlates well with the mechanical results, showing that the green composite pieces processed with PS-co-GMA were slightly more rigid but also less tough. Finally, in Figure 4f, the fracture surface of the green composite piece processed with MLO is shown. One can observe that the multi-functionalized vegetable oil caused a remarkable change in the fracture surface morphology. In particular, the MLO incorporation promoted the formation of a smooth surface with considerable number of small cavities homogeneously distributed along the PLA matrix, causing a porous-like surface. Similar morphologies were reported in previous studies for MLO-containing PLA materials [22,60]. Other authors described a similar phenomenon for plasticizers due to they form a separated spherical plasticizer-rich phase that result in the creation of holes during fracture [82–84]. One can also observe that the gaps existing around fibers were shorter compared with those observed in the uncompatibilized PLA/FF piece, but larger than those attained in the other compatibilized composite pieces. This indicates that MLO mainly induced a matrix modification and it did not effectively enhance the interfacial adhesion as much as the other compatibilizers, which is in agreement with the mechanical analysis described above.

3.4. Thermal Properties of the Green Composite Pieces

Figure 5 shows the DSC thermograms during the first heating of the neat PLA piece and the PLA/FF pieces. The average values of the main thermal transitions obtained from the DSC curves are gathered in Table 3. The enthalpies associated with the cold crystallization and the melting processes are also reported in the table. A step change in the base lines in the 60–70 °C range can be observed, which corresponds to the T_g of PLA. For the neat PLA sample, the mean T_g value was located at 67.3 °C. Then, the exothermic peaks located between 110 °C and 130 °C can be attributed to the cold crystallization temperature (T_{cc}) of PLA, which was 114.7 °C for the neat piece of PLA. Finally, at higher temperatures, in the range of 150–180 °C, the endothermic peaks represent the melting process of the crystalline PLA domains. In the thermogram of the neat PLA sample, one can observe the presence of two overlapped peaks. The first one appeared at 167.7 °C and the second one at 174.1 °C. This double-peak phenomenon is ascribed to crystal reorganization upon melting in polyesters, by which imperfect crystals melt and the amorphous regions are ordered into spherulites with thicker lamellar thicknesses that thereafter melt at higher temperatures [59,85].

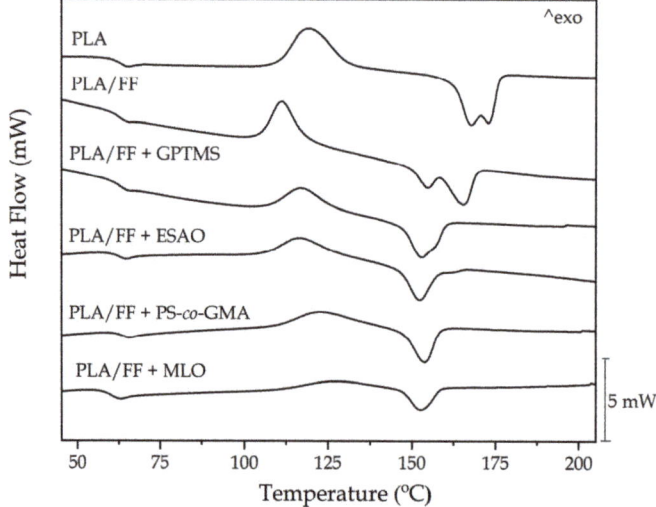

Figure 5. Differential scanning calorimetry (DSC) thermograms corresponding to the polylactide (PLA)/flaxseed fiber (FF) pieces compatibilized with (3-glycidyloxypropyl) trimethoxysilane (GPTMS), epoxy-based styrene-acrylic oligomer (ESAO), poly(styrene-co-glycidyl methacrylate) (PS-co-GMA), and maleinized linseed oil (MLO).

Table 3. Thermal properties of the polylactide (PLA)/flaxseed fiber (FF) pieces compatibilized with (3-glycidyloxypropyl) trimethoxysilane (GPTMS), epoxy-based styrene-acrylic oligomer (ESAO), poly(styrene-co-glycidyl methacrylate) (PS-co-GMA), and maleinized linseed oil (MLO) in term of: glass transition temperature (T_g), cold crystallization temperature (T_{cc}), melting temperature (T_m), cold crystallization enthalpy (ΔH_{cc}), melting enthalpy (ΔH_m), and degree of crystallinity (X_c).

Piece	T_g (°C)	T_{cc} (°C)	T_m (°C)	ΔH_{cc} (J/g)	ΔH_m (J/g)	X_c (%)
PLA	67.3 ± 0.1	114.7 ± 0.5	167.7 ± 0.6/174.1 ± 0.4	28.61 ± 0.2	33.13 ± 0.2	4.86 ± 0.4
PLA/FF	67.0 ± 0.3	109.0 ± 0.4	154.3 ± 0.5/ 164.9 ± 0.4	19.50 ± 0.4	22.47 ± 0.2	3.99 ± 0.2
PLA/FF + GPTMS	64.7 ± 0.4	110.3 ± 0.4	151.3 ± 0.6	15.55 ± 0.5	19.88 ± 0.4	5.81 ± 0.5
PLA/FF + ESAO	64.0 ± 0.4	111.7 ± 0.6	151.0 ± 0.7	13.88 ± 0.6	17.48 ± 0.6	4.89 ± 0.3
PLA/FF + PS-co-GMA	64.3 ± 0.6	118.3 ± 0.2	153.2 ± 0.4	13.47 ± 0.4	18.32 ± 0.5	6.58 ± 0.4
PLA/FF + MLO	61.7 ± 0.2	124.5 ± 0.4	152.3 ± 0.4	10.48 ± 0.6	13.97 ± 0.4	4.92 ± 0.4

As can be observed in the thermograms, the incorporation of the alkalized FFs produced a slight reduction of the T_g value, which remained at 67 °C. The cold crystallization process, however, occurred at a lower temperature, that is, 109 °C. This reduction of the T_{cc} value indicates that FF could act as external nuclei for the formation of PLA crystallites. Indeed, the nucleating effect of lignocellulosic particles on PLA has been widely reported [86,87]. It is also worth noting that the presence of FFs also slightly reduced the percentage of crystallinity, which suggests that the crystallites formed were less perfect due to the presence of the fillers. In this regard, Huang et al. [88] reported a similar phenomenon as a result of the interfacial interaction of the fillers with the matrix that could favor crystal nucleation but reduce crystal growth. This phenomenon was further correlated with the observed reduction in the T_m values, that is, 154.3 and 164.9 °C, since less perfect crystals underwent the melting process. One should also mention that double melting continued to be observable, confirming the absence or low interaction between the fillers and the PLA matrix.

In all the compatibilized PLA/FF composites, the T_g values remained in the 64–65 °C range, with the exception of MLO. In the case of the green composite piece processed with the multi-functionalized vegetable oil, T_g decreased to 61.7 °C. This observation further supports that MLO plasticized the PLA matrix by reducing the interaction between the biopolymer chains and increasing their mobility [84].

This phenomenon was well evidenced by Ferri et al. [60], who showed that the reduction of the T_g values fully correlated with the amount of MLO added. With regard to the cold crystallization process, an increase in the T_{cc} values was observed for all the compatibilized formulations when compared with the uncompatibilized green composite. This result points out that the compatibilizers could partially occupy the free volume between FFs and the PLA molecules and restricted the nucleation of the fillers. Furthermore, in the case of the green composite samples processed with ESAO, PS-co-GMA and, even, MLO, this crystallinity restriction can be related to the formation of chain-extended and/or branched structures of PLA with a higher impairment to crystallize [74,89]. Some authors have additionally concluded that this reaction can create cross-linked structures that could lead to difficulties in the chains mobility and, hence, cold crystallization is delayed [50,74,90]. In the case of the PLA/FF + MLO piece, showing the highest T_{cc} value, that is, 124.5 °C, the newly formed spherical-shape phases dispersed in the PLA matrix could further contribute to delay crystallization of PLA. One can also observe that all the compatibilizers suppressed the double-melting peak phenomenon and PLA melted in a single peak in the 151–154 °C range. This result suggests that all the crystalline structures presented similar lamellae thicknesses, being lower than that of PLA in the pieces without compatibilizer due to the effect of those on the PLA chains [53]. The degree of crystallinity of PLA in the green composite pieces slightly increased when the fibers were silanized or processed with the reactive compatibilizers. This effect has been related to a higher fiber-to-matrix adhesion but it could also be due to an improvement of M_w by a chain extension and/or prevention of random chain scission reactions (e.g., hydrolysis) by which more mass of biopolymer crystallized in the mold [59].

Figure 6 shows the TGA curves of the FF, the neat PLA piece, and the green composite pieces uncompatibilized and compatibilized by the different methods. The thermal stability values extracted from the TGA curves are summarized in Table 4, which gathers the temperature required for a loss of weight of 5% ($T_{5\%}$) that is representative for the onset of degradation, the degradation temperatures (T_{deg1} and T_{deg2}), and the amount of residual mass at 700 °C. Figure 6a shows the mass loss as a function of temperature, whereas, in Figure 6b, their respective DTG curves are included.

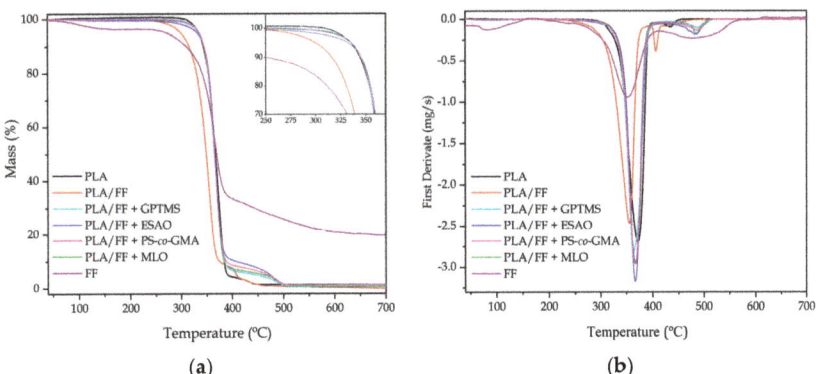

Figure 6. (a) Thermogravimetric analysis (TGA) curves with inset zooming the onset of degradation; and (b) first derivate thermogravimetric (DTG) curves corresponding to the polylactide (PLA)/flaxseed fiber (FF) pieces compatibilized with (3-glycidyloxypropyl) trimethoxysilane (GPTMS), epoxy-based styrene-acrylic oligomer (ESAO), poly(styrene-co-glycidyl methacrylate) (PS-co-GMA), and maleinized linseed oil (MLO).

Table 4. Main thermal degradation parameters of the polylactide (PLA)/flaxseed fiber (FF) pieces compatibilized with (3-glycidyloxypropyl) trimethoxysilane (GPTMS), epoxy-based styrene-acrylic oligomer (ESAO), poly(styrene-co-glycidyl methacrylate) (PS-co-GMA), and maleinized linseed oil (MLO) in term of: temperature required for a loss of weight of 5% ($T_{5\%}$), degradation temperatures (T_{deg1} and T_{deg2}), and residual mass at 700 °C.

Piece	$T_{5\%}$ (°C)	T_{deg1} (°C)	T_{deg2} (°C)	Residual Mass (%)
FF	255.3 ± 0.4	316.6 ± 0.4	419.8 ± 0.9	20.6 ± 0.2
PLA	334.2 ± 0.7	373.3 ± 0.4	410.3 ± 1.1	1.5 ± 0.3
PLA/FF	279.1 ± 0.6	340.6 ± 0.7	368.4 ± 1.1	3.4 ± 0.3
PLA/FF + GPTMS	337.7 ± 0.4	375.8 ± 1.1	411.6 ± 0.8	4.2 ± 0.4
PLA/FF + ESAO	332.2 ± 0.3	381.2 ± 0.9	412.7 ± 0.9	4.1 ± 0.3
PLA/FF + PS-co-GMA	331.6 ± 0.9	383.9 ± 1.2	411.2 ± 1.3	4.4 ± 0.2
PLA/FF + MLO	337.6 ± 0.5	381.7 ± 0.8	409.6 ± 1.2	3.6 ± 0.5

One can see that thermal degradation of the alkali-pretreated FF took place in three main steps. Initially, at a temperature close to 100 °C, a slight mass loss of ~3.5% was observed, which corresponds with the moisture and remaining solvent retained in the fibers. Following the TGA curve, a change in the slope of the curve was observed when temperature reached ~255 °C. This represents a mass loss of ~60% and it finalized at ~390 °C. Previous studies reporting the thermal decomposition of lignocellulosic particles have attributed this mass loss to the oxidation of their main constituents, primarily in the amorphous regions (hemicellulose), followed by cellulose and lignin. This phenomenon is normally referred as the "active pyrolysis zone" and it is followed by a second degradation step, denominated "passive pyrolysis zone", where the oxidation of the residual char occurs [91]. In the present FF curve, this pyrolysis step was observable in the 400–600 °C range. The TGA curve of FF is totally in agreement with others reports focused on lignocellulosic fibers [92]. For temperatures above 600 °C, thermal degradation continued but no significant weight losses were detected, indicating that the entire organic component were already pyrolyzed. The residual mass at the end of the analysis was over 20%, which can be related to the high mineral content present in flax [53].

Neat PLA showed a typical thermal degradation for a biopolyester based on two steps. In the range from nearly 335 °C to 390 °C, around 95% of the mass was lost. This peak was sharp and intense due to the chain-scission reaction through breakage of its ester groups [93]. After this, a second low-intense degradation stage took place, which was more detectable in the DTG graph. This corresponds with the fully oxidation of the char residue produced by the formation of some complexes, C = C bonds, aromatic species, etc. during the first stage [94]. After the incorporation of the alkalized FFs, the thermal stability of PLA was reduced. Both main thermal stability values, that is, $T_{5\%}$ and T_{deg1}, were reduced by approximately 55 °C and 33 °C, respectively. The lower thermal stability of green composites has mainly been explained in terms of the low thermal stability of the lignocellulosic fillers and also to the remaining moisture, which is extremely difficult to remove [53]. Moreover, FF increased the intensity of the second degradation peak centered at ~410 °C, which can be related to the passive pyrolysis of the lignocellulosic fillers. In relation to the compatibilized PLA/FF samples, all coupling strategies yielded a similar positive effect on the thermal stability of the green composite. In particular, the $T_{5\%}$ values ranged from 330 °C to 340 °C, while T_{deg1} was in the 375–385 °C range. Therefore, the use of different compatibilizers resulted in green composites with thermal stability values in the same range or even higher than those observed for neat PLA. This thermal enhancement can be related to the improved interfacial adhesion achieved in the green composite due to the chemical interaction established between the lignocellulosic fillers and the biopolymer matrix. As in the mechanical properties, the highest improvement was observed for the green composite processed with ESAO, although the thermal stability of all the compatibilized samples was relatively similar and close to that of neat PLA.

3.5. Thermomechanical Properties of the Green Composite Pieces

As shown in Figure 7, the thermomechanical behavior of the PLA and PLA/FF pieces was studied by the variation of the storage modulus (G′) and the damping factor (tanδ) as a function of temperature. Figure 7a gathers the evolution of G′ with temperature. For all the PLA pieces, it can be observed a relatively high stiffness at low temperatures and then a sharply decrease of the G′ values between 50 °C and 80 °C due to the temperature reached the alpha (α)-relaxation of the biopolymer. With the temperature increment, in the 85–100 °C range, the cold crystallization phenomenon occurred, as previously discussed during the DSC analysis. Table 5 shows the G′ values of the storage modulus measured at 40 °C, 75 °C, and 110 °C, being these temperatures representative of the stored elastic energy of the material in the three mentioned states, that is, amorphous glassy, amorphous rubber, and semi-crystalline. One can observe that all the green composite pieces presented higher values of G′ than the neat PLA piece. In particular, at 40 °C, the G′ of the PLA/FF piece was 1080.3 MPa, which represents almost the double of the unfilled PLA, that is, 566.5 MPa. This is consistent with the above-reported effect of mechanical reinforcement of PLA by the lignocellulosic fibers. Similar remarkable increases in stiffness for natural filler-reinforced polymer composites have been reported previously [22,48,56]. The reinforcing effect was more pronounced after the glass transition region, in which the G′ value measured at 75 °C increased from 2.2. MPa, in the neat PLA piece, to 13.1 MPa, in the PLA/FF piece. This points out the high effect of relative hard fillers in a rubber-like matrix. The G′ value also showed a nearly three-fold increased, from 47.7 MPa to 154.6 MPa, at 110 °C, after cold crystallization, when the alkali-pretreated FF were incorporated into PLA. One can also notice in the DMTA curves that, as also observed during the DSC analysis, the fillers presence also promoted the occurrence of the cold crystallization phenomenon at lower temperatures.

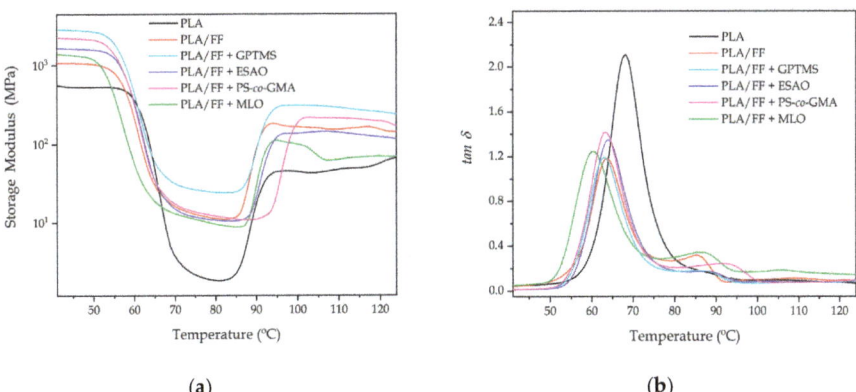

Figure 7. Evolution as a function of temperature of the (**a**) storage modulus and (**b**) dynamic damping factor (*tan δ*) of the polylactide (PLA)/flaxseed fiber (FF) pieces compatibilized with (3-glycidyloxypropyl) trimethoxysilane (GPTMS), epoxy-based styrene-acrylic oligomer (ESAO), poly(styrene-*co*-glycidyl methacrylate) (PS-*co*-GMA), and maleinized linseed oil (MLO).

Table 5. Thermomechanical properties of the polylactide (PLA)/flaxseed fiber (FF) pieces compatibilized with (3-glycidyloxypropyl) trimethoxysilane (GPTMS), epoxy-based styrene-acrylic oligomer (ESAO), poly(styrene-co-glycidyl methacrylate) (PS-co-GMA), and maleinized linseed oil (MLO) in term of: storage modulus (G') measured at 40 °C, 75 °C, and 110 °C and glass transition temperature (T_g).

Piece	T_g (°C)	G' (MPa)		
		40 °C	75 °C	110 °C
PLA	68.4 ± 1.3	566.5 ± 8.2	2.2 ± 0.7	47.7 ± 0.5
PLA/FF	63.8 ± 0.8	1080.3 ± 9.5	13.1 ± 0.5	154.6 ± 0.5
PLA/FF + GPTMS	63.1 ± 1.1	2908.8 ± 7.8	26.7 ± 0.7	293.8 ± 0.7
PLA/FF + ESAO	64.2 ± 1.2	1683.7 ± 7.3	11.9 ± 0.8	141.7 ± 0.5
PLA/FF + PS-co-GMA	63.4 ± 0.6	2293.4 ± 6.4	13.9 ± 0.8	214.2 ± 0.6
PLA/FF + MLO	60.8 ± 0.9	1437.6 ± 9.1	11.2 ± 0.7	63.6 ± 0.6

The use of the different compatibilizers also yielded higher G' values at low temperatures, with the exception of MLO due to its secondary role as plasticizer. This observation confirms that all the here-tested compatibilization strategies successfully increased the adhesion of the alkali-pretreated FFs to the PLA matrix in the green composites. The highest G' values were attained for the green composite pieces processed with the alkalized FFs that were further pretreated with GPTMS. In particular, their G' value was nearly 3 GPa at 40 °C, being around three times higher than that of the PLA/FF pieces and representing an increment of ~500% compared with the neat PLA pieces. This trend was also kept at higher temperatures, showing G' values of 26.7 MPa and 293.8 MPa at 75 °C and 110 °C, respectively. Some authors have concluded that silanization of lignocellulosic fibers can induce a noticeable increment in the mechanical rigidity of the polymer composites, though it also reduces the elastic limit, especially when the fibers are subjected to alkaline pretreatment [70,79]. This combination of surface pretreatments can notably modify and/or remove some substances present on the fiber surfaces (e.g., hemicellulose, cellulose, and lignin), which can favor the interaction with the polymer molecules, but it can also cause damage in the fiber due to an excess of delignification [95]. This statement is in agreement with the FESEM images shown above, in which one can observe that the fibers were rugged due the decomposition of its constituent compounds and rougher by the presence of the silane layer. Therefore, the structural modification of the fibers played a key role in the thermomechanical properties of the green composites This explanation further supports both the morphological properties, which showed that the fibers were highly adhered to the biopolyester matrix, but also the fact that the mechanical strength and toughness was lower when compared with ESAO and PS-co-GMA since in the latter cases the fiber morphology was better preserved. The slight difference of the pieces in terms of their mechanical and thermomechanical performance can also be related to the type of experiment since in the tensile test the effort was uniaxial, while in the DMTA tests the pieces were subjected to a combination of shear-torsion stresses and the contribution of the fiber-to-adhesion strength was higher. It is also worth noting that both reactive compatibilizers, in particular PS-co-GMA, delayed cold crystallization of PLA due to the above-described restriction in chain mobility caused by the formation of a macromolecular structure with higher M_w.

Figure 7b represents the evolution of tanδ versus temperature. The peaks of the graphs represent the change in the thermomechanical behavior when the α-relaxation of the biopolymer is reached, which relates to its T_g. As shown in Table 5, the peak for the neat PLA piece was located at 68.4 °C, which is similar to the T_g value reported by DSC. One can also observe that the green composite pieces showed lower peak values for α-relaxation, centered at ~63 °C, being also very similar to the values of T_g attained by DSC. In the case of the green composite processed with MLO, this value was 60.8 °C due to the plasticizing effect of the multi-functionalized vegetable oil. It should also be noted that all green composite samples had α-relaxation peaks with lower intensities than the neat PLA sample. This observation points out that the relaxation of the PLA chains was partially suppressed by the presence of the alkalized FFs and, hence, lower number of PLA chains underwent glass transition [44]. This fact also correlates well the crystallinity results obtained by DSC that showed that PLA developed slightly higher crystallinity in the compatibilized green composite pieces.

4. Conclusions

Four different routes were tested to improve the overall performance of injection-molded pieces of green composites made of PLA and short FFs. These strategies included a silanization pretreatment of alkalized FFs with GPTMS and the use of three reactive compatibilizers added during melt extrusion, that is, ESAO, PS-co-GMA, and MLO. The morphological analysis showed that the highest interfacial adhesion was obtained in the green composite pieces containing the fibers pretreated with GPTMS due to the glycidyl-silane reactive coupling agent was able to react with the –OH terminal groups present in both cellulose and PLA. Although the mechanical and thermal performance of the green composite pieces was improved, the fiber–biopolymer interface was not strong enough to modify the fracture energy and the improvement was relatively low. This effect was related to the fiber morphology damage that occurred due to an excess of delignification during silanization. In relation to the two petroleum derived reactive additives, that is, ESAO and PS-co-GMA, it was observed that these led to the highest mechanical resistance and toughness improvement and also the highest thermal stability, showing the multi-functional oligomer the most optimal properties. Since the morphological analysis revealed that the filler–matrix was relatively low, the enhancement attained was mainly ascribed to a matrix modification by the formation of a chain-extended, branched or even cross-linked structure through the functional groups present in both additives that can react with –OH groups of PLA. Regarding the multi-functionalized vegetable oil, the green composite processed with MLO also showed improved mechanical and thermal properties, showing the highest ductility. However, it was also observed that it plasticized the PLA matrix and, as a result, the mechanical and thermomechanical properties were not as high as those attained by means of the other strategies. Despite their relatively low properties, the MLO-containing green composites are still great candidates for more flexible applications and it is also the only solution fully based on natural raw materials. According to the here-attained results, one can conclude that the compatibilizer of highest potential for green composites is ESAO since it allows to produce, at very low contents, high-performing bio-based and biodegradable materials in terms of mechanical resistance, toughness, and thermal stability.

Author Contributions: Conceptualization, S.T.-G. and Á.A.; methodology, O.F. and D.L.; validation, D.G.-S., Á.A., and S.R.-L.; formal analysis, J.I.-M. and D.L.; investigation, D.G.S.; data curation, D.L., J.I.-M., and S.R.-L.; writing—original draft preparation, Á.A. and D.G.-S.; writing—review and editing, O.F. and S.T.-G.; supervision, Á.A. and O.F.; and project administration, O.F. and S.T.-G. All authors have read and agreed to the published version of the manuscript.

Funding: This research work was funded by the Spanish Ministry of Science and Innovation (MICI) project numbers RTI2018-097249-B-C21 and MAT2017-84909-C2-2-R.

Acknowledgments: D.L. and J.I.-M. thanks UPV for the grant received through the PAID-01-18 and PAID-01-19 (SP2019001) programs, respectively. S.T.-G. is recipient of a Juan de la Cierva contract (IJCI-2016-29675) from MICIU while S.R.J. is funded through the Santiago Gisolía contract (GRISOLIAP/2019/132) from Generalitat Valenciana (GVA). The microscopy services of UPV are also acknowledged for their help in collecting and analyzing the FESEM images.

Conflicts of Interest: The authors declare no conflict of interest.

References

1. Garkhail, S.K.; Heijenrath, R.W.H.; Peijs, T. Mechanical properties of natural-fibre-mat-reinforced thermoplastics based on flax fibres and polypropylene. *Appl. Compos. Mater.* **2000**, *7*, 351–372. [CrossRef]
2. Khandelwal, V.; Sahoo, S.K.; Kumar, A.; Manik, G. Electrically conductive green composites based on epoxidized linseed oil and polyaniline: An insight into electrical, thermal and mechanical properties. *Compos. Part B Eng.* **2018**, *136*, 149–157. [CrossRef]

3. Mittal, G.; Dhand, V.; Rhee, K.Y.; Park, S.-J.; Lee, W.R. A review on carbon nanotubes and graphene as fillers in reinforced polymer nanocomposites. *J. Ind. Eng. Chem.* **2015**, *21*, 11–25. [CrossRef]
4. Reddy, M.M.; Vivekanandhan, S.; Misra, M.; Bhatia, S.K.; Mohanty, A.K. Biobased plastics and bionanocomposites: Current status and future opportunities. *Prog. Polym. Sci.* **2013**, *38*, 1653–1689. [CrossRef]
5. Jha, K.; Kataria, R.; Verma, J.; Pradhan, S. Potential biodegradable matrices and fiber treatment for green composites: A review. *AIMS Mater. Sci.* **2019**, *6*, 119–138. [CrossRef]
6. Mohanty, A.K.; Wibowo, A.; Misra, M.; Drzal, L.T. Effect of process engineering on the performance of natural fiber reinforced cellulose acetate biocomposites. *Compos. Part A Appl. Sci. Manuf.* **2004**, *35*, 363–370. [CrossRef]
7. Ghanbarzadeh, B.; Almasi, H. Biodegradable polymers. In *Biodegradation-Life of Science*; InTech Publications: Rijeka, Croatia, 2013; pp. 141–186.
8. Madhavan Nampoothiri, K.; Nair, N.R.; John, R.P. An overview of the recent developments in polylactide (PLA) research. *Bioresour. Technol.* **2010**, *101*, 8493–8501. [CrossRef]
9. Corrado, A.; Polini, W. Measurement of high flexibility components in composite material by touch probe and force sensing resistors. *J. Manuf. Process.* **2019**, *45*, 520–531. [CrossRef]
10. Chen, J.; Gao, X. Directional dependence of electrical and thermal properties in graphene-nanoplatelet-based composite materials. *Results Phys.* **2019**, *15*, 102608. [CrossRef]
11. Omrani, E.; Menezes, P.L.; Rohatgi, P.K. State of the art on tribological behavior of polymer matrix composites reinforced with natural fibers in the green materials world. *Eng. Sci. Technol. Int. J.* **2016**, *19*, 717–736. [CrossRef]
12. Eichhorn, S.J.; Baillie, C.A.; Zafeiropoulos, N.; Mwaikambo, L.Y.; Ansell, M.P.; Dufresne, A.; Entwistle, K.M.; Herrera-Franco, P.J.; Escamilla, G.C.; Groom, L.; et al. Review: Current international research into cellulosic fibres and composites. *J. Mater. Sci.* **2001**, *36*, 2107–2131. [CrossRef]
13. Harada, J.; de Souza, A.G.; de Macedo, J.R.; Rosa, D.S. Soil culture: Influence of different natural fillers incorporated in biodegradable mulching film. *J. Mol. Liq.* **2019**, *273*, 33–36. [CrossRef]
14. Bledzki, A.K.; Gassan, J. Composites reinforced with cellulose based fibres. *Prog. Polym. Sci.* **1999**, *24*, 221–274. [CrossRef]
15. Liu, W.; Misra, M.; Askeland, P.; Drzal, L.T.; Mohanty, A.K. 'Green' composites from soy based plastic and pineapple leaf fiber: Fabrication and properties evaluation. *Polymer* **2005**, *46*, 2710–2721. [CrossRef]
16. Hoffmann, R.; Morais, D.; Braz, C.; Haag, K.; Wellen, R.; Canedo, E.; de Carvalho, L.; Koschek, K. Impact of the natural filler babassu on the processing and properties of PBAT/PHB films. *Compos. Part A Appl. Sci. Manuf.* **2019**, *124*, 105472. [CrossRef]
17. Kapatel, P.M. Investigation of green composite: Preparation and characterization of alkali-treated jute fabric-reinforced polymer matrix composites. *J. Nat. Fibers* **2019**. [CrossRef]
18. Srivastava, K.R.; Singh, M.K.; Mishra, P.K.; Srivastava, P. Pretreatment of banana pseudostem fibre for green composite packaging film preparation with polyvinyl alcohol. *J. Polym. Res.* **2019**, *26*, 95. [CrossRef]
19. Quiles-Carrillo, L.; Montanes, N.; Lagaron, J.M.; Balart, R.; Torres-Giner, S. On the use of acrylated epoxidized soybean oil as a reactive compatibilizer in injection-molded compostable pieces consisting of polylactide filled with orange peel flour. *Polym. Int.* **2018**, *67*, 1341–1351. [CrossRef]
20. Torres-Giner, S.; Hilliou, L.; Melendez-Rodriguez, B.; Figueroa-Lopez, K.J.; Madalena, D.; Cabedo, L.; Covas, J.A.; Vicente, A.A.; Lagaron, J.M. Melt processability, characterization, and antibacterial activity of compression-molded green composite sheets made of poly(3-hydroxybutyrate-*co*-3-hydroxyvalerate) reinforced with coconut fibers impregnated with oregano essential oil. *Food Packag. Shelf Life* **2018**, *17*, 39–49. [CrossRef]
21. Melendez-Rodriguez, B.; Torres-Giner, S.; Aldureid, A.; Cabedo, L.; Lagaron, J.M. Reactive melt mixing of poly(3-hydroxybutyrate)/rice husk flour composites with purified biosustainably produced poly(3-hydroxybutyrate-*co*-3-hydroxyvalerate). *Materials* **2019**, *12*, 2152. [CrossRef]
22. Quiles-Carrillo, L.; Montanes, N.; Sammon, C.; Balart, R.; Torres-Giner, S. Compatibilization of highly sustainable polylactide/almond shell flour composites by reactive extrusion with maleinized linseed oil. *Ind. Crops Prod.* **2018**, *111*, 878–888. [CrossRef]

23. Montava-Jordà, S.; Quiles-Carrillo, L.; Richart, N.; Torres-Giner, S.; Montanes, N. Enhanced interfacial adhesion of polylactide/poly(ε-caprolactone)/walnut shell flour composites by reactive extrusion with maleinized linseed oil. *Polymers* **2019**, *11*, 758. [CrossRef] [PubMed]
24. Brett, C.T.; Waldron, K.W. *Physiology and Biochemistry of Plant Cell Walls*; Chapman & Hall: London, UK, 1996; Volume 2.
25. Rong, M.Z.; Zhang, M.Q.; Liu, Y.; Yang, G.C.; Zeng, H.M. The effect of fiber treatment on the mechanical properties of unidirectional sisal-reinforced epoxy composites. *Compos. Sci. Technol.* **2001**, *6*, 1437–1447. [CrossRef]
26. Bos, H.L.; Donald, A.M. In situ ESEM study of the deformation of elementary flax fibres. *J. Mater. Sci.* **1999**, *34*, 3029–3034. [CrossRef]
27. Wang, H.H.; Drummond, J.G.; Reath, S.M.; Hunt, K.; Watson, P.A. An improved fibril angle measurement method for wood fibers. *Wood Sci. Technol.* **2001**, *34*, 493–503. [CrossRef]
28. Scaffaro, R.; Maio, A.; Lopresti, F. Physical properties of green composites based on poly-lactic acid or Mater-Bi® filled with Posidonia Oceanica leaves. *Compos. Part A Appl. Sci. Manuf.* **2018**, *112*, 315–327. [CrossRef]
29. Hughes, M.; Carpenter, J.; Hill, C. Deformation and fracture behaviour of flax fibre reinforced thermosetting polymer matrix composites. *J. Mater. Sci.* **2007**, *42*, 2499–2511. [CrossRef]
30. Garkhail, S.K.; Heijenrath, R.W.H.; Peijs, T. Natural-fibre-mat-reinforced thermoplastic composites based on flax fibres and polypropylene. *Appl. Compos. Mater.* **1996**, *7*, 351–372. [CrossRef]
31. Diederichsen, A.; Raney, J.P. Seed colour, seed weight and seed oil content in *Linum usitatissimum* accessions held by Plant Gene Resources of Canada. *Plant Breed.* **2006**, *125*, 372–377. [CrossRef]
32. FAOSTAT. *Flax and Linseed Crops*; FAOSTAT: Roma, Italy, 2017.
33. Sharma, H.S.S.; Faughey, G.; Lyons, G. Comparison of physical, chemical, and thermal characteristics of water-, dew-, and enzyme-retted flax fibers. *J. Appl. Polym. Sci.* **1999**, *74*, 139–143. [CrossRef]
34. Charlet, K.; Jernot, J.P.; Gomina, M.; Bréard, J.; Morvan, C.; Baley, C. Influence of an Agatha flax fibre location in a stem on its mechanical, chemical and morphological properties. *Compos. Sci. Technol.* **2009**, *69*, 1399–1403. [CrossRef]
35. Maity, S.; Gon, D.P.; Paul, P. A review of flax nonwovens: Manufacturing, properties, and applications. *J. Nat. Fibers* **2014**, *11*, 365–390. [CrossRef]
36. Hearle, J.W.; Morton, W.E. *Physical Properties of Textile Fibres*; Elsevier: Amsterdam, The Netherlands, 2008.
37. Yan, L.; Chouw, N.; Jayaraman, K. Flax fibre and its composites—A review. *Compos. Part B Eng.* **2014**, *56*, 296–317. [CrossRef]
38. Herrera-Franco, P.J.; Valadez-González, A. Mechanical properties of continuous natural fibre-reinforced polymer composites. *Compos. Part A Appl. Sci. Manuf.* **2004**, *35*, 339–345. [CrossRef]
39. Muñoz-Vélez, M.F.; Hidalgo-Salazar, M.A.; Mina-Hernández, J.H. Fique fiber an alternative for reinforced plastics. Influence of surface modification. *Biotecnol. Sect. Agropecu. Agroind.* **2014**, *12*, 60–70.
40. Valadez-Gonzalez, A.; Cervantes-Uc, J.M.; Olayo, R.; Herrera-Franco, P.J. Effect of fiber surface treatment on the fiber–matrix bond strength of natural fiber reinforced composites. *Compos. Sci. Technol.* **1999**, *30*, 309–320. [CrossRef]
41. Tang, Q.; Wang, F.; Guo, H.; Yang, Y.; Du, Y.; Liang, J.; Zhang, F. Effect of coupling agent on surface free energy of organic modified attapulgite (OAT) powders and tensile strength of OAT/ethylene-propylene-diene monomer rubber nanocomposites. *Powder Technol.* **2015**, *270*, 92–97. [CrossRef]
42. Tang, Q.; Wang, F.; Liu, X.; Tang, M.; Zeng, Z.; Liang, J.; Guan, X.; Wang, J.; Mu, X. Surface modified palygorskite nanofibers and their applications as reinforcement phase in cis-polybutadiene rubber nanocomposites. *Appl. Clay Sci.* **2016**, *132*, 175–181. [CrossRef]
43. Faruk, O.; Bledzki, A.K.; Fink, H.-P.; Sain, M. Biocomposites reinforced with natural fibers: 2000–2010. *Prog. Polym. Sci.* **2012**, *37*, 1552–1596. [CrossRef]
44. Torres-Giner, S.; Montanes, N.; Fenollar, O.; García-Sanoguera, D.; Balart, R. Development and optimization of renewable vinyl plastisol/wood flour composites exposed to ultraviolet radiation. *Mater. Des.* **2016**, *108*, 648–658. [CrossRef]
45. Pothan, L.A.; Thomas, S. Polarity parameters and dynamic mechanical behaviour of chemically modified banana fiber reinforced polyester composites. *Compos. Sci. Technol.* **2003**, *63*, 1231–1240. [CrossRef]

46. Valadez-Gonzalez, A.; Cervantes-Uc, J.M.; Olayo, R.; Herrera-Franco, P.J. Chemical modification of henequén fibers with an organosilane coupling agent. *Compos. Part B Eng.* **1998**, *30*, 321–331. [CrossRef]
47. España, J.M.; Samper, M.D.; Fages, E.; Sánchez-Nácher, L.; Balart, R. Investigation of the effect of different silane coupling agents on mechanical performance of basalt fiber composite laminates with biobased epoxy matrices. *Polym. Compos.* **2013**, *34*, 376–381. [CrossRef]
48. García-García, D.; Carbonell, A.; Samper, M.D.; García-Sanoguera, D.; Balart, R. Green composites based on polypropylene matrix and hydrophobized spend coffee ground (SCG) powder. *Compos. Part B Eng.* **2015**, *78*, 256–265. [CrossRef]
49. Nyambo, C.; Mohanty, A.K.; Misra, M. Effect of maleated compatibilizer on performance of PLA/Wheat straw-based green composites. *Macromol. Mater. Eng.* **2011**, *296*, 710–718. [CrossRef]
50. Jorda, M.; Montava-Jorda, S.; Balart, R.; Lascano, D.; Montanes, N.; Quiles-Carrillo, L. Functionalization of Partially Bio-Based Poly(Ethylene Terephthalate) by Blending with Fully Bio-Based Poly(Amide) 10,10 and a Glycidyl Methacrylate-Based Compatibilizer. *Polymers* **2019**, *11*, 1331. [CrossRef] [PubMed]
51. Jaszkiewicz, A.; Meljon, A.; Bledzki, A. Bledzki. mechanical and thermomechanical properties of PLA/man-made cellulose green composites modified with functional chain extenders—A comprehensive study. *Polym. Compos.* **2018**, *39*, 1716–1723. [CrossRef]
52. Carbonell-Verdu, A.; Bernardi, L.; Garcia-Garcia, D.; Sanchez-Nacher, L.; Balart, R. Development of environmentally friendly composite matrices from epoxidized cottonseed oil. *Eur. Polym. J.* **2015**, *63*, 1–10. [CrossRef]
53. Agüero, Á.; Lascano, D.; Garcia-Sanoguera, D.; Fenollar, O.; Torres-Giner, S. Valorization of linen processing by-products for the development of injection-molded green composite pieces of polylactide with improved performance. *Sustainability* **2020**, *12*, 652. [CrossRef]
54. Quiles-Carrillo, L.; Montanes, N.; Garcia-Garcia, D.; Carbonell-Verdu, A.; Balart, R.; Torres-Giner, S. Effect of different compatibilizers on injection-molded green composite pieces based on polylactide filled with almond shell flour. *Compos. Part B Eng.* **2018**, *147*, 76–85. [CrossRef]
55. Sever, K.; Sarikanat, M.; Seki, Y.; Erkan, G.; Erdoğan, Ü.H. The mechanical properties of γ-methacryloxypropyltrimethoxy silane-treated jute/polyester composites. *J. Compos. Mater.* **2010**, *44*, 1913–1924. [CrossRef]
56. Aguero, A.; Quiles-Carrillo, L.; Jorda-Vilaplana, A.; Fenollar, O.; Montanes, N. Effect of different compatibilizers on environmentally friendly composites from poly(lactic acid) and diatomaceous earth. *Polym. Int.* **2019**, *68*, 893–903. [CrossRef]
57. Quiles-Carrillo, L.; Boronat, T.; Montanes, N.; Balart, R.; Torres-Giner, S. Injection-molded parts of fully bio-based polyamide 1010 strengthened with waste derived slate fibers pretreated with glycidyl- and amino-silane coupling agents. *Polym. Test.* **2019**, *77*, 105875. [CrossRef]
58. Samper, M.D.; Petrucci, R.; Sánchez-Nacher, L.; Balart, R.; Kenny, J.M. Effect of silane coupling agents on basalt fiber-epoxidized vegetable oil matrix composite materials analyzed by the single fiber fragmentation technique. *Polym. Compos.* **2015**, *36*, 1205–1212. [CrossRef]
59. Torres-Giner, S.; Montanes, N.; Boronat, T.; Quiles-Carrillo, L.; Balart, R. Melt grafting of sepiolite nanoclay onto poly(3-hydroxybutyrate-*co*-4-hydroxybutyrate) by reactive extrusion with multi-functional epoxy-based styrene-acrylic oligomer. *Eur. Polym. J.* **2016**, *84*, 693–707. [CrossRef]
60. Ferri, J.M.; Garcia-Garcia, D.; Montanes, N.; Fenollar, O.; Balart, R. The effect of maleinized linseed oil as biobased plasticizer in poly(lactic acid)-based formulations. *Polym. Int.* **2017**, *66*, 882–891. [CrossRef]
61. Torres-Giner, S.; Gimeno-Alcañiz, J.V.; Ocio, M.J.; Lagaron, J.M. Optimization of electrospun polylactide-based ultrathin fibers for osteoconductive bone scaffolds. *J. Appl. Polym. Sci.* **2011**, *122*, 914–925. [CrossRef]
62. Foulk, J.A.; Akin, D.E.; Dodd, R.B. Harvesting and processing of flax in the USA. In *Recent Progress in Medicinal Plants*; Studium Press, LLC: Houston, TX, USA, 2006; p. 15.
63. Foulk, J.; Akin, D.; Dodd, R.; Ulven, C. Production of Flax Fibers for Biocomposites. In *Cellulose Fibers: Bio- and Nano-Polymers Composites*; Springer-Verlag Berlin Heidelberg: Heidelberg, Germany, 2011; pp. 61–95.
64. Bos, H.L.; Müssig, J.; van den Oever, M.J.A. Mechanical properties of short-flax-fibre reinforced compounds. *Compos. Part A Appl. Sci. Manuf.* **2006**, *37*, 1591–1604. [CrossRef]
65. Baley, C. Analysis of the flax fibres tensile behaviour and analysis of the tensile stiffness increase. *Compos. Part A Appl. Sci. Manuf.* **2002**, *33*, 939–948. [CrossRef]

66. Ray, D.; Sarkar, B.K. Characterization of alkali-treated jute fibers for physical and mechanical properties. *J. Appl. Polym. Sci.* **2000**, *80*, 1013–2001. [CrossRef]
67. Keener, T.J.; Stuart, R.K.; Brown, T.K. Maleated coupling agents for natural fibre composites. *Compos. Part A Appl. Sci. Manuf.* **2004**, *35*, 357–362. [CrossRef]
68. Lu, J.Z.; Wu, Q.; McNabb, H.S. Chemical coupling in wood fiber and polymer composites: A review of coupling agents and treatments. *Wood Fiber Sci.* **2000**, *32*, 88–104.
69. Lu, J.Z.; Wu, Q.; Negulescu, I.I. The influence of maleation on polymer adsorption and fixation, wood surface wettability, and interfacial bonding strength in wood-PVC composites. *Wood Fiber Sci.* **2002**, *34*, 434–459.
70. Asim, M.; Jawaid, M.; Abdan, K.; Ishak, M.R. Effect of Alkali and silane treatments on mechanical and fibre-matrix bond strength of kenaf and pineapple leaf fibres. *J. Bionic Eng.* **2016**, *13*, 426–435. [CrossRef]
71. Lopattananon, N.; Panawarangkul, K.; Sahakaro, K.; Ellis, B. Performance of pineapple leaf fiber–natural rubber composites: The effect of fiber surface treatments. *J. Appl. Polym. Sci.* **2006**, *102*, 1974–1984. [CrossRef]
72. Puglia, D.; Monti, M.; Santulli, C.; Sarasini, F.; De Rosa, I.M.; Kenny, J.M. Effect of alkali and silane treatments on mechanical and thermal behavior of Phormium tenax fibers. *Fibers Polym.* **2013**, *14*, 423–427. [CrossRef]
73. Jaszkiewicz, A.; Bledzki, A.K.; van der Meer, R.; Franciszczak, P.; Meljon, A. How does a chain-extended polylactide behave?: A comprehensive analysis of the material, structural and mechanical properties. *Polym. Bull.* **2014**, *71*, 1675–1690. [CrossRef]
74. Anakabe, J.; Zaldua Huici, A.M.; Eceiza, A.; Arbelaiz, A. The effect of the addition of poly(styrene-*co*-glycidyl methacrylate) copolymer on the properties of polylactide/poly(methyl methacrylate) blend. *J. Appl. Polym. Sci.* **2016**, *133*, 43935. [CrossRef]
75. Sun, S.; Zhang, M.; Zhang, H.; Zhang, X. Polylactide toughening with epoxy-functionalized grafted acrylonitrile-butadiene-styrene particles. *J. Appl. Polym. Sci.* **2011**, *122*, 2992–2999. [CrossRef]
76. Battegazzore, D.; Abt, T.; Maspoch, M.L.; Frache, A. Multilayer cotton fabric bio-composites based on PLA and PHB copolymer for industrial load carrying applications. *Compos. Part B Eng.* **2019**, *163*, 761–768. [CrossRef]
77. Battegazzore, D.; Frache, A.; Abt, T.; Maspoch, M.L. Epoxy coupling agent for PLA and PHB copolymer-based cotton fabric bio-composites. *Compos. Part B Eng.* **2018**, *148*, 188–197. [CrossRef]
78. Bocqué, M.; Voirin, C.; Lapinte, V.; Caillol, S.; Robin, J.-J. Petro-based and bio-based plasticizers: Chemical structures to plasticizing properties. *J. Polym. Sci. Part A Polym. Chem.* **2016**, *54*, 11–33. [CrossRef]
79. Nor Azowa, I.; Kamarul Arifin, H.; Abdan, K. Effect of fiber treatment on mechanical properties of kenaf fiber-ecoflex composites. *J. Reinf. Plast. Compos.* **2009**, *29*, 2192–2198. [CrossRef]
80. Móczó, J.; Pukánszky, B. Polymer micro and nanocomposites: Structure, interactions, properties. *J. Ind. Eng. Chem.* **2008**, *14*, 535–563. [CrossRef]
81. Mortazavian, S.; Fatemi, A. Fatigue behavior and modeling of short fiber reinforced polymer composites: A literature review. *Int. J. Fatigue* **2015**, *70*, 297–321. [CrossRef]
82. Lourdin, D.; Bizot, H.; Colonna, P. "Antiplasticization" in starch-glycerol films? *J. Appl. Polym. Sci.* **1997**, *63*, 1047–1053. [CrossRef]
83. Rizzuto, M.; Mugica, A.; Zubitur, M.; Caretti, D.; Müller, A.J. Plasticization and anti-plasticization effects caused by poly(lactide-*ran*-caprolactone) addition to double crystalline poly(L-lactide)/poly(ε-caprolactone) blends. *CrystEngComm* **2016**, *18*, 2014–2023. [CrossRef]
84. Sanyang, M.; Sapuan, S.; Jawaid, M.; Ishak, M.; Sahari, J. Effect of plasticizer type and concentration on tensile, thermal and barrier properties of biodegradable films based on sugar palm (*Arenga pinnata*) Starch. *Polymers* **2015**, *7*, 1106–1124. [CrossRef]
85. Cacciotti, I.; Mori, S.; Cherubini, V.; Nanni, F. Eco-sustainable systems based on poly(lactic acid), diatomite and coffee grounds extract for food packaging. *Int. J. Biol. Macromol.* **2018**, *112*, 567–575. [CrossRef]
86. Perinović, S.; Andričić, B.; Erceg, M. Thermal properties of poly(L-lactide)/olive stone flour composites. *Thermochim. Acta* **2010**, *510*, 97–102. [CrossRef]
87. Pilla, S.; Gong, S.; O'Neill, E.; Rowell, R.M.; Krzysik, A.M. Polylactide-pine wood flour composites. *Polym. Eng. Sci.* **2008**, *48*, 578–587. [CrossRef]
88. Huang, Y.; Liu, Y.; Zhao, C. Morphology and properties of PET/PA-6/E-44 blends. *J. Appl. Polym. Sci.* **1998**, *69*, 1505–1515. [CrossRef]
89. Corre, Y.-M.; Duchet, J.; Reignier, J.; Maazouz, A. Melt strengthening of poly(lactic acid) through reactive extrusion with epoxy-functionalized chains. *Rheol. Acta* **2011**, *50*, 613–629. [CrossRef]

90. Ojijo, V.; Ray, S.S. Super toughened biodegradable polylactide blends with non-linear copolymer interfacial architecture obtained via facile in-situ reactive compatibilization. *Polymer* **2015**, *80*, 1–17. [CrossRef]
91. Mansaray, K.G.; Ghaly, A.E. Thermogravimetric analysis of rice husks in an air atmosphere. *Energy Sources* **2007**, *20*, 653–663. [CrossRef]
92. Wielage, B.; Lampke, T.; Marx, G.; Nestler, K.; Starke, D. Thermogravimetric and differential scanning calorimetric analysis of natural fibres and polypropylene. *Thermochim. Acta* **1999**, *337*, 169–177. [CrossRef]
93. Sánchez-Jiménez, P.E.; Pérez-Maqueda, L.A.; Perejón, A.; Criado, J.A. Generalized kinetic master plots for the thermal degradation of polymers following a random scission mechanism. *J. Phys. Chem. A* **2010**, *114*, 7868–7876. [CrossRef]
94. Zhan, J.; Song, L.; Nie, S.; Hu, Y. Combustion properties and thermal degradation behavior of polylactide with an effective intumescent flame retardant. *Polym. Degrad. Stabil.* **2009**, *94*, 291–296. [CrossRef]
95. Mishra, S.; Mohanty, A.K.; Drzal, L.T.; Misra, M.; Parija, S.; Nayak, S.K.; Tripathy, S.S. Studies on mechanical performance of biofibre/glass reinforced polyester hybrid composites. *Compos. Sci. Technol.* **2003**, *63*, 1377–1385. [CrossRef]

© 2020 by the authors. Licensee MDPI, Basel, Switzerland. This article is an open access article distributed under the terms and conditions of the Creative Commons Attribution (CC BY) license (http://creativecommons.org/licenses/by/4.0/).

Article

Thermomechanical and Morphological Properties of Poly(ethylene terephthalate)/Anhydrous Calcium Terephthalate Nanocomposites

Franco Dominici [1], Fabrizio Sarasini [2], Francesca Luzi [1], Luigi Torre [1] and Debora Puglia [1,*]

[1] Civil and Environmental Engineering Department, University of Perugia, Strada di Pentima 4, 05100 Terni, Italy; franco.dominici@unipg.it (F.D.); francesca.luzi@unipg.it (F.L.); luigi.torre@unipg.it (L.T.)
[2] Department of Chemical Engineering Materials Environment, Sapienza-Università di Roma and UdR INSTM, Via Eudossiana 18, 00184 Roma, Italy; fabrizio.sarasini@uniroma1.it
* Correspondence: debora.puglia@unipg.it; Tel.: +39-0744-492916

Received: 1 December 2019; Accepted: 21 January 2020; Published: 30 January 2020

Abstract: Calcium terephthalate anhydrous salts (CATAS), synthetized by reaction of terephthalic acid with metal (Ca) oxide were incorporated at different weight contents (0–30 wt. %) in recycled Poly(ethylene terephthalate) (rPET) by melt processing. Their structure, morphology, thermal and mechanical properties (tensile and flexural behavior) were investigated. Results of tensile strength of the different formulations showed that when the CATAS content increased from 0.1 to 0.4 wt. %, tangible changes were observed (variation of tensile strength from 65.5 to 69.4 MPa, increasing value for E from 2887 up to 3131 MPa, respectively for neat rPET and rPET_0.4CATAS). A threshold weight amount (0.4 wt. %) of CATAS was also found, by formation at low loading, of a rigid amorphous fraction at the rPET/CATAS interface, due to the aromatic interactions (π–π conjugation) between the matrix and the filler. Above the threshold, a restriction of rPET/CATAS molecular chains mobility was detected, due to the formation of hybrid mechanical percolation networks. Additionally, enhanced thermal stability of CATAS filled rPET was registered at high content (T_{max} shift from 426 to 441 °C, respectively, for rPET and rPET_30CATAS), essentially due to chemical compatibility between terephthalate salts and polymer molecules, rich in stable aromatic rings. The singularity of a cold crystallization event, identified at the same loading level, confirmed the presence of an equilibrium state between nucleation and blocking effect of amorphous phase, basically related to the characteristic common terephthalate structure of synthetized Ca–Metal Organic Framework and the rPET matrix.

Keywords: recycled poly(ethylene terephthalate); rPET; Calcium terephthalate salts; high performance nanocomposites

---

1. Introduction

Disposal of waste materials has become an urgent problem that can be solved by following two general paths: re-using the materials disposed, for suitable applications, as received, or recycling the waste to produce regenerated materials that could be applied again in the same or in another industrial field [1]. For instance, waste poly(ethylene terephthalate) (PET) plastic waste is neither biodegradable nor compostable, and due to its wide use worldwide substantial disposal problems are created. Recycling has emerged as the most practical method to deal with this issue, especially with products such as recycled PET (rPET) beverage bottles or fibers for textiles [2].

On the other hand, although PET is a recyclable polyester, its post-consumer reuse is limited by the weakening of its physical-mechanical properties, due to the reduction of molecular weight and viscosity, derived from the cleavage of the chains during processing [3]. As a solution to this problem, extruded blends with satisfactory properties based on virgin PET, with marginal proportions of rPET, were developed and used at an industrial level. However, it is necessary to increase the use of recycled PET in these mixed formulations to reduce costs and contribute to compliance with increasingly stringent regulations on plastic waste disposal.

To achieve the result of using increasing fractions of rPET without adding virgin material, it is mandatory to increase the mechanical properties of the polymer with the incorporation of low cost, commercially available nanoparticles/nanofillers [4]. There are many examples available in the literature dealing with the selection of unmodified or chemically modified layered nanoclays [5–7].

However, even if it was shown that substantial increase of mechanical properties can be achieved by high MMT (montmorillonite) content and optimum strength and stiffness at low level [8], the main limit of the use of organo-modified nanoclays relies in the decomposition of alkylammonium ether that can influence the PET degradation during the processing. A recent paper investigated the use of graphene nanoplatelets [9], demonstrating how the higher-order structure of PET can be tuned by considering different graphene loading. The authors interestingly found that the presence of a significant rigid amorphous fraction (RAF) in the nanocomposites increased with an increase in the graphene loading, while at low loadings the stiffening effect was observed to be quite small in the glassy state region. At a value of 2 wt. %, graphene formed a mechanical percolation network with the RAF of PET and the PET chains were geometrically restricted by the incorporation of graphene, resulting in an unexpectedly higher modulus of nanocomposites both below and above T_g.

The effect of the nanometric calcium terephthalate anhydrous salts (CATAS) was studied in analogy to the above mentioned PET/GNP (polyethylene terephthalate/graphene nanoparticles) nanocomposites work. The idea is to design a circular use of PET, using terephthalic acid (TPA) recovered from the hydrolysis of PET for the preparation of terephthalate salts of various metals such as aluminum, magnesium, sodium, calcium potassium, etc. [10,11]. The production of these salts with nano-sized lamellar structure can be obtained with appropriate control of the reaction parameters [12,13]. Many different terephthalate salts, or metal–organic framework (MOFs) may be obtained by using sodium, potassium, aluminum, magnesium, calcium and other metals salts [14,15]. They are hybrid materials, composed of metal nodes and coordinating organic linkers, arranged in highly regular motifs that lead to materials exhibiting ultra-high surface areas. Applications are therefore proposed that use this porosity for reversible host–guest behavior, for example, in gas storage, catalysis and drug delivery. Attempts were made to develop existing MOFs and explore new applications using known functionalities, and introducing flexibility, defects and stimuli responsive behavior [16]. The formation of nanocomposites by combining MOFs with more processable materials such as polymers, not only engages with the theme of new materials discovery, but also offers new solutions for the manufacturing of robust bulk structures.

A few examples are available in the scientific literature on the use of these salts in thermoplastic nanocomposites, as in the case of applications that consider terephthalate salts in batteries [17–19] or gas separation applications, by taking into advantage of their defined porosity, high surface area, and catalytic activity. A recent review covers various topics of MOF/polymer research, where MOF/polymer hybrid and composite materials are highlighted, encompassing a range of material classes, such as bulk materials, membranes, and dispersed materials [20].

On the other hand, the chemical compatibility between the terephthalate salts and the polymer molecules, rich in stable aromatic rings, suggested the possibility of investigating the role of these salts in the thermomechanical behavior of recycled PET.

The objective of the present investigation was to study the structure of rPET_CATAS nanocomposites and the effect of nanofillers introduction on mechanical properties. The study was conducted by selecting two levels of loadings (0.1–1 wt. % and 2–30 wt. %), aiming to correlate the internal structure with the bulk thermomechanical performance of recycled PET. We specifically aimed to study these properties from the different perspective of evaluating not only the dispersibility at the macrolevel, but also checking the higher-order structure of the polymeric matrix in presence of the selected nanofillers.

2. Materials and Methods

Polyethylene terephthalate (I. V., 0.65 dl/g, moisture content 1.3%, bulk density 250–350 kg m^{-3}) was supplied by Extremadura Torrepet (Torremejia, Spain). The recycling process adopted by the company intended to recycle post-consumer food grade poly(ethylene terephthalate) containers to produce recycled PET pellets. A four steps process consisted of in house treatment of PET bottles into hot caustic, drying and extrusion of flakes into pellets under vacuum, continuous feeding to a reactor for crystallization at high temperature, under atmospheric pressure for a predefined residence time, after that decontamination by solid state polymerisation process (SSP) is considered. A nanometric metal organic framework consisting of Calcium ions as metal clusters coordinated to terephthalic acid as organic ligand was synthesized in our laboratories to be used as filler in the nanocomposites. Calcium oxide, terephthalic acid, ammonia and water chemical reagents were supplied by Sigma Aldrich (Milan, Italy). The synthesis method is briefly described in the next section.

2.1. Preparation of Terephthalate Salts

Powdered terephthalic acid (TPA) was solubilized in an alkaline solution of water and ammonia with stirring at 80 °C for 1 h. Then, calcium oxide was added in a stoichiometric proportion to the acid according to Reaction (1):

$$C_8H_6O_4 + CaO + 2 H_2O \rightarrow C_8H_4O_4Ca \bullet 3 H_2O \qquad (1)$$

The solution was kept under stirring at 80 °C for 30 min, after that insoluble calcium terephthalate precipitated forming a whitish deposit on the bottom. The solution was filtered, and the solid residue subjected to appropriate washings to eliminate unwanted reaction residues. At the end of the synthesis, calcium terephthalate trihydrate is obtained. To eliminate both moisture and water-bound molecules, the product was kept in a vacuum oven for 2 h at 190 °C. Finally, anhydrous calcium terephthalate (CATAS) was pulverized and sieved to eliminate any conglomerates of salts.

2.2. Production of Composite Materials

Composite materials were manufactured using the melt mixing method by blending the recycled polyethylene terephthalate matrix with the calcium terephthalate salts as shown in Table 1. A co-rotating twin-screw extruder, Microcompounder 5 & 15cc by DSM (DSM explorer 5&15 CC Micro Compounder, Xplore Instruments BV, Sittard, The Netherlands) was used for the mixing at 90 rpm for 60 s by setting a temperature profile of 250–255–265 °C in the three heating zones from top to die. A Micro Injection Molding Machine 10cc, Xplore line by DSM (DSM explorer 5&15 CC Micro Compounder, Xplore Instruments BV, Sittard, The Netherlands) was used to produce the samples according to the standards for flexural and tensile tests. An adequate pressure/time profile was used for the injection of each type of samples, while barrel and mold temperatures were set at 280 and 120 °C, respectively.

Table 1. Developed formulations based on rPET and CATAS.

Sample Name	rPET [%] wt.	CATAS [%] wt.
rPET	100	—
rPET_0.1CATAS	99.9	0.1
rPET_0.25CATAS	99.75	0.25
rPET_0.4CATAS	99.6	0.4
rPET_0.5CATAS	99.5	0.5
rPET_1CATAS	99	1
rPET_2CATAS	98	2
rPET_3CATAS	97	3
rPET_10CATAS	90	10
rPET_20CATAS	80	20
rPET_30CATAS	70	30

2.3. Characterization of Terephthalate Salts and rPET/CATAS Nanocomposites

Morphological characterization of salts and rPET/CATAS nanocomposites was carried out using a field emission scanning electron microscope (FESEM) Supra 25 by Zeiss (Oberkochen, Germany) taking micrographs with an accelerating voltage of 5 kV at different magnifications. Previously, the samples were gold sputtered to provide electric conductivity.

X-ray diffraction (XRD) analysis was performed with a diffractometer X'Pert PRO by Philips (Malvern Panalytical Ltd., Malvern, UK) (CuKα radiation = 1.54060 Å) at room temperature. XRD patterns were collected in the range of 2θ = 10–80° with a step size of 0.02° scan and a time per step of 34 s.

Thermal characteristics of rPET/CATAS nanocomposites were investigated with a temperature-modulated differential scanning calorimeter (MDSC) Q200 by TA Instruments (TA Instrument, Q200, New Castle, DE, USA). A heating scan between 0 and 300 °C was performed at a rate of 10 °C/min in nitrogen flow at 60 mL/min. A three phases model was adopted to describe the higher-order structure of rPET/CATAS nanocomposites. It was demonstrated that sometimes ΔC_p at T_g in semicrystalline polymers results inconsistent with the amorphous fraction (X_a), if calculated as complement to one of the crystalline fraction (X_c). In this case, it is possible to calculate the amorphous fraction with the variation of thermal capacity at T_g and a total rigid fraction ($X_f = 1 - X_a$) that consists of the crystalline phase and another fraction of a phase which remains rigid beyond the glass transition region. Since this rigid phase cannot be detected by ΔC_p at T_g that is not properly amorphous phase and it does not have a crystalline structure, it was defined rigid amorphous phase X_{raf}:

$$X_a = \frac{\Delta C_{P-meas}^{\Delta T_g} - w_{CATAS} * \Delta C_{P-CATAS}^{\Delta T_g}}{\Delta C_{P-aPET}^{\Delta T_g}} = \frac{\Delta C_{P-PET}^{\Delta T_g}}{\Delta C_{P-aPET}^{\Delta T_g}} = 1 - X_f \quad (2)$$

$$X_a = 1 - (X_c + X_{raf}) \quad (3)$$

where $\Delta C_{P-meas}^{\Delta T_g}$, $\Delta C_{P-CATAS}^{\Delta T_g}$, $\Delta C_{P-aPET}^{\Delta T_g}$ are ΔC_p at T_g, respectively, for measured nanocomposites, CATAS and 100% amorphous neat PET, while w_{CATAS} is the weight fraction of the salts in the nanocomposites. It is assumed that $\Delta C_{P-aPET}^{\Delta T_g} = 0.405$ J g^{-1} K^{-1} and heat of melting of 100% crystalline PET is $\Delta H_{m-cPET}^0 = 140$ J g^{-1}. [21].

Thermogravimetric analysis (TGA) on nanocomposites samples was carried out using Seiko Exstar 6300 from Seiko instruments (Seiko, Tokyo, Japan). Thermograms obtained from TGA analysis provides the information related to the thermal stability of material. About 10–15 mg of the sample was subjected for TGA analysis, performed on the samples between room temperature and 900 °C with heating rate of 10 °C/min in nitrogen atmosphere.

A universal electronic dynamometer LR30K by LLOYD Instruments (Lloyd Instrument, Segens worth West, Foreham, UK) was used to carry out tensile tests by setting a cross-head speed of 5 mm min^{-1}, in accordance with ASTM D638 standard. Three points bending tests were carried out by setting a span between supports of 64 mm and loading at a deflection rate of 2 mm min^{-1}, according to the standard ASTM D790. The analysis of mechanical tests data was carried out by using the Nexigen software (Elis (Electronic Instruments & Systems) s.r.l., Rome, Italy).

Dynamic mechanical analysis of the rPET-based nanocomposites was performed with an Ares N2 rheometer (Rheometric Scientific, Reichelsheim, Germany), by testing samples of about 4 × 10 × 40 mm, gripped with a gap of about 20 mm and tested with rectangular torsion at a frequency of 2π rad/s with a strain of 0.05%, with a temperature ramp of 3 °C/min applied in the range from 30 to 180 °C. Storage (G') and loss moduli (G") as a function of the temperature were determined.

3. Results and Discussion

3.1. Characterization of Calcium Terephthalate Salts

The structural characteristics of terephthalate calcium salts are shown in Figure 1a [22–24]. The reaction between terephthalic acid and calcium oxide in water was carried out to produce calcium terephthalate trihydrate salts (CATS) [25]. The production process required an appropriate monitoring of reaction conditions, by acting on the kinetics and using appropriate additives to obtain the nanometric structures. Some attempts were made to optimize the characteristics of the fillers. Lastly, insoluble calcium terephthalate trihydrate salts were obtained by their precipitation in water. A heat treatment at 200 °C for 1 h in a vacuum oven transforms the trihydrate salts (Figure 1b) into the corresponding anhydrous calcium terephthalate (CATAS) salts (Figure 1c): this step was necessary, being the presence of bound water harmful for the melt blending process of the nanocomposites, due to hydrolysis phenomena and formation of cavities within the samples.

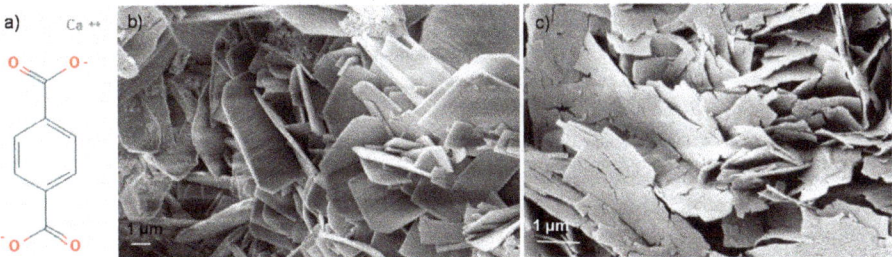

Figure 1. Chemical structure of calcium terephthalate salts (**a**) and FESEM morphology in their hydrated (**b**) and anhydrous state (**c**).

Figure 1c shows that after the dehydration (T = 200 °C), the anhydrous salts have cracks with rough surface, revealing that volume contraction happens in the dehydration process [26].

The X-ray diffraction pattern of CATAS shows diffraction peaks at 18.8°, 21.6°, 25.7°, 26.9°, and 31.2°. According to the report by Mou et al., this structure can be indexed with a space group of C2/c (n. 15) as monoclinic crystal system [17] (Figure 2a). More precisely, the process involves rotation of the ligand, disconnection, and protonation of the carboxylate group on one side and partial disconnection and reformation of Ca–O bond on the other side of the ligand. Mazaj proposes a reversible hydration/dehydration mechanism, which through the breaking of the bonds between Ca^{2+} and carboxylate groups, rotating of ligand and re-coordination of COO^- groups to Ca^{2+} centers. The crystal-to-crystal transformations are driven by the tendencies to change the bonding modes between COO^- and Ca^{2+} with the change of Ca^{2+} coordination number. Thus, the decrease of Ca^{2+} coordination number, which is usually a consequence of activation, leads to a structural rearrangement, with expansion or contraction of the pores [27].

TG and DTG plots for CATS and anhydrous CATAS (Figure 2b,c) showed multiple decomposition steps. In the case of CATS, the first peak at a temperature lower than 100 °C refers to the loss of hygroscopic water and any residual volatile additives present during the reaction. The second peak, at a temperature between 100 °C and 170 °C, refers to the loss of molecular water since a weight loss of 22% (moisture free weight) corresponds to three water molecules per formula unit. The decomposition of the acid salts involved the breaking of the carboxyl groups, the formation of carbonates, and the release of gaseous products. The subsequent mass losses are due to thermal destruction of organic ligands over 500 °C. Figure 2c shows that in the case of anhydrous salts, the main weight loss can be detected between 500 and 800 °C that corresponds to the decomposition of the organic ligands [25,28]. Evaluation of thickness distribution showed that the average size for anhydrous salts was set at 50 ± 5 nm.

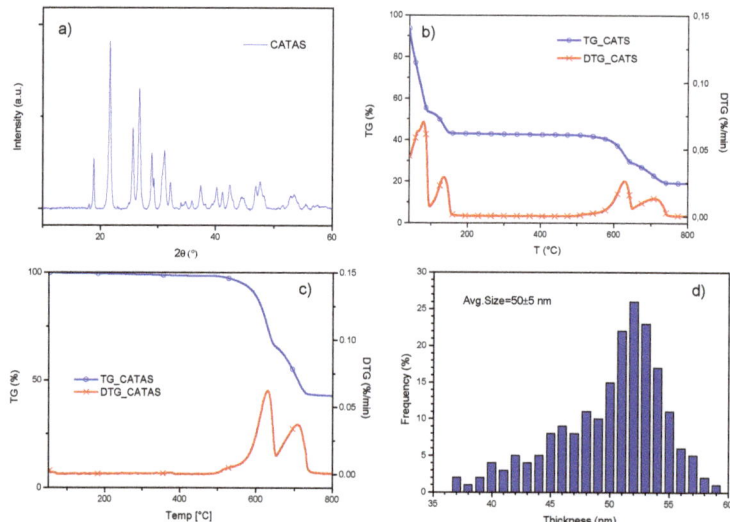

Figure 2. XRD profile for CATAS (**a**); TG/DTG curves for CATS (**b**); and CATAS nanofillers (**c**); thickness distribution for CATAS (**d**).

3.2. Characteristics of rPET/CATAS Nanocomposites

3.2.1. DMTA (Dynamic Mechanical Thermal Analysis)

To evaluate the influence of CATAS on viscoelastic behavior of rPET matrix, dynamic mechanical thermal analysis of high loaded rPET nanocomposites was performed. The variation of the storage moduli (G′) and tan δ (G″/G′) with temperature are presented in Figure 3a,b, respectively. The storage modulus behavior in Figure 3a shows that the rPET/CATAS nanocomposites possessed higher storage modulus compared to the reference rPET below and above the T_g. A gradual increase in storage modulus is observed with increase in CATAS content in rPET matrix (Table 2). Among the nanocomposite samples, rPET_30CATAS had the highest storage modulus compared to reference rPET and other nanocomposite samples. The adsorption of the CATAS onto the macromolecular chains of rPET leads to a constraint in the chains movement and therefore, an improved storage modulus [29], while a sharp decrease in the moduli was observed at the glass–rubber transition at around 90 °C [4]. Regarding the glass temperatures (T_g), results of tan δ from the DMTA tests can be seen in Figure 3b and Table 2. Compared to rPET, the nanocomposites had the maximum of the peak at slightly higher temperature with increasing content of CATAS temperatures. Thus, the increased value of T_g resulting on the addition of nanometric salts confirms that the mobility of polymeric chains was slightly reduced, which in turn affected the flexibility of the samples. On the other hand, the peak values of tan δ increased with an increase in CATAS up to 10% wt., where the influence of the elastic component starts to prevail on the viscous one: at higher filler contents (20–30 wt. %), a decrease in the tan δ is observed, mainly due to a combination of viscoelastic and physical behavior. In details, a prevalent rigid amorphous fraction (X_{raf}), formed in the interfacial region between rPET and CATAS, justifies the increase in elastic modulus, so that the nanofiller is able to restrict the molecular movement of the amorphous rPET chains both in the interfacial region and in the amorphous region. This restriction is also evident in the registered G′ values when measured at 150 °C (after Tg) (Table 2): the limited decrease in G′ after the Tg can be mainly related to the CATAS physical interactions that when present in large amount physically constrains the mobility of the polymeric chains.

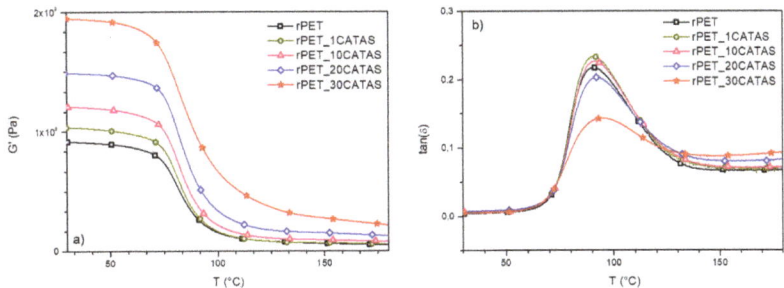

Figure 3. G' (a) and tan δ curve (b) from DMTA tests for rPET and rPET/CATAS nanocomposites.

Table 2. Calculated values of tan δ and G' at 50 and 150 °C for the rPET/CATAS nanocomposites.

MATERIAL	G' @ T = 50 °C ($\times 10^9$ Pa)	G' @ T = 150 °C ($\times 10^8$ Pa)	T_g (at tan δ peak) (°C)
rPET	0.89	0.61	90.41
rPET_1CATAS	1.00	0.68	90.48
rPET_10CATAS	1.19	0.96	90.89
rPET_20CATAS	1.47	1.54	91.29
rPET_30CATAS	1.92	2.76	93.43

3.2.2. Mechanical Characterization (Flexural and Tensile Tests)

The influence of CATAS presence on tensile and flexural properties of rPET nanocomposites was also investigated. Stress-strain curves for flexural and tensile tests of representative samples are respectively reported in Figure 4a,b, while the variations of the tensile and flexural parameters of all the rPET_CATAS nanocomposites are included in Figure 4c,d. Complementary information on performed tests are also summarized in Table 3. The results of tensile strength of the different formulations showed that when the CATAS content increased from 0.1 to 0.4 wt. %, tangible changes were observed (from 65.5 to 69.4 MPa) and this improvement can be attributed to strong interaction between the salts and the polymer matrix owing to uniform dispersion of nanoparticles, as further confirmed by following FESEM analysis. Besides, amounts of nanoparticles higher than 0.4 wt. % negatively affects the mechanical properties of the nanocomposites, due to agglomeration of nanoparticles [30]. The possible formation of mechanical percolation networks with the RAF of rPET at interfacial regions with confined rPET matrix could justify these results above 0.4 wt. % of CATAS loading. As already observed by Aoyama et al. for grapheme nanoparticles [9], a mechanical percolation threshold can be estimated at these low loading level (between 1 and 2 wt. % in the case of GNP): a rigid amorphous fraction of rPET in nanocomposites is formed at the rPET/CATAS interface, due to the aromatic interactions between the matrix and the filler, both of which contain a π–π conjugation. Molecular chains of rPET in nanocomposites are clearly confined, in compared with neat rPET: below 0.4% wt. of CATAS loading, the restriction of rPET chains is mostly limited to the interfacial region (i.e., X_{raf}), therefore, the stiffness enhancement effect is dictated by simple mixing rule (increasing value for E from 2887 up to 3131 MPa, respectively for neat rPET and rPET_0.4CATAS). Above 0.4 wt. %, the restriction of molecular chains of rPET/CATAS is in turn due to the formation of hybrid mechanical percolation networks of CATAS with the X_{raf} of rPET that geometrically restricts the mobility of rPET chains in rPET. This effect was visible also in the registered values for strain at break, where a visible decrease was noted up to the same level of CATAS amount (0.4 wt. %), after that the values showed an increase, due to the absence of X_{raf} fraction between 0.4 and 2 wt. % of CATAS. When the loading was clearly too high (2–30 wt. %), the rPET nanocomposites became extremely fragile.

Similar trend in terms of mechanical performance was found for the nanocomposites in flexural tests. Below the mechanical percolation threshold, estimated between 1% and 2% for these nanostructured systems, the flexural strength remains constant. In particular, a threshold value of 0.4% CATAS for the formation of a mechanical percolation network is conceivable; this model justifies the improvements in the mechanical characteristics of the rPET_0.4CATAS system. The deformation reaches values two times higher than the matrix, maintaining a good flexural strength. However, flexural strength decreases for composites with higher nanofiller loadings (beyond 0.4 wt. %), which could be due to agglomeration of CATAS and reduced interfacial interaction between rPET and nanofiller [31].

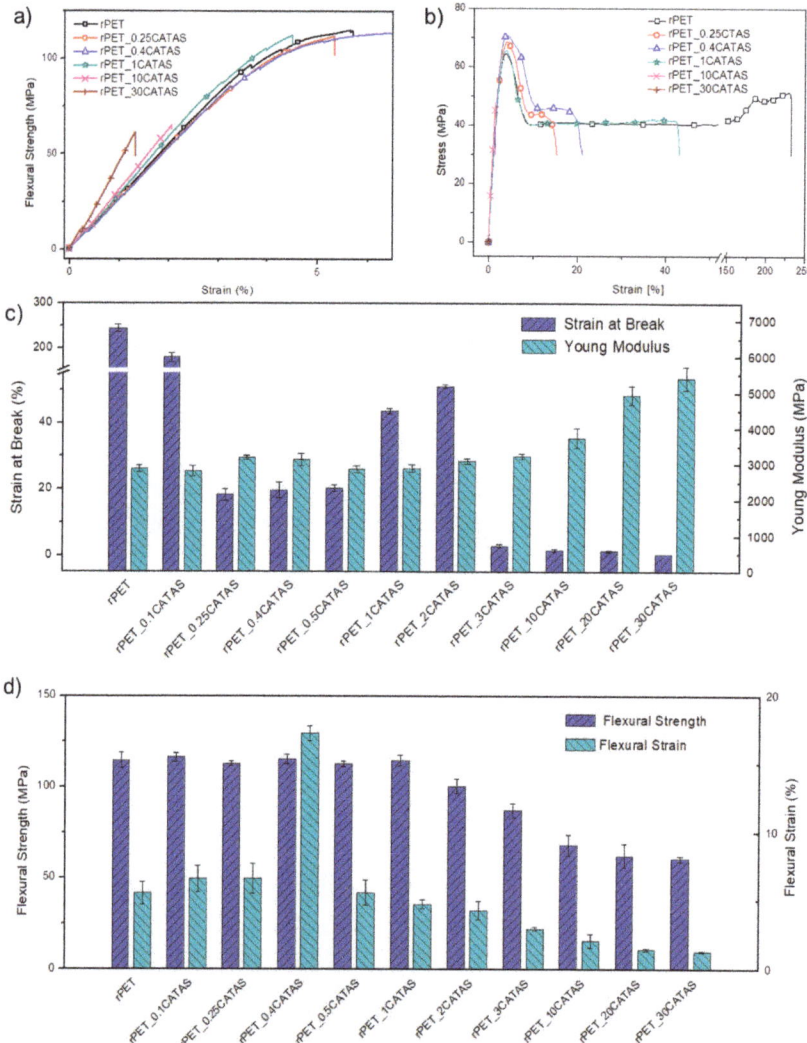

Figure 4. Stress-strain curves from flexural (**a**) and tensile tests (**b**), variation of tensile modulus and strain at break (**c**), variation of flexural stress and strain (**d**) for rPET_CATAS nanocomposites at the different loading levels.

This result can be justified again by the formation of aromatic interactions between the matrix and the filler. At this threshold the effect of CATAS is optimal and an exceptional performance is obtained. At values of reinforcement higher than the 2 wt. % CATAS threshold, the excess filler produces block of molecular flows (due to the mechanical effect at the filler/matrix interface and due to the increasing presence of the rigid amorphous phase, which blocks the structure of the composite limiting the deformability of the samples). In high-filling samples an increase in flexural modulus (see Figure 4a) is obtained but, when the deformability limit becomes excessive, the fragile behavior causes premature failure of the samples.

Table 3. Tensile parameters of rPET_CATAS nanocomposites at the different loading levels.

MATERIAL	Young's Modulus (MPa)		Stress (MPa)		Strain at Maximum Stress (%)		Stress at Break (MPa)		Strain at Break (%)	
rPET	2887	±178	65.5	±3.6	3.99	±0.21	41.0	±5.2	244.2	±16.9
rPET_0.1CATAS	2819	±283	61.7	±4.2	3.68	±0.23	41.7	±3.1	178.9	±20.7
rPET_0.25CATAS	3186	±115	68.9	±2.3	3.98	±0.28	41.9	±1.9	18.3	±3.6
rPET_0.4CATAS	3131	±342	69.4	±1.7	3.89	±0.27	42.4	±3.2	19.7	±4.5
rPET_0.5CATAS	2867	±201	66.8	±0.1	3.90	±0.28	39.6	±0.5	20.2	±2.3
rPET_1CATAS	2898	±206	67.4	±2.5	3.78	±0.26	43.2	±1.7	43.6	±1.6
rPET_2CATAS	3099	±145	64.8	±1.4	3.12	±0.38	39.4	±0.5	51.0	±0.9
rPET_3CATAS	3225	±148	61.9	±8.8	2.91	±0.36	61.8	±8.8	2.9	±1.1
rPET_10CATAS	3751	±549	45.5	±14.1	1.60	±0.42	45.5	±14.1	1.6	±0.7
rPET_20CATAS	4943	±517	36.5	±10.2	1.34	±0.51	36.5	±10.3	1.3	±0.5
rPET_30CATAS	5408	±645	0.3	0.0	0.04	0.01	0.3	0.0	0.3	±0.1

3.2.3. XRD Analysis

The XRD patterns of recycled PET and some representative rPET/CATAS nanocomposites (Figure 5) indicate semi-crystalline triclinic structure for rPET matrix.

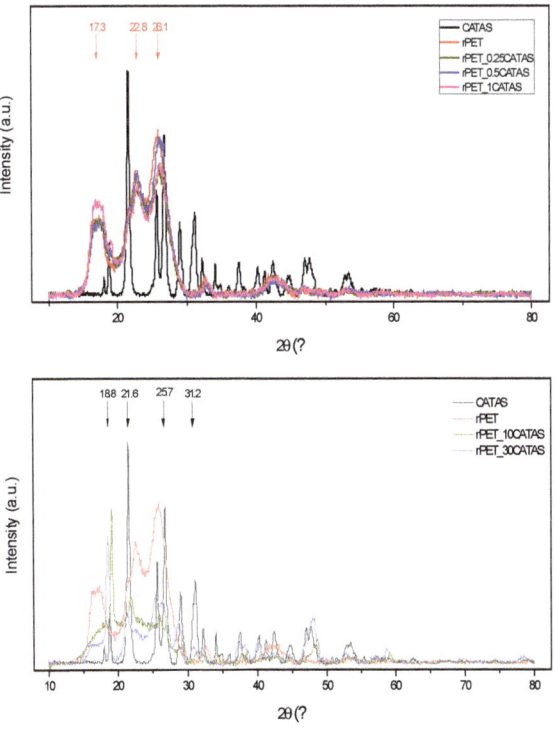

Figure 5. XRD diffraction curves for rPET and rPET/CATAS nanocomposites at low and high loadings.

Semi-crystalline PET typically shows characteristic crystalline XRD peaks at 2θ = 17.3, 21.7, 22.8, 26.1 and 32.5°, corresponding to the (010), (111), (110), (100), (021), (002) and (101) crystal planes [32]. Although some minor peaks are observed as the amount of CATAS increased in the sample, the intensities of the peaks for rPET decreased in presence of increasing amount of salts, while in parallel the signals related to the CATAS increased, indicating that polymer and salts maintained their original structure [33].

3.2.4. Morphological Analysis

The morphology of the samples was investigated by using FESEM analysis, performed on the fractured surfaces of the specimens after the tensile test. The SEM images of unreinforced rPET and rPET_CATAS nanocomposite samples at different loading levels are presented in Figure 6. It can be seen a gradual transition from ductile fractured surface of rPET to essentially fragile morphology of rPET_10CATAS, while at intermediate fraction (1 wt. % CATAS) the failure behavior changes, becoming partially brittle. This observation is consistent with the increasing content of amorphous fraction at higher concentrations: this fraction can act to prematurely fracture the composite, reducing the stiffness of the composite, due to CATAS role as stress concentration sites, facilitating crack initiation [34].

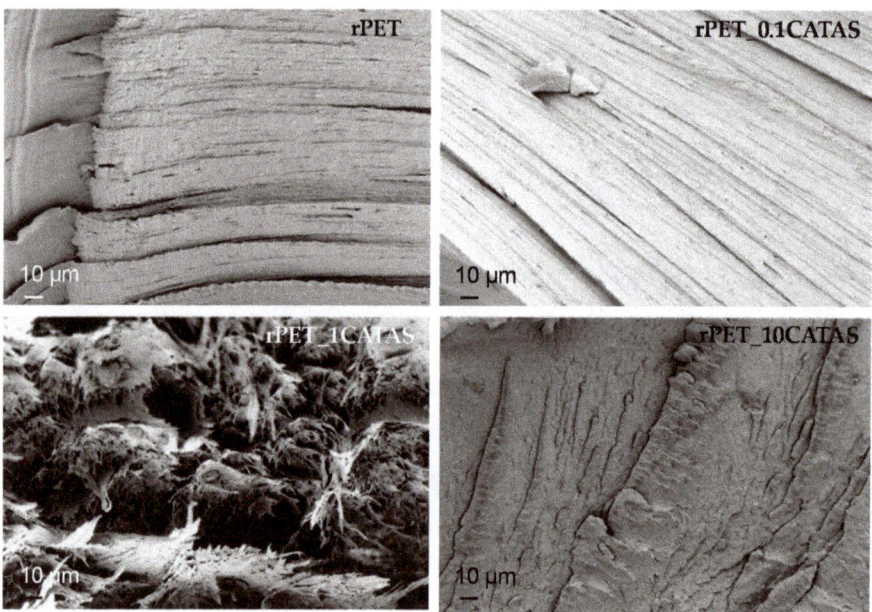

Figure 6. FESEM images of fractured surfaces for rPET and rPET/CATAS nanocomposites.

3.2.5. Thermal Characterization

TGA: The TG weight loss curves and first derivative TG curves for reference rPET and the nanocomposite samples under nitrogen atmosphere are presented in Table 4 and Figure 7a,b. Neat rPET exhibited a single-step decomposition profile: the main degradation process during TGA analysis under nitrogen atmosphere is due to the combined degradation processes—polymer chain degradation through end group initiated mechanism and the thermal degradation of the products formed during polymer chain degradation [35]. The thermal decomposition of polymer is leading to thermally stable cross-linked carbonaceous species which are not undergoing any further decomposition due to the presence of inert nitrogen atmosphere and hence the large amounts of residue at the end of TGA analysis. The addition of different amount of CATAS enhanced the thermal stability of the matrix to some extent (peak temperature shifted from 426 °C for neat rPET to 427 and 441 °C for rPET_3CATAS and rPET_30CATAS), respectively: nevertheless, all the formulations presented a degradation curve similar to the matrix, with a superposition, in the thermal degradation pattern, due to the concomitant decomposition of the salts [36]. The values of residual mass at the end of the tests, measured as 13.7%, 12.3% and 23.6% for neat rPET, rPET_3CATAS and rPET_30CATAS, respectively, are in line with the expected values, according to the registered weight loss for CATAS (43% at the end of the test) already measured in Figure 2b. The decreased residual mass for the rPET_3CATAS system at the end of the test, in comparison with rPET, can be justified by considering that in addition to polymer decomposition, the presence of the alkaline earth metal-based MOF catalyzed the thermal degradation of the neat matrix, lowering the values for final mass residue.

Table 4. TGA parameters of rPET_CATAS nanocomposites at three representative loading levels (0, 3 and 30 wt. %).

Material	T_{on} (°C)	$T_{20\%}$ (°C)	T_{max} (°C)	Residual Mass (%) (at 900 °C)
rPET	350	417	426	13.7
rPET_3CATAS	350	418	427	12.3
rPET_30CATAS	335	439	441	23.6

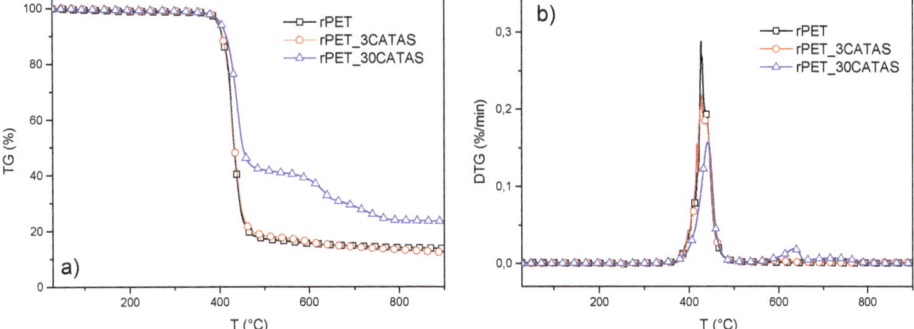

Figure 7. TG curve (a) and DTG curve (b) for rPET and rPET/CATAS nanocomposites at 3 and 30 wt. % of CATAS.

DSC: DSC heating curves for neat rPET and rPET_CATAS nanocomposites, reported in Figure 8a (low loading) and Figure 8b (high loading), registered the presence of three main events during the heating scan: a stepwise endothermic changes for glass transition at around 76 °C, an exothermic peak for cold crystallization at 117 °C, and one endothermic peak for crystal melting at 250 °C [9]. The peak heat flow for cold crystallization with very low quantity of CATAS nanocomposites become larger compared with that of neat rPET, indicating that CATAS play the role of nucleating agents for crystallization in rPET_0.1CATAS and rPET_0.25CATAS nanocomposite samples. Furthermore, the peak shapes of PET-based nanocomposites became wider, indicating that the nanocomposite cold crystallization process takes a longer time to complete than for neat rPET. This crystallization peak is negligible in rPET_0.4CATAS and becomes again visible in rPET_0.5CATAS as in rPET_0.25CATAS. Despite the variability observed across samples in terms of the cold crystallization behavior, it was observed for all of the samples that the crystallinity level is almost the same for all the nanocomposites with a slight increase at higher filler content.

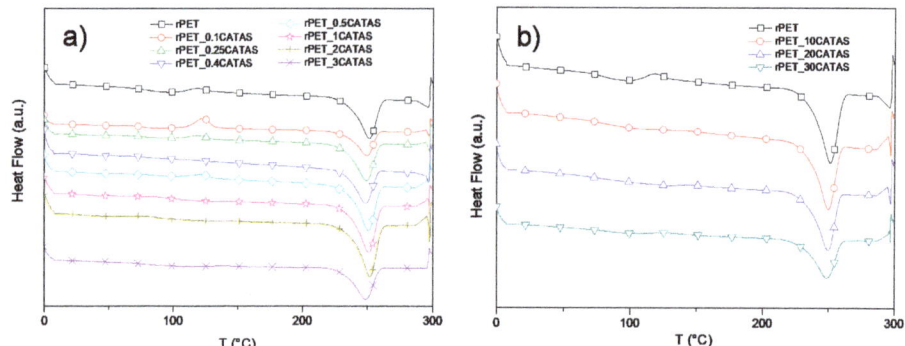

Figure 8. DSC heating scan for rPET and rPET/CATAS nanocomposites at low (**a**) and high (**b**) loading levels.

The presence of CATAS, in addition to the nucleating effect, produces a blocking effect of the amorphous phase causing the increase of the fraction of the rigid amorphous phase. At the lower percentages of filler, the nucleating effect is prevalent on the blocking so that the X_{raf} remains lower than 12%. It was hypothesized that the singularity of the rPET_0.4CATAS composite consists of a condition of equilibrium between the nucleating effect of the crystalline phase and the blocking effect of the amorphous phase, coordinating the terephthalate structure of the salts with the terephthalic chains of the matrix. This situation produces a slight increase of X_c with a reduction of X_{raf} to values close to rPET neat. The progressive increase of filler in the subsequent formulation bring to a prevalence of the blocking effect and then to a progressive increase of X_{raf}. The formation of an important fraction of a rigid amorphous phase limits the formation of crystals as opposed phenomena, hindering the peak of cold crystallization. In addition, during the heating, melting and reconstitution of crystallites reach to an equilibrium between the phases that is influenced by many factors as time, temperature, heating rate, etc. In our systems, a three-phase model show the phase ratio reported in Figure 9 that helps to give a comprehensive explanation of the mechanical results [37,38].

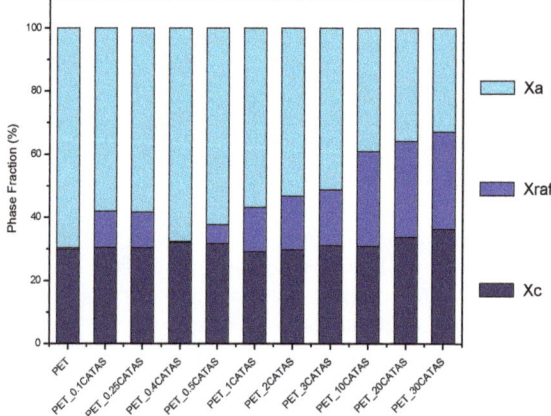

Figure 9. Evolution of different phase fractions for rPET and rPET_CATAS content.

As already observed, flexural test at a speed rate of 2 mm/min shows a large strain for rPET and rPET_0.4CATAS that indeed have a very similar phases ratio with the addition of a possible interfacial slipping effect of the filler in the composite. Tensile test carried out at higher speed (5 mm/min) shows an increase in strength at low amount of CATAS, but it is necessary a larger percentage of filler to obtain the slipping effect at this strain rate. The blocking effect of X_{raf} show its effect even on moduli that increased progressively with the CATAS as soon as the composite become fragile at higher amounts of filler.

4. Conclusions

Preparation, by melt extrusion, of CATAS reinforced rPET matrix for nanocomposites loaded at different levels was reported. Results from thermomechanical characterization indicated that CATAS content can be tailored, opening the use of these materials in unexplored industrially relevant applications that require high strength and thermal stability at high temperatures.

In detail, it was observed that when the CATAS content increased from 0.1 to 0.4 wt. %, an increase for tensile strength and elastic moduli was observed. Nevertheless, a threshold weight amount (0.4 wt. %) of CATAS was also found, by formation at low loading, of a rigid amorphous fraction at the rPET/CATAS interface, that restricted the mobility, as shown by G' values measured at temperature below and above Tg temperature. When the threshold was surpassed, a restriction of rPET/CATAS molecular chains mobility was also detected, due to the formation, of mechanical percolation networks in addition to physical restrictions (hydrogen and π-π bonding). Additionally, enhanced thermal stability of CATAS filled rPET was registered at a high content due to chemical compatibility between synthetized Ca–Metal Organic Framework and the rPET matrix, rich in stable aromatic rings.

Given the chemical affinity of Ca-based salts and the rPET matrix and the simplicity of the processing methodology necessary to make these systems, this work can pave the way for the facile preparation of similar mixed-matrix materials.

Author Contributions: Conceptualization, F.D.; Data curation, F.L.; Investigation, F.D., F.S., F.L. and D.P.; Supervision, L.T.; Writing–original draft, F.D., F.S. and D.P.; Writing–review & editing, D.P. All authors have read and agreed to the published version of the manuscript.

Funding: This research received no external funding.

Conflicts of Interest: The authors declare no conflict of interest.

References

1. Allen, R.D. Waste PET: A Renewable Resource. *Joule* **2019**, *3*, 910. [CrossRef]
2. Thomas, S.; Rane, A.V.; Kanny, K.; Abitha, V.K.; Thomas, M.G. *Recycling of Polyethylene Terephthalate Bottles*; William Andrew: Norwich, NY, USA, 2018.
3. Palkopoulou, S.; Joly, C.; Feigenbaum, A.; Papaspyrides, C.D.; Dole, P. Critical review on challenge tests to demonstrate decontamination of polyolefins intended for food contact applications. *Trends Food Sci. Technol.* **2016**, *49*, 110–120. [CrossRef]
4. Chowreddy, R.R.; Nord-Varhaug, K.; Rapp, F. Recycled Poly(ethylene Terephthalate)/Clay Nanocomposites: Rheology, Thermal and Mechanical Properties. *J. Polym. Environ.* **2019**, *27*, 37–49. [CrossRef]
5. Pegoretti, A.; Kolarik, J.; Peroni, C.; Migliaresi, C. Recycled poly(ethylene terephthalate)/layered silicate nanocomposites: Morphology and tensile mechanical properties. *Polymer* **2004**, *45*, 2751–2759. [CrossRef]
6. Tsai, T.-Y.; Bunekar, N.; Wu, T.-C. The Role of Organomodified Different LDHs on Recycled Poly(ethylene terephthalate) Nanocomposites. *J. Chin. Chem. Soc.* **2017**, *64*, 851–859. [CrossRef]
7. Kráčalík, M.; Mikešová, J.; Puffr, R.; Baldrian, J.; Thomann, R.; Friedrich, C. Effect of 3D structures on recycled PET/organoclay nanocomposites. *Polym. Bull.* **2007**, *58*, 313–319. [CrossRef]
8. Zare, Y. Recent progress on preparation and properties of nanocomposites from recycled polymers: A review. *Waste Manag.* **2013**, *33*, 598–604. [CrossRef]
9. Aoyama, S.; Ismail, I.; Park, Y.T.; Macosko, C.W.; Ougizawa, T. Higher-Order Structure in Amorphous Poly(ethylene terephthalate)/Graphene Nanocomposites and Its Correlation with Bulk Mechanical Properties. *ACS Omega* **2019**, *4*, 1228–1237. [CrossRef]

10. Dyosiba, X.; Ren, J.; Musyoka, N.M.; Langmi, H.W.; Mathe, M.; Onyango, M.S. Preparation of value-added metal-organic frameworks (MOFs) using waste PET bottles as source of acid linker. *Sustain. Mater. Technol.* **2016**, *10*, 10–13. [CrossRef]
11. Doan, V.D.; Do, T.L.; Ho, T.M.T.; Le, V.T.; Nguyen, H.T. Utilization of waste plastic pet bottles to prepare copper-1, 4-benzenedicarboxylate metal-organic framework for methylene blue removal. *Sep. Sci. Technol.* **2020**, *55*, 444–455. [CrossRef]
12. Jones, P.G.; Ossowski, J.; Kus, P.; Dix, I. Three Crystal Structures of Terephthalic Acid Salts of Simple Amines. *Z. Für Nat. B* **2009**, *64*, 865–870. [CrossRef]
13. Kaduk, J.A. Terephthalate salts: Salts of monopositive cations. *Acta Crystallogr. Sect. B Struct. Sci.* **2000**, *56*, 474–485. [CrossRef] [PubMed]
14. Panasyuk, G.P.; Azarova, L.A.; Budova, G.P.; Izotov, A.D. Synthesis and characterization of ammonium terephthalates. *Inorg. Mater.* **2002**, *38*, 385–389. [CrossRef]
15. Panasyuk, G.P.; Khaddaj, M.; Privalov, V.I.; Miroshnichenko, I.V. Poly(ethylene terephthalate) Transformations during Autoclaving in Water Vapor. *Plast. Massy* **2002**, *2*, 27–31.
16. Hou, J.; Ashling, C.W.; Collins, S.M.; Krajnc, A.; Zhou, C.; Longley, L.; Johnstone, D.; Chater, P.A.; Li, S.; Coudert, F. Metal-organic framework crystal-glass composites. *Nat. Commun.* **2019**, *10*, 1–10. [CrossRef]
17. Mou, C.; Wang, L.; Deng, Q.; Huang, Z.; Li, J. Calcium terephthalate/graphite composites as anode materials for lithium-ion batteries. *Ionics* **2015**, *21*, 1893–1899. [CrossRef]
18. Sachdeva, S.; Koper, S.J.H.; Sabetghadam, A.; Soccol, D.; Gravesteijn, D.J.; Kapteijn, F.; Sudhölter, E.J.R.; Gascon, J.; De Smet, L.C.P.M. Gas phase sensing of alcohols by Metal Organic Framework–polymer composite materials. *ACS Appl. Mater. Interfaces* **2017**, *9*, 24926–24935. [CrossRef]
19. Stock, N.; Biswas, S. Synthesis of metal-organic frameworks (MOFs): Routes to various MOF topologies, morphologies, and composites. *Chem. Rev.* **2011**, *112*, 933–969. [CrossRef]
20. Schmidt, B.V. Metal-Organic Frameworks in Polymer Science: Polymerization Catalysis, Polymerization Environment, and Hybrid Materials. *Macromol. Rapid Commun.* **2019**, *41*. [CrossRef]
21. Wunderlich, B. *Thermal Analysis of Polymeric Materials*; Springer Science & Business Media: Berlin/Heidelberg, Germany, 2005.
22. Matsuzaki, T.; Iitaka, Y. The crystal structure of calcium terephthalate trihydrate. *Acta Crystallogr. Sect. B Struct. Crystallogr. Cryst. Chem.* **1972**, *28*, 1977–1981. [CrossRef]
23. Alavi, M.A.; Morsali, A. Alkaline-earth metal carbonate, hydroxide and oxide nano-crystals synthesis methods, size and morphologies consideration. *Nanocrystal* **2011**, 237–262. [CrossRef]
24. Mazaj, M.; Mali, G.; Rangus, M.; Žunkovič, E.; Kaučič, V.; Zabukovec Logar, N. Spectroscopic studies of structural dynamics induced by heating and hydration: A case of calcium-terephthalate metal–organic framework. *J. Phys. Chem. C* **2013**, *117*, 7552–7564. [CrossRef]
25. Dominici, F.; Puglia, D.; Luzi, F.; Sarasini, F.; Rallini, M.; Torre, L. A Novel Class of Cost Effective and High Performance Nanocomposites Based on Terephthalate Salts Reinforced Polyether Ether Ketone. *Polymers* **2019**, *11*, 2097. [CrossRef]
26. Liang, P.-C.; Liu, H.-K.; Yeh, C.-T.; Lin, C.-H.; Zima, V.T.Z. Supramolecular Assembly of Calcium Metal–Organic Frameworks with Structural Transformations. *Cryst. Growth Des.* **2011**, *11*, 699–708. [CrossRef]
27. Wang, L.; Mou, C.; Sun, Y.; Liu, W.; Deng, Q.; Li, J. Structure-property of metal organic frameworks calcium terephthalates anodes for lithium-ion batteries. *Electrochim. Acta* **2015**, *173*, 235–241. [CrossRef]
28. Mazaj, M.; Zabukovec Logar, N.A. Phase Formation Study of Ca-Terephthalate MOF-Type Materials. *Cryst. Growth Des.* **2015**, *15*, 617–624. [CrossRef]
29. Howarth, A.J.; Liu, Y.; Li, P.; Li, Z.; Wang, T.C.; Hupp, J.T.; Farha, O.K. Chemical, thermal and mechanical stabilities of metal–organic frameworks. *Nat. Rev. Mater.* **2016**, *1*, 15018. [CrossRef]
30. Ahani, M.; Khatibzadeh, M.; Mohseni, M. Preparation and characterization of Poly(ethylene terephthalate)/hyperbranched polymer nanocomposites by melt blending. *Nanocomposites* **2016**, *2*, 29–36. [CrossRef]
31. Golpour, M.; Pakizeh, M. Preparation and characterization of new PA-MOF/PPSU-GO membrane for the separation of KHI from water. *Chem. Eng. J.* **2018**, *345*, 221–232. [CrossRef]
32. Strain, I.N.; Wu, Q.; Pourrahimi, A.M.; Hedenqvist, M.S.; Olsson, R.T.; Andersson, R.L. Electrospinning of recycled PET to generate tough mesomorphic fibre membranes for smoke filtration. *J. Mater. Chem. A* **2015**, *3*, 1632–1640. [CrossRef]

33. Kim, J.; Choi, J.; Soo Kang, Y.; Won, J. Matrix effect of mixed-matrix membrane containing CO2-selective MOFs. *J. Appl. Polym. Sci.* **2016**, *133*. [CrossRef]
34. Shabafrooz, V.; Bandla, S.; Allahkarami, M.; Hanan, J.C. Graphene/polyethylene terephthalate nanocomposites with enhanced mechanical and thermal properties. *J. Polym. Res.* **2018**, *25*, 256. [CrossRef]
35. Wang, X.-S.; Li, X.-G.; Yan, D. Thermal decomposition kinetics of poly (trimethylene terephthalate). *Polym. Degrad. Stab.* **2000**, *69*, 361–372. [CrossRef]
36. Cacho-Bailo, F.; Téllez, C.; Coronas, J. Interactive thermal effects on metal–organic framework Polymer composite membranes. *Chem. A Eur. J.* **2016**, *22*, 9533–9536. [CrossRef] [PubMed]
37. Chung, S.-C.; Hahm, W.-G.; Im, S.-S.; Oh, S.-G. Poly(ethylene terephthalate)(PET) nanocomposites filled with fumed silicas by melt compounding. *Macromol. Res.* **2002**, *10*, 221–229. [CrossRef]
38. Qiu, G.; Tang, Z.L.; Huang, N.X.; Gerking, L. Dual melting endotherms in the thermal analysis of Poly(ethylene terephthalate). *J. Appl. Polym. Sci.* **1998**, *69*, 729–742. [CrossRef]

© 2020 by the authors. Licensee MDPI, Basel, Switzerland. This article is an open access article distributed under the terms and conditions of the Creative Commons Attribution (CC BY) license (http://creativecommons.org/licenses/by/4.0/).

Article

Development of Injection-Molded Polylactide Pieces with High Toughness by the Addition of Lactic Acid Oligomer and Characterization of Their Shape Memory Behavior

Diego Lascano [1,2], Giovanni Moraga [1], Juan Ivorra-Martinez [1], Sandra Rojas-Lema [1,2], Sergio Torres-Giner [3,*], Rafael Balart [1], Teodomiro Boronat [1] and Luis Quiles-Carrillo [1]

1 Technological Institute of Materials (ITM), Universitat Politècnica de València (UPV), Plaza Ferrándiz y Carbonell 1, 03801 Alcoy, Spain; dielas@epsa.upv.es (D.L.); giovannimoraga92@gmail.com (G.M.); juaivma@epsa.upv.es (J.I.-M.); sanrole@epsa.upv.es (S.R.-L.); rbalart@mcm.upv.es (R.B.); tboronat@dimm.upv.es (T.B.); luiquic1@epsa.upv.es (L.Q.-C.)
2 Escuela Politécnica Nacional, 17-01-2759 Quito, Ecuador
3 Novel Materials and Nanotechnology Group, Institute of Agrochemistry and Food Technology (IATA), Spanish National Research Council (CSIC), Calle Catedrático Agustín Escardino Benlloch 7, 46980 Paterna, Spain
* Correspondence: storresginer@iata.csic.es; Tel.: +34-963-900-022

Received: 11 November 2019; Accepted: 11 December 2019; Published: 14 December 2019

Abstract: This work reports the effect of the addition of an oligomer of lactic acid (OLA), in the 5–20 wt% range, on the processing and properties of polylactide (PLA) pieces prepared by injection molding. The obtained results suggested that the here-tested OLA mainly performs as an impact modifier for PLA, showing a percentage increase in the impact strength of approximately 171% for the injection-molded pieces containing 15 wt% OLA. A slight plasticization was observed by the decrease of the glass transition temperature (T_g) of PLA of up to 12.5 °C. The OLA addition also promoted a reduction of the cold crystallization temperature (T_{cc}) of more than 10 °C due to an increased motion of the biopolymer chains and the potential nucleating effect of the short oligomer chains. Moreover, the shape memory behavior of the PLA samples was characterized by flexural tests with different deformation angles, that is, 15°, 30°, 60°, and 90°. The obtained results confirmed the extraordinary effect of OLA on the shape memory recovery (R_r) of PLA, which increased linearly as the OLA loading increased. In particular, the OLA-containing PLA samples were able to successfully recover over 95% of their original shape for low deformation angles, while they still reached nearly 70% of recovery for the highest angles. Therefore, the present OLA can be successfully used as a novel additive to improve the toughness and shape memory behavior of compostable packaging articles based on PLA in the new frame of the Circular Economy.

Keywords: PLA; OLA; impact modifier; shape memory; packaging applications

1. Introduction

Polylactide (PLA) is a linear thermoplastic biodegradable polyester that can be obtained from starch-rich materials by fermentation to give lactide, which is polymerized at the industrial scale by ring-opening polymerization (ROP) [1]. PLA is a sound candidate to substitute some plastic commodities such as polypropylene (PP) or polystyrene (PS) in packaging applications [2–6], electronics [7], automotive [8], agriculture [9], textile, consumer goods, 3D printing applications [10–13], biomedical devices [14,15], pharmaceutical carriers [16,17], etc. The main advantage of PLA over other biopolymers is its relatively low cost and overall balanced properties and processability [18,19], resulting in compostable articles [20]. In 2018, PLA production represented 10.3% of the worldwide production capacity of bioplastics, reaching nearly 220,000 tons/year and it is estimated a growth around 60% by 2023 [21].

Although PLA is a very versatile biopolymer, it results in extremely brittle materials with very low elongation at break and low toughness [22]. Many research works have been focused on overcoming or, at least, minimizing the intrinsic brittleness of PLA materials. One possible strategy is copolymerization with long aliphatic monomers, as suggested by Zhang et al. [23]. This is the case of the new type of polyester amide (PEA) copolymers developed by Zou et al. [24] combining poly(L-lactic acid) (PLLA) and poly(butylene succinate) (PBS) flexible segments. A similar approach was developed by Lan et al. [25] based on PLA-co-PBS copolymers. Nevertheless, the most widely used approach is blending with other biopolymers due to its cost effectiveness. For instance, Garcia-Campo et al. [26,27] reported interesting toughening effects on ternary blends composed of PLA, poly(3-hydroxybutyrate) (PHB), and different rubbery polymers such as PBS, poly(butylene succinate-co-adipate) (PBSA) or poly(ε-caprolactone) (PCL). Recently, Sathornluck et al. [28] developed binary blends of PLA with epoxidized natural rubber (ENR), which can positively contribute to improving toughness due to the effect of the rubber phase finely dispersed in a brittle PLA matrix. Su et al. [29] and Zhang et al. [30] have also reported different approaches to overcome the intrinsic brittleness of PLA by adding different contents of PBS using reactive compatibilizers. Fortelny et al. [31] recently reported, however, that it is possible to obtain "super-toughened" PLA formulations by blending PLA with poly(ε-caprolactone) (PCL) without any compatibilizer but using the appropriate PCL particle size. Addition of modified or unmodified particles is another way to overcome the intrinsic PLA brittleness. For instance, Wang et al. [32] improved the PLA toughness by means of silanized helical carbon nanotubes (CNTs). Similarly, Li et al. [33] reported the positive effect of cellulose nanofibers (CNFs) and a Surlyn® ionomer to improve the interfacial interaction between CNFs and PLA, leading to a remarkable increase in toughness. The work carried out by Gonzalez-Ausejo et al. [34] reported a clear improvement in toughness by using sepiolite nanoclays as gas barrier and compatibilizer in PLA/poly(butylene adipate-co-terephthalate) (PBAT) blends.

Even though copolymerization and blending are some of the most used procedures to overcome the intrinsic brittleness of PLA, plasticization is another interesting approach. Plasticizers have been widely used in PLA formulations to reduce fragility by decreasing the glass transition temperature (T_g). Plasticizers contribute to an increase in ductile properties, such as elongation at break, but mechanical resistant properties are also negatively decreased. Therefore, the use of plasticizers not always provides enhanced toughness since it depends on both ductile and resistant properties. For instance, Tsou et al. [35] improved toughness of PLA by using an adipate ester plasticizer. It has also been reported the positive effect of combining two plasticizers: one solid plasticizer, namely poly(ethylene glycol) (PEG), and a liquid plasticizer derived from soybean oil [36]. Indeed, many research works have been focused on using environmentally friendly plasticizers, which can contribute to improving toughness without compromising the overall biodegradability of PLA. In this regard, Kang et al. [37] described the usefulness of cardanol obtained from cashew nutshell liquid (CNSL) to obtain a 2.6-fold increase in toughness over neat PLA. Carbonell-Verdu et al. [38] developed a new series of dual plasticizer and compatibilizer additives derived from cottonseed oil subjected to epoxidation and/or maleinization as an environmentally friendly solution instead of using conventional epoxy-styrene acrylic oligomers. Other similar multi-functionalized vegetable oils have demonstrated good compatibilizing effects on PLA [39,40] and also on its blends with other biopolymers [41]. Ferri et al. [42] recently reported a remarkable increase in toughness of neat PLA by using maleinized linseed oil (MLO) as a bio-based plasticizer, which yielded a slight plasticization that overlapped with chain extension, branching, and some cross-linking. Tributyl citrate (TBC) has also been proposed as an effective plasticizer for PLA by Notta-Cuvier et al. [43], who reported the synergistic effect of TBC in PLA formulations containing halloysite nanotubes (HNTs).

Oligomers of lactic acid (OLAs) can also provide plasticization to PLA and PLA-based materials [44]. As reported by Burgos et al. [45], OLAs can contribute to a significant decrease in T_g, though their role as impact modifiers require further research. Therefore, this study focuses on the effect of the addition of a new type of OLA on PLA pieces prepared by injection molding. In particular, the effect of varying the OLA content on the mechanical, thermal, thermomechanical properties and the morphology of PLA was reported. Finally, the shape memory behavior of the pieces was analyzed and related to the OLA dispersion within the PLA matrix.

2. Experimental

2.1. Materials

PLA was supplied by NatureWorks LLC (Minnetonka, MN, USA) as Ingeo™ Biopolymer 6201D. This PLA grade contains 2 mol% D-lactic acid. It was supplied in pellet form and it had a density of 1.24 g cm^{-3} whereas its melt flow index (MFI) was 20 g/10 min, measured at 210 °C and 2.16 kg. OLA was kindly supplied by Condensia Química S.A. (Barcelona, Spain) as Glyplast OLA 2. It was provided in a liquid form with a viscosity of 90 mPa.s at 40 °C. According to the manufacturer, it has an ester content >99%, a density of 1.10 g cm^{-3}, a maximum acid index of 2.5 mg KOH g^{-1}, and a maximum water content of 0.1 wt%.

2.2. Manufacturing of OLA-Containing PLA Pieces

As PLA is very sensitive to moisture, the biopolyester pellets were dried at 60 °C for 24 h. The OLA content varied in the 0–20 wt% at weight steps of 5 wt%. The terminology used for the formulations was "PLA-OLA x%", where x represents the weight fraction of OLA in PLA. All the compositions were compounded in a twin-screw co-rotating extruder Coperion ZS-B 18 (Stuttgart, Germany) equipped with a main hopper in which the PLA pellets were fed and a side feeder for liquids to feed OLA. The liquid feeder was heated at 50 °C during compounding to decrease the OLA viscosity and allow efficient mixing. The rotating speed was set to 150 rpm while the temperature profile was modified according to the formulation as shown in Table 1. These temperatures were selected due to the change in viscosity produced by the OLA addition in order to optimize the processing conditions for each formulation. After extrusion, the different strands were cooled in air and then pelletized using an air-knife unit.

Table 1. Temperature profile of the seven barrels in the twin-screw extruder during the compounding of the polylactide (PLA)/oligomer of lactic acid (OLA) formulations.

PLA (wt%)	OLA (wt%)	T (°C) Zone 1	T (°C) Zone 2	T (°C) Zone 3	T (°C) Zone 4	T (°C) Zone 5	T (°C) Zone 6	T (°C) Zone 7
100	0	145	150	160	170	175	180	180
95	5	145	150	160	170	175	180	180
90	10	145	150	160	170	170	175	175
85	15	145	150	155	155	160	160	170
80	20	145	150	155	155	160	160	160

The compounded pellets were stored in an air-circulating oven at 60 °C for 24 h to avoid moisture gain. Standard samples for characterization were thereafter obtained in an injection molding machine Meteor 270/75 from Mateu & Solé (Barcelona, Spain). The temperature profile of the four barrels was programmed as indicated in Table 2. Similar to the extrusion process, it was necessary to optimize the temperature profile for each formulation due to the drop in the melt viscosity after the OLA addition.

Table 2. Temperature profile of the four barrels and maximum pressure in the injection-molding machine during the manufacturing of polylactide (PLA)/oligomer of lactic acid (OLA) pieces.

PLA (wt%)	OLA (wt%)	T (°C) Zone 1	T (°C) Zone 2	T (°C) Zone 3	T (°C) Zone 4	Maximum Pressure (Bar)
100	0	175	180	185	190	93
95	5	175	180	185	190	90
90	10	160	165	170	175	118
85	15	150	155	165	170	120
80	20	145	150	155	165	125

2.3. Mechanical Characterization

Mechanical properties of the OLA-containing PLA pieces were obtained in tensile conditions as indicated by ISO 527-1:1996. A universal testing machine ELIB 30 from S.A.E. Ibertest (Madrid, Spain) was used. At least six different samples were tested at room temperature using a load cell of 50 kN. The cross-head speed rate was set to 10 mm min^{-1}. As recommended by the standard, the tensile modulus (E_t), tensile strength at break (σ_b), and elongation at break (%ε_b) were determined and averaged. The toughness was estimated using the Charpy method following the guidelines of ISO 179-1:2010 with a 6-J pendulum from Metrotec (San Sebastián, Spain). Unnotched rectangular samples with dimensions 80 × 10 mm^2 and a thickness of 4 mm were tested. The impact strength was obtained from testing five different samples and calculated as the absorbed-energy per unit area (kJ m^{-2}). The Shore D hardness was determined in a 673-D durometer from J. Bot Instruments (Barcelona, Spain) according to ISO 868:2003. Ten different measurements were collected from randomly selected zones and various samples were tested to obtain the average values.

2.4. Microscopy

The morphology of the fracture surfaces of the OLA-containing PLA pieces was studied by field emission scanning electron microscopy (FESEM) after the impact tests. A ZEISS Ultra 55 FESEM microscope from Oxford Instruments (Abingdon, UK) operating at 2 kV was used to collect the FESEM images at 1000×. To avoid electrical charging during observation, samples were previously coated with an ultrathin gold-palladium alloy in an EMITECH SC7620 sputter-coater from Quorum Technologies Ltd. (East Sussex, UK) in an argon atmosphere.

2.5. Thermal Characterization

Thermal properties of the OLA-containing PLA pieces were obtained by differential scanning calorimetry (DSC) and thermogravimetric analysis (TGA). DSC characterization was performed on a Mettler-Toledo DSC calorimeter DSC821e (Schwerzenbach, Switzerland). To carry out the DSC runs, small specimens of each composition with an average weight of 5–7 mg were placed in standard aluminum crucibles with a total volume of 40 mL and sealed with a cap. Then, the samples were subjected to a three-step temperature program. A first heating cycle from 30 °C to 200 °C was followed by a cooling step down to −60 °C and, finally, a second heating cycle from −60 °C to 350 °C was applied. The heating and cooling rates were set to 10 °C min^{-1} whereas the selected atmosphere was nitrogen at a flow-rate of 30 mL min^{-1}. The resultant DSC curves allowed obtaining T_g, the cold crystallization peak temperature (T_{cc}), and the melting peak temperature (T_m). Besides, the cold crystallization (ΔH_{cc}) and melting enthalpies (ΔH_m) were obtained from the integration of the corresponding peaks. The maximum degree of crystallinity (%χ_{cmax}) was calculated as indicated in Equation (1):

$$\%\chi_{cmax} = \frac{\Delta H_m}{\Delta H_m^0} \cdot \frac{100}{w} \qquad (1)$$

where w (g) stands for the weight fraction of PLA and ΔH_m^o (J g^{-1}) represents the theoretical melting enthalpy of a fully crystalline PLA polymer, which is close to 93.7 J g^{-1} [46–49].

The effect of the OLA addition on the thermal stability of PLA was studied by TGA in a TGA/SDTA 851 thermobalance from Mettler-Toledo (Schwerzenbach, Switzerland). The temperature sweep was scheduled from 30 °C to 700 °C at a heating rate of 10 °C min^{-1} in air atmosphere. Samples with an average weight comprised between 5 and 7 mg were placed on standard alumina crucibles and sealed with the corresponding cap. The onset degradation temperature, which was assumed at a weight loss of 5 wt% ($T_{5\%}$), and the maximum degradation rate temperature (T_{deg}) were obtained. All DSC and TGA tests were run in triplicate to obtain reliable results.

2.6. Viscosity Characterization

To study the influence of the OLA addition on the viscosity of PLA, cylindrical disks sizing 40 mm diameter and 5 mm thickness were manufactured by hot-press molding in a BUEHLER SimpliMet 1000 (Lake Bluff, IL, USA) at a temperature of 165 °C and a pressure of 7.5 ton. Parallel-plate oscillatory rheometry (OR) was conducted in an AR-G2 rheometer from TA Instruments (New Castle, DE, USA) to obtain the evolution of the complex viscosity ($|\eta^*|$) as a function of the angular frequency. The selected isothermal temperature was 200 °C and the angular frequency varied in the 100–0.01 rad s^{-1} range. The maximum shear strain (γ) was set to 1% and the tests were carried out in air atmosphere in triplicate.

2.7. Thermomechanical Characterization

The effect of the OLA addition on the thermomechanical behavior of PLA was carried out by dynamic mechanical thermal analysis (DMTA) in a DMA 1 from Mettler-Toledo (Schwerzenbach, Switzerland) working in single cantilever flexural conditions. Rectangular samples with dimensions of 10 × 7 mm^2 and a thickness of 4 mm were subjected to a dynamic heating program from −30 °C to 130 °C at a constant heating rate of 2 °C min^{-1}. The maximum deflection in the free edge was set to 10 μm and the selected frequency was 1 Hz. The storage modulus (E') and the dynamic damping factor (tan δ) were collected as a function of increasing temperature.

The dimensional stability of the OLA-containing PLA pieces was studied by thermomechanical analysis (TMA) in a Q400 thermomechanical analyzer from TA Instruments (New Castle, DE, USA). The applied force was set to 0.02 N and the temperature program was scheduled from −30 to 120 °C in air atmosphere (50 mL min^{-1}) at a constant heating rate of 2 °C min^{-1}. The coefficient of linear thermal expansion (CLTE) of the PLA pieces, both below and above T_g, was determined from the change in dimensions versus temperature.

2.8. Characterization of the Shape Memory Behavior

The flexural method was used to evaluate the extension of the shape memory behavior. To this end, samples were compression-molded into sheets in the hot-press molding with a thickness of 0.4–0.5 mm. Two different parameters were then calculated to analyze the shape memory behavior in flexural conditions, namely the shape memory recovery (R_r) and stability ratio (R_f) [50,51]. The procedure is described as follows. In the first stage, the sheet specimens were deformed at a particular angle θ_f. For this, the samples were heated at 65 °C and forced to adapt between two aluminum sheets with different angles (15°, 30°, 60°, and 90°) to form a sandwich structure: aluminum/PLA sheet/aluminum. The sandwich was then clamped to allow a permanent deformation to the desired angle. Thereafter, the sandwich was cooled down to 14 °C to achieve its permanent shape after a slight temporary recovery of θ_t. Finally, the specimen was heated above its T_g in an air circulating oven at 65 °C for 3 min. After this, the final angle, that is, θ_f, was measured. At least three different sheets were tested for each composition to obtain reliable values. The same procedure has been used to characterize the shape recovery behavior in several polymer systems [52,53]. The values of R_f and R_r were calculated using the following equations:

$$\%R_f = \frac{(\pi - \theta_t)}{(\pi - \theta_f)}. \qquad (2)$$

$$\%R_r = \frac{(\pi - \theta_f)/(\pi - \theta_r)}{(\pi - \theta_f)}. \tag{3}$$

3. Results and Discussion

3.1. Effect of OLA on the Mechanical Properties of PLA

Table 3 gathers the main results obtained after the mechanical characterization of the PLA pieces with different OLA contents. As expected, the progressive addition of OLA led to lowering σ_b values from 64.6 MPa, for the neat PLA, down to 37.4 MPa, for the PLA pieces containing 20 wt% OLA. This decreasing tendency was almost linear as it can be seen in the table for the other compositions. This tendency on mechanical strength is the typical a plasticizer produces on the base polymer. Some other OLA additives have demonstrated a similar effect on mechanical properties by increasing remarkably ductility, mainly in the film form, as reported by Burgos et al. [45]. In the latter work, it was reported an increase in ε_b from 4% to 315% with an OLA content of 25 wt%, but it is worthy to note the OLA previously used was designed for plasticization of PLA films. Such dramatic increase in the ε_b was not observed when using OLA in this work since the primary use of this type of OLA was to improve impact strength, as indicated by the supplier and it will be discussed further.

Table 3. Mechanical properties of the polylactide (PLA) pieces with different weight contents of oligomer of lactic acid (OLA) in terms of: tensile modulus (E_t), strength at break (σ_b), elongation at break (ε_b), impact strength, and Shore D hardness.

Piece	Tensile Test			Impact Strength (kJ m^{-2})	Shore D Hardness
	E_t (MPa)	σ_b (MPa)	ε_b (%)		
PLA	2251 ± 45	64.6 ± 1.1	7.9 ± 0.1	25.7 ± 2.7	78.8 ± 0.9
PLA-OLA 5%	2272 ± 94	52.0 ± 2.1	6.8 ± 0.3	30.4 ± 3.6	81.2 ± 1.4
PLA-OLA 10%	2300 ± 82	42.1 ± 1.7	5.6 ± 0.2	54.2 ± 4.8	80.9 ± 2.3
PLA-OLA 15%	2400 ± 58	41.4 ± 1.0	5.3 ± 0.3	69.7 ± 5.2	80.9 ± 0.9
PLA-OLA 20%	2101 ± 91	37.4 ± 2.2	5.0 ± 0.2	38.4 ± 5.7	80.5 ± 0.4

Regarding mechanical ductility, the neat PLA piece showed a very low value of 7.9% and the addition of OLA did not promote an increase in ε_b, but a slight decrease down to values of 5%. It is not usual that a plasticizer promotes a decrease in ductility since the typical effect of a plasticizer is a decrease in the tensile resistant properties (σ_b and E_t) and an increase in ductile properties (ε_b). Nevertheless, it has been reported that some plasticizers promote a clear plasticization that is detectable by a decrease in T_g while no improvement in ductility occurs. This atypical behavior was reported by Ambrosio-Martín et al. [54] in PLA films blended with different synthesized OLAs. A ε_b value of 5.25% was reported for neat PLA, while the addition of 25 wt% of a purified OLA yielded a ε_b of 2.52%. It was also reported a slight increase in E_t and an apparent decrease in σ_b, in a similar way as obtained in this work. It was concluded that, although there is clear evidence of the mechanical plasticization of OLA-containing PLA films, they were not more deformable, which is also in agreement with the work performed by Courgneau et al. [55]. Concerning E_t, the neat PLA piece was characterized by a value of nearly 2.2 GPa and the values remained in the 2.2–2.4 GPa range after the addition of OLA. Thus, the main effect of this type of OLA on the tensile mechanical properties was a remarkable decrease in σ_b, which representative for some plasticization, but also a slight decrease in ε_b. Interestingly, as it can also be seen in Table 3, the addition of OLA successfully increased the impact strength of PLA. Neat PLA showed an impact strength of 25.7 kJ m^{-2}, which indicates a brittle behavior, and the only addition of 5 wt% OLA provided a slight increase in the impact strength to 30.4 kJ m^{-2}. Nevertheless, the most remarkable changes were obtained for OLA additions of 10 wt% and 15 wt%, showing impact strength values of 54.2 kJ m^{-2} and 69.7 kJ m^{-2}, respectively. Therefore, the PLA piece with 15 wt% OLA presented the maximum impact strength with a percentage increase of approximately 171%

with regard to the neat PLA. It is also worthy to mention that the PLA piece containing 20 wt% OLA showed a decrease in impact strength in comparison with the other OLA-containing PLA pieces, thus suggesting certain OLA saturation in the PLA matrix. In this regard, Fortunati et al. [56] reported the use of isosorbide diester (ISE) as plasticizer for PLA. It was observed plasticizer saturation at 20 wt% ISE and this was attributed to a limitation of T_g decrement. In addition, a noticeable decrease in ε_b was observed once the plasticizer saturation was achieved. Furthermore, Ferri et al. [57] reported the plasticization of PLA by fatty acid esters, observing a remarkable decrease in impact strength above 5 parts per one hundred parts (phr) of PLA. Accordingly, a remarkable decrease in T_g was attained for contents of up to 5 phr whereas, above this, the T_g values did not change in a noticeable way, corroborating the relationship between the impact and thermal properties. Regarding hardness, one can observe that the Shore D values remained nearly constant after the OLA addition, showing values in the 78–82 range. Therefore, the most important feature this OLA can potentially provide to the mechanical properties of PLA is a remarkable improvement in impact strength while the elasticity can be slightly improved and ductility reduced. This particular mechanical behavior could be ascribed to an increase in the sample crystallinity and also to the presence of soft domains of OLA dispersed within the PLA matrix, simultaneously improving impact strength and reducing flexibility.

As previously indicated, one of the most widely used strategies to improve toughness in PLA-based formulations is blending with rubber-like polymers such as PCL [58], PBS [29,59], or PBAT [60,61]. In these immiscible blends, the energy absorption is related to presence of finely dispersed rubber-like small polymer droplets embedded in the brittle PLA matrix. In some cases, a synergistic effect can be found when different reactive or non-reactive compatibilizers are used. In this work, OLA has the same chemical structure than PLA, thus leading to miscibility without the need of compatibilizers. In this regard, Burgos et al. [45] have reported the similarity between the solubility parameters of both PLA and OLA, which plays an essential role in miscibility. According to this, Figure 1 gathers the FESEM images corresponding to the fracture surfaces from the impact tests of the PLA pieces with the different OLA loadings. Figure 1a shows the fracture surface of the neat PLA piece. As one can observe in this micrograph, the surface was smooth with multiple microcracks presence, which is an indication of a brittle behavior. As opposite, Figure 1b shows that the fracture surface morphology of the PLA piece with 5 wt% OLA changed noticeably. Phase separation could not be detected due to the high chemical affinity between PLA and OLA and the microcracks were not observed but, in contrast, macrocracks were produced. Therefore, the presence of OLA seems to inhibit microcrack formation and growth and, therefore, the cracks could grow to a greater extent thus leading to a rougher surface that is responsible for higher energy absorption during impact. Figure 1c,d show the FESEM images corresponding to the fracture surfaces of the PLA pieces with 10 wt% and 15 wt% OLA, respectively. In these images, the above-mentioned effect was more intense then showing rougher surfaces that are related to enhanced energy absorption. Finally, Figure 1e shows that the PLA piece containing 20 wt% OLA presented a similar fracture surface than the other pieces and, despite there was a clear loss of toughness, its morphology did not allow identifying phase separation.

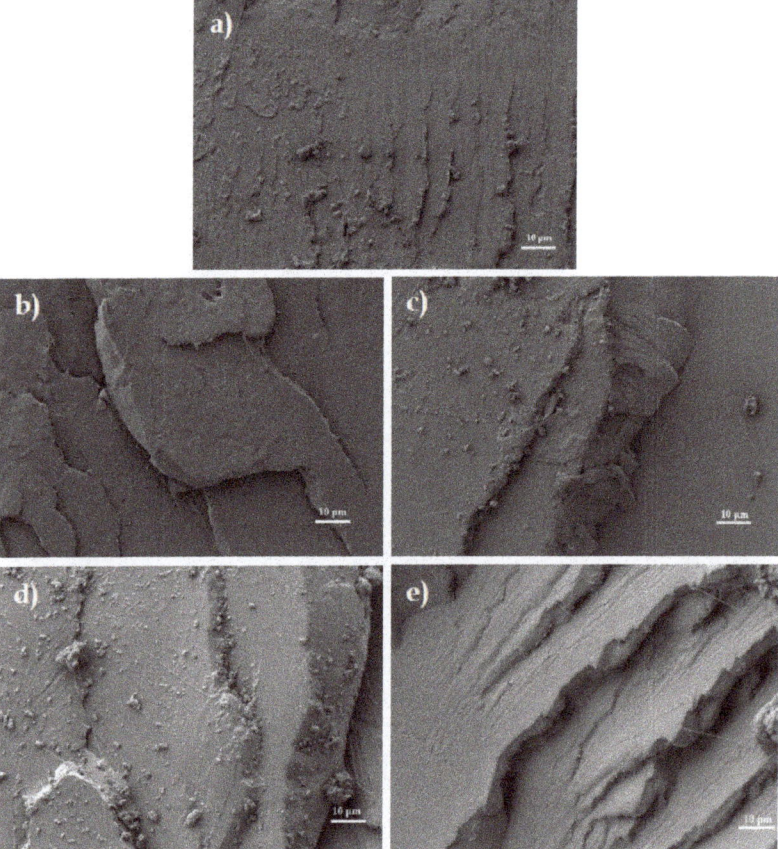

Figure 1. Field emission scanning electron microscopy (FESEM) images of the fracture surfaces of the of the polylactide (PLA) pieces with different weight contents of oligomer of lactic acid (OLA): (**a**) 0 wt%; (**b**) 5 wt%; (**c**) 10 wt%; (**d**) 15 wt%; and (**e**) 20 wt%. Images were taken at 1000× and scale markers are of 10 μm.

3.2. Effect of OLA on the Thermal and Rheological Properties of PLA

Figure 2 shows a comparative graph with the characteristic DSC thermograms during first heating corresponding to the neat PLA piece and the PLA pieces with different OLA loadings. Table 4 summarizes the main thermal values obtained from the thermograms. One can observe that T_g of neat PLA was located close to 63 °C. Then, the PLA sample cold crystallized indicating that the biopolyester chains could not crystallize in the injection mold. The process of cold crystallization was characterized by a peak at 109.8 °C. Finally, the melting process was defined by a peak temperature of 170.9 °C in which the whole crystalline fraction melted. The effect of the OLA addition on the thermal properties was remarkable. Concerning T_g, a clear decreasing tendency can be observed. In particular, T_g was reduced after the OLA addition down to 50.8 °C, thus indicating plasticization. Ambrosio-Martín et al. [54] reported a decrease in T_g with addition of OLA to PLA films from 60 °C to 27.7 °C, concluding that this fact is directly related to the mechanical properties of OLA and it depends on the synthesis procedure [62]. In the present work, the T_g value of PLA decreased after the addition of OLA but this reduction was much lower than other reported to other plasticizers. For instance, Ljungberg et al. [63] reported a T_g decrease of 30 °C with 15 wt% addition of different plasticizers such as triacetin, tributyl citrate (TBC), triethyl citrate (TEC), acetyl tributyl citrate (ATBC), and acetyl triethyl citrate (ATEC).

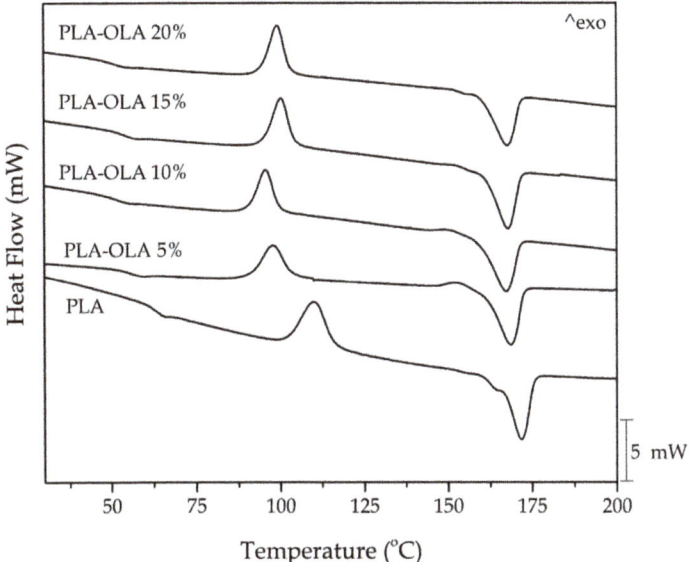

Figure 2. Differential scanning calorimetry (DSC) thermograms corresponding to the polylactide (PLA) pieces with different weight contents of oligomer of lactic acid (OLA).

Table 4. Thermal properties of the polylactide (PLA) pieces with different weight contents of oligomer of lactic acid (OLA) in terms of: glass transition temperature (T_g), cold crystallization temperature (T_{CC}), cold crystallization enthalpy (ΔH_{CC}), melting temperature (T_m), melting enthalpy (ΔH_m), and degree of crystallinity (χ_{cmax}).

Piece	T_g (°C)	T_{cc} (°C)	ΔH_{cc} (J g^{-1})	T_m (°C)	ΔH_m (J g^{-1})	χ_{cmax} (%)
PLA	63.3 ± 1.5	109.8 ± 3.8	28.6 ± 0.8	170.9 ± 1.7	33.4 ± 1.6	35.6 ± 1.7
PLA-OLA 5%	55.9 ± 2.4	97.8 ± 2.1	22.5 ± 5.1	168.2 ± 2.4	46.2 ± 4.5	51.9 ± 4.7
PLA-OLA 10%	53.9 ± 1.9	100.1 ± 2.5	21.9 ± 1.7	167.1 ± 3.0	46.3 ± 4.8	54.9 ± 5.3
PLA-OLA 15%	51.3 ± 0.4	96.7 ± 2.1	26.3 ± 3.8	166.5 ± 2.1	42.4 ± 2.1	53.2 ± 2.5
PLA-OLA 20%	50.8 ± 3.2	99.1 ± 2.4	25.0 ± 2.5	166.9 ± 1.8	34.3 ± 3.2	45.8 ± 4.0

Furthermore, the added OLA induced an internal lubricating effect that shifted the cold crystallization of PLA to lower temperatures due to an increase of chain mobility. Then T_{cc} lowered to values in the range of 96–101 °C with the different OLA loadings. Other authors suggested a specific nucleating effect provided by the short length OLA molecules, which are more readily to pack the PLA macromolecular structure thus favoring the cold crystallization process [64]. In addition, a small and broad exothermic peak was seen in the PLA sample processed with OLA, particularly noticeable at the lowest OLA contents. This exothermic peak is related to a pre-melt crystallization just before melting. In this regard, one can consider that the presence of OLA promoted the formation of different crystallites [63]. As reported by Maróti et al. [65] and also Maiza et al. [66], this peak has been observed in neat PLA depending on the heating rate and the applied thermal cycle. One can also observe a slight decrease in the T_m value with the increasing OLA content. Similar findings have been reported by Burgos et al. [62] in PLA films with different OLAs.

In addition to the characteristic values of T_g, T_{cc}, and, T_m, the enthalpies corresponding to the cold crystallization and melting processes, that is, ΔH_{cc} and ΔH_m, respectively, were collected.

The maximum degree of crystallinity, that is, χ_{cmax}, which does not consider the amount of crystals formed during cold crystallization, was 35.6% for the neat PLA. Then, χ_{cmax} increased up to values of around 50% for the compositions containing 5–15 wt% OLA, while slightly lower values of crystallinity were obtained for the composition containing 20 wt% OLA, that is, 45.8%. This increase in crystallinity can be related to the plasticizing effect of OLA, as earlier reported by Burgos et al. [62] in PLA formulations with 15 wt% of different OLAs. This latter study also reported a decrease in the T_m value of approximately 5 °C.

Regarding thermogravimetric characterization, Figure 3 shows the mass versus temperature (Figure 3a) and the first derivative (DTG) versus temperature (Figure 3b) curves for all the PLA pieces. The main results of the thermal decomposition of PLA-OLA blends are summarized in Table 5. One can observe that the neat PLA was much more thermally stable than the toughened PLA formulations with the different OLA loadings. As the OLA content increased, the characteristic TGA curves in Figure 3a shifted to lower temperatures, thus indicating a decrease in thermal stability. DTG curves were very useful to determine the maximum degradation rate temperature (T_{deg}), which was seen as peaks in Figure 3b. It can be seen in the graph that there was a clear decreasing tendency of T_{deg} with increasing OLA content. Furthermore, the residual mass for all PLA formulations with OLA was almost the same, being below 1 wt%.

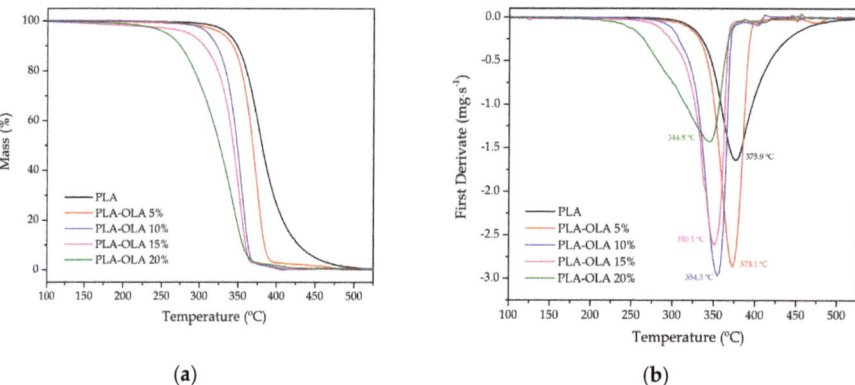

(a) (b)

Figure 3. (a) Thermogravimetric analysis (TGA) and (b) first derivate thermogravimetric (DTG) curves corresponding to the polylactide (PLA) pieces with different weight contents of oligomer of lactic acid (OLA).

Table 5. Main thermal parameters of the polylactide (PLA) pieces with different weight contents of oligomer of lactic acid (OLA) in terms of: onset temperature of degradation ($T_{5\%}$), degradation temperature (T_{deg}), and residual mass at 700 °C.

Piece	$T_{5\%}$ (°C)	T_{deg} (°C)	Residual Mass (%)
PLA	336.1 ± 2.7	375.9 ± 1.9	0.11 ± 0.02
PLA-OLA 5%	327.2 ± 1.2	373.1 ± 2.3	0.57 ± 0.08
PLA-OLA 10%	310.1 ± 3.2	354.3 ± 3.1	0.65 ± 0.12
PLA-OLA 15%	296.5 ± 3.8	350.1 ± 1.8	0.39 ± 0.08
PLA-OLA 20%	254.8 ± 2.8	344.8 ± 2.6	0.53 ± 0.07

Therefore, OLA induced a remarkable reduction in the thermal stability of PLA, as similarly described by Ambrosio-Martín et al. [54]. In this regard, Burgos et al. [62] also reported individual TGA characterization of different OLAs with their corresponding thermal degradation parameters. A variation in the onset degradation temperature was observed from 179 °C to 214 °C. The lower thermal stability was related to the lower T_g values. Moreover, it was reported a maximum degradation rate temperature ranging from 259 °C to 291 °C. These characteristic degradation temperatures are remarkably lower than those of the neat PLA herein studied. As the OLA content in PLA pieces increased, both the $T_{5\%}$ and the T_{max} values showed a clear decreasing tendency. This decrease was more pronounced in the case of $T_{5\%}$, which varied from 336.1 °C, for the neat PLA piece, to 254.8 °C, for the PLA piece containing 20 wt% OLA.

As indicated previously, OLA provided a lubricating effect, which could potentially lead to a subsequent decrease in the viscosity. Figure 4 shows the effect of the different OLA loadings on the |η*| values of the PLA sheets as a function of the angular frequency. As it can be seen, the effect of the angular frequency on the complex viscosity was not highly pronounced but it was possible to detect a decreasing tendency of |η*| with increasing the OLA content. This confirms that OLA lowers the viscosity of the PLA melt during processing and it can therefore potentially act as a processing aid. In fact, as indicated previously, it was necessary to adjust the thermal profile for optimum processing when the OLA content was modified.

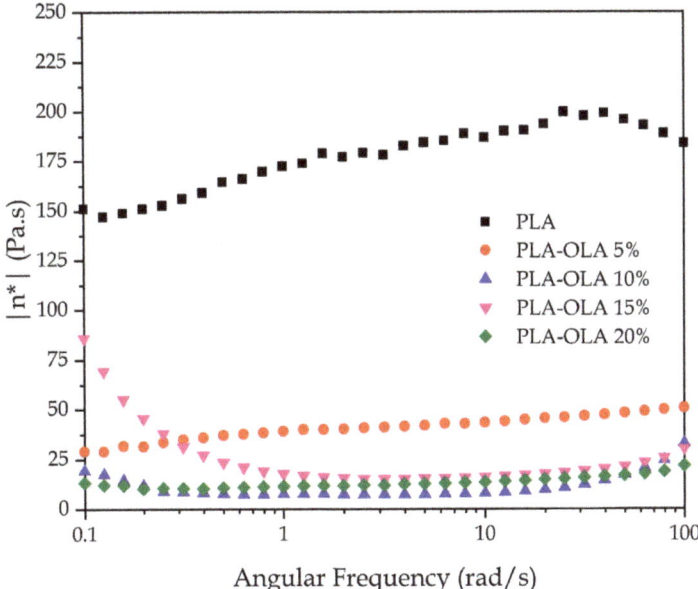

Figure 4. A comparative plot of the complex viscosity (|η*|) of the polylactide (PLA) sheets with different weight contents of oligomer of lactic acid (OLA) at a constant temperature of 200 °C as a function of increasing angular frequency.

3.3. Effect of OLA on the Thermomechanical Properties of PLA

The results described above indicated a definite improvement in the PLA toughness by using OLA as an impact modifier. Mechanical characterization showed a decrease in mechanical strength while ductility was also slightly reduced. DMTA allows characterization of mechanical properties in dynamic conditions (sinusoidal applied stress) as a function of a heating cycle. Figure 5 shows the DMTA curves for the neat PLA piece and the PLA pieces containing different loadings of OLA. The results of the thermomechanical properties obtained by DMTA are summarized in Table 6. The variation of the E' values of the neat PLA (Figure 5a) showed a dramatic drop between 50 °C and 70 °C, which is representative of the α-relaxation process of the PLA chains as the glass transition region was surpassed. In particular, a three-fold decrease in E' was observed. As the OLA loading increased, the E' curves shifted to lower temperatures thus indicating a decrease in T_g, as previously observed by DSC analysis. In particular, the E' values decreased from 1500 MPa, for the neat PLA piece, to 1197 MPa, for the

PLA piece blended with 20 wt% OLA at 30 °C. As reported by other authors, both nucleating agents and plasticizers play a crucial role in the DMTA behavior of PLA [67,68]. Although the characteristic E' curves showed a decrease in T_g, more accurate values can be obtained by determining the peak maximum of *tan δ*, as observed in Figure 5b. Neat PLA showed a T_g of 68.2 °C and the T_g values decreased progressively as the OLA loading increased, reaching a minimum value of 49.4 °C for the PLA piece containing 20 wt% OLA. These results are in total agreement with the above-described results obtained during the DSC characterization.

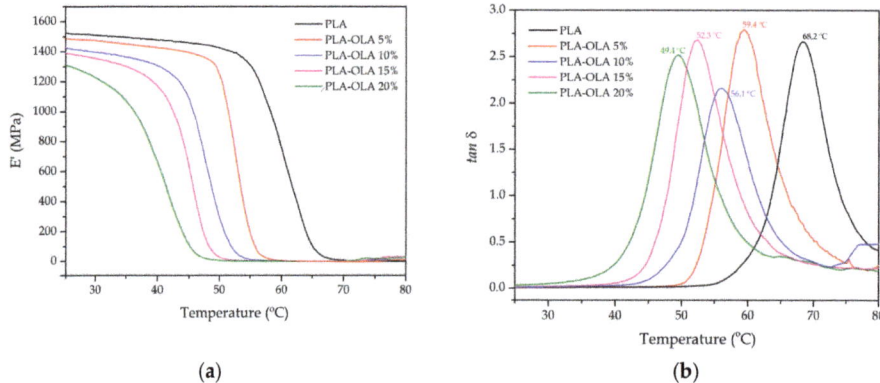

Figure 5. Evolution as a function of temperature of the (**a**) storage modulus (E') and (**b**) dynamic damping factor (*tan δ*) of the polylactide (PLA) pieces with different weight contents of oligomer of lactic acid (OLA).

Table 6. Main thermomechanical parameters of the polylactide (PLA) pieces with different weight contents of oligomer of lactic acid (OLA) in terms of: storage modulus (E') measured at 30 °C and 70 °C, glass transition temperature (T_g), and coefficient of linear thermal expansion (CLTE) below and above T_g.

Piece	DMTA			TMA	
	E' at 30 °C (MPa)	E' at 70 °C (MPa)	T_g * (°C)	CLTE below T_g (µm m^{-1} K^{-1})	CLTE above T_g (µm m^{-1} K^{-1})
PLA	1500 ± 55	4.9 ± 0.3	68.2 ± 1.2	79.9 ± 3.5	155.4 ± 6.2
PLA-OLA 5%	1467 ± 49	2.4 ± 0.2	59.4 ± 1.4	87.1 ± 4.0	166.4 ± 7.9
PLA-OLA 10%	1389 ± 52	4.6 ± 0.4	56.1 ± 0.7	88.9 ± 3.8	173.8 ± 6.2
PLA-OLA 15%	1343 ± 38	5.2 ± 0.6	52.3 ± 0.9	90.1 ± 1.9	179.9 ± 7.1
PLA-OLA 20%	1197 ± 47	9.9 ± 1.1	49.4 ± 0.8	91.6 ± 3.9	184.5 ± 6.8

* T_g was obtained as the peak maximum of the dynamic damping factor (*tan δ*).

Another attractive thermomechanical property is the effect of temperature on the dimensional stability of the PLA-based materials with different OLA loadings. TMA is a very useful technique to determine the CLTE values, which is a crucial property related to dimensional stability in terms of temperature exposition. Table 6 gathers these coefficients for the neat PLA piece and the PLA pieces containing different amount of OLA. As one can observe in the table, two different CLTE values were determined, one corresponding to the slope below T_g and another one corresponding to that above T_g. Plasticization can be observed by seeing the CLTE both below and above T_g. A slight increasing tendency was obtained, thus, indicating more ductility. The CLTE below T_g changed from 79.9 µm m^{-1} K^{-1} to 91.6 µm m^{-1} K^{-1}. The maximum change was therefore 11.7 µm m^{-1} K^{-1}, which is a very narrow range, typical of values below T_g then indicating excellent dimensional stability. Above T_g, the maximum change was 29.1 µm m^{-1} K^{-1}, which is in accordance with the typical plastic thermomechanical behavior above T_g.

3.4. Effect of OLA on the Shape Memory Behavior of PLA

Initially, the shape memory behavior was studied qualitatively by introducing the sheet specimens into a glass tube at room temperature, as observed in Figure 6a, and remained inside for 5 min to retain the shape. Then, the crimped PLA sheets were immersed in a water bath at 70 °C, above the biopolyester's T_g, and allowed to recover their shape. As can it be seen in Figure 5b, flat sheet shapes were obtained for the PLA materials containing >10 wt% OLA loadings in a short period of 4–10 s, therefore giving support to the significant effect of OLA on the shape memory behavior of PLA. Similar results, under the same conditions, were reported in the development of poly(L-lactide-co-ε-caprolactone) (PLACL), which showed recovery times of approximately 20 s [69]. Another study was focused on PLA/thermoplastic polyurethane (TPU) blends in which the recovery times at 70 °C were very similar to those obtained in this work, that is, 7–12 s [50].

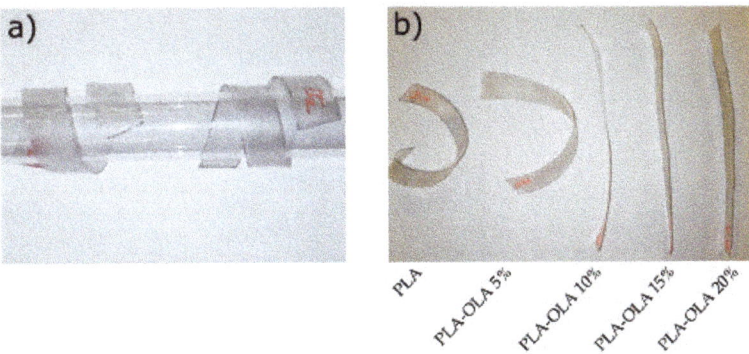

Figure 6. Photographs of the qualitative study of the shape memory recovery capacity of the polylactide (PLA) sheets with different weight contents of oligomer of lactic acid (OLA): (**a**) initial deformation of the sheets by introducing them into a glass tube and (**b**) recovered shape of the sheets after heating at 70 °C.

In addition to the qualitative characterization shown above, a quantitative study was carried out to evaluate the influence of the OLA impact modifier on the shape memory behavior of PLA. Figure 7 shows the evolution of R_r for different angles as the OLA content was increased. As it can be seen in the plot, the neat PLA sheet showed limited shape recovery properties and, obviously, it highly depended on the deformation angle. This shape memory recovery was close to 77% for a deformation angle of 90° and it was remarkably lower with more aggressive deformations. For instance, PLA could only recover 55.5% when the initial deformation angle was 15°. As the OLA loading increased, the ability of PLA to recover its initial flat shape (angle of 180°) increased considerably. It is worthy to note the effect of the addition of 20 wt% OLA, which yielded an almost constant increase in R_r of approximately 20% for all the tested angles. Therefore, for this OLA loading, the shape recovery ability of PLA remarkably improved thus leading to an exciting shape memory behavior. As reported by Leonés et al. [70], a good shape memory behavior can be attained in electrospun PLA-based fibers containing different OLA loadings in the 10–30 wt% range.

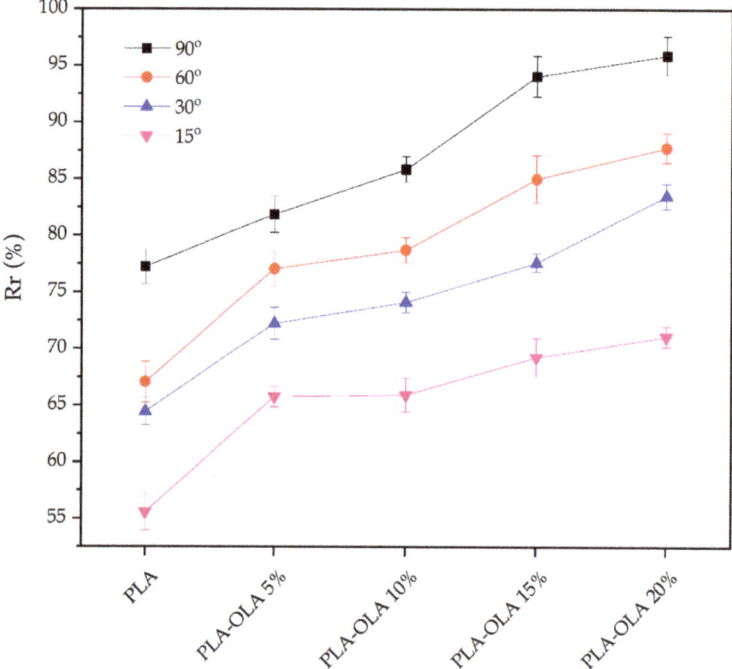

Figure 7. Evolution of the percentage of shape memory recovery (%R_r) of polylactide (PLA) sheets with different weight contents of oligomer of lactic acid (OLA) at different initial deformation angles: 15°, 30°, 60°, and 90°.

The R_f value is representative for the dimensional stability below T_g, after an initial deformation. As it can be seen in Table 7, R_f was very high for all the systems, thus indicating excellent dimensional stability after the first deformation. In general, as the deformation angle was lower, for instance 90°, the stability ratio was higher, showing recovery values over 98%. Alternatively, if the initial deformation angle was very aggressive, for instance 15°, a slight decrease in the stability ratio can be observed down to values of nearly 85%. Anyway, these stability ratios can be considered high for all compositions and angles. In this regard, Jing et al. [50] reported lower stability ratios in PLA/TPU blends, which showed less dimensional stability than the materials developed herein. Therefore, one can conclude indicating that this type of OLA is also an exceptional additive for shape memory recovery as shown by the high %R_r and %R_f values obtained.

Table 7. Variation of the percentage of stability ratio (%R_f) after different initial deformation angles (θ_f) for the polylactide (PLA) sheets with different weight contents of oligomer of lactic acid (OLA).

Sheet	R_f (%)			
	$\theta_f = 90°$	$\theta_f = 60°$	$\theta_f = 30°$	$\theta_f = 15°$
PLA	99.6 ± 1.3	98.9 ± 2.1	99.6 ± 1.3	88.2 ± 1.1
PLA-OLA 5%	98.7 ± 1.2	98.1 ± 1.3	88.2 ± 1.3	87.2 ± 2.5
PLA-OLA 10%	98.4 ± 1.5	94.2 ± 2.4	96.8 ± 1.5	86.7 ± 1.3
PLA-OLA 15%	99.3 ± 1.4	98.1 ± 1.8	97.8 ± 2.1	85.2 ± 3.6
PLA-OLA 20%	91.4 ± 2.3	96.5 ± 2.4	98.4 ± 1.4	84.9 ± 1.6

4. Conclusions

The positive effect of OLA to improve PLA toughness was evaluated in this study. The addition of 15 wt% OLA provided an increase in the impact strength from 25.7 kJ m^{-2} to almost 70 kJ m^{-2}, thus showing an extraordinary effect on toughness. Furthermore, the OLA impact modifier also provided a mechanical plasticization as observed by a decrease in the tensile strength though a slight reduction in ductility was also noticed. This plasticizing effect was observable by DSC in which the characteristic T_g of the neat PLA was reduced from 63.3 °C to 50.8 °C for the PLA piece with 20 wt% OLA. Moreover, since OLA was less thermally stable than PLA, a decrease in the onset degradation temperature, reported as $T_{5\%}$, was observed as the OLA loading increased. Nevertheless, the $T_{5\%}$ value corresponding to the PLA piece with the highest OLA loading was still high, that is, 254.8 °C, which successfully allows processing these blends without thermal degradation. Another exciting feature that this type of OLA can provide to PLA is the improvement of its shape memory behavior. In particular, the shape memory recovery parameter, that is, R_r, was very high compared to other PLA-based blends or plasticized PLA systems, thus showing the extraordinary effect of this OLA on the shape memory recovery. As a general conclusion, the here-studied OLA additive represents an interesting technical and environmentally friendly solution to improve the intrinsic brittleness of PLA and it also contributes to somewhat plasticization that allows enhanced shape memory recovery properties. The resultant toughened PLA materials can be of high interest for the development of compostable packaging articles, such as food trays and films, or disposable articles, such as cutlery and straws.

Author Contributions: Conceptualization, R.B. and D.L.; methodology, G.M., S.R.-L.; validation, L.Q.-C., T.B. and S.T.-G.; formal analysis, J.I.-M. and R.B.; investigation, D.L.; data curation, G.M., D.L., J.I.-M., and S.R.-L.; writing—original draft preparation, L.Q.-C. and T.B.; writing—review and editing, R.B., S.T.-G.; supervision, T.B., L.Q.-C., and D.L.; project administration, S.T.-G. and R.B.

Funding: This research work was funded by the Spanish Ministry of Science, Innovation, and Universities (MICIU) project numbers RTI2018-097249-B-C21 and MAT2017-84909-C2-2-R.

Acknowledgments: L.Q.-C. wants to thank Generalitat Valenciana (GVA) for his FPI grant (ACIF/2016/182) and the Spanish Ministry of Education, Culture, and Sports (MECD) for his FPU grant (FPU15/03812). D.L. thanks Universitat Politècnica de València (UPV) for the grant received through the PAID-01-18 program. S.T.-G. is recipient of a Juan de la Cierva contract (IJCI-2016-29675) from MICIU. S.R.-L. is recipient of a Santiago Grisolía contract (GRISOLIAP/2019/132) from GVA. J.I.-M. wants to thank UPV for an FPI grant PAID-01-19 (SP2019001). Microscopy services of UPV are acknowledged for their help in collecting and analyzing the microscopy images. Authors also thank Condensia Química S.A. for kindly supplying Glyoplast OLA 2.

Conflicts of Interest: The authors declare no conflict of interest.

References

1. Dijkstra, P.J.; Du, H.; Feijen, J. Single site catalysts for stereoselective ring-opening polymerization of lactides. *Polym. Chem.* **2011**, *2*, 520–527. [CrossRef]
2. Quiles-Carrillo, L.; Montanes, N.; Lagaron, J.M.; Balart, R.; Torres-Giner, S. Bioactive multilayer polylactide films with controlled release capacity of gallic acid accomplished by incorporating electrospun nanostructured coatings and interlayers. *Appl. Sci.* **2019**, *9*, 533. [CrossRef]
3. Radusin, T.; Torres-Giner, S.; Stupar, A.; Ristic, I.; Miletic, A.; Novakovic, A.; Lagaron, J.M. Preparation, characterization and antimicrobial properties of electrospun polylactide films containing A*llium ursinum* L. extract. *Food Packag. Shelf Life* **2019**, *21*, 100357. [CrossRef]
4. Scarfato, P.; Di Maio, L.; Milana, M.R.; Giamberardini, S.; Denaro, M.; Incarnato, L. Performance properties, lactic acid specific migration and swelling by simulant of biodegradable poly(lactic acid)/nanoclay multilayer films for food packaging. *Food Addit. Contam. Part A* **2017**, *34*, 1730–1742. [CrossRef]
5. Scarfato, P.; Di Maio, L.; Incarnato, L. Recent advances and migration issues in biodegradable polymers from renewable sources for food packaging. *J. Appl. Polym. Sci.* **2015**, *132*, 42597. [CrossRef]
6. Tawakkal, I.S.M.A.; Cran, M.J.; Miltz, J.; Bigger, S.W. A review of poly(lactic acid)-based materials for antimicrobial packaging. *J. Food Sci.* **2014**, *79*, R1477–R1490. [CrossRef]

7. Paula, K.T.; Gaal, G.; Almeida, G.F.B.; Andrade, M.B.; Facure, M.H.M.; Correa, D.S.; Ruil, A.; Rodrigues, V.; Mendonca, C.R. Femtosecond laser micromachining of polylactic acid/graphene composites for designing interdigitated microelectrodes for sensor applications. *Opt. Laser Technol.* **2018**, *101*, 74–79. [CrossRef]
8. Jeoung, S.K.; Ha, J.U.; Ko, Y.K.; Kim, B.R.; Yoo, S.E.; Lee, K.D.; Lee, S.N.; Lee, P.C. Aerobic biodegradability of polyester/polylactic acid composites for automotive NVH parts. *Int. J. Precis. Eng. Manuf.* **2014**, *15*, 1703–1707. [CrossRef]
9. Finkenstadt, V.L.; Tisserat, B. Poly(lactic acid) and osage orange wood fiber composites for agricultural mulch films. *Ind. Crop. Prod.* **2010**, *31*, 316–320. [CrossRef]
10. Chang, Y.C.; Chen, Y.; Ning, J.L.; Hao, C.; Rock, M.; Amer, M.; Feng, S.; Falahati, M.; Wang, L.J.; Chen, R.K.; et al. No such thing as trash: A 3D-printable polymer composite composed of oil-extracted spent coffee grounds and polylactic acid with enhanced impact toughness. *ACS Sustain. Chem. Eng.* **2019**, *7*, 15304–15310. [CrossRef]
11. Gao, Y.; Li, Y.; Hu, X.R.; Wu, W.D.; Wang, Z.; Wang, R.G.; Zhang, L.Q. Preparation and properties of novel thermoplastic vulcanizate based on bio-based polyester/polylactic acid, and its application in 3D printing. *Polymers* **2017**, *9*, 694. [CrossRef] [PubMed]
12. Kumar, S.; Singh, R.; Singh, T.P.; Batish, A. Investigations of polylactic acid reinforced composite feedstock filaments for multimaterial three-dimensional printing applications. *Proc. Inst. Mech. Eng. Part C J. Eng. Mech. Eng. Sci.* **2019**, *233*, 5953–5965. [CrossRef]
13. Matos, B.D.M.; Rocha, V.; da Silva, E.J.; Moro, F.H.; Bottene, A.C.; Ribeiro, C.A.; Dias, D.D.; Antonio, S.G.; do Amaral, A.C.; Cruz, S.A.; et al. Evaluation of commercially available polylactic acid (PLA) filaments for 3d printing applications. *J. Therm. Anal. Calorim.* **2019**, *137*, 555–562. [CrossRef]
14. Bayer, I.S. Thermomechanical properties of polylactic acid-graphene composites: A state-of-the-art review for biomedical applications. *Materials* **2017**, *10*, 748. [CrossRef]
15. Pierchala, M.K.; Makaremi, M.; Tan, H.L.; Pushpamalar, J.; Muniyandy, S.; Solouk, A.; Lee, S.M.; Pasbakhsh, P. Nanotubes in nanofibers: Antibacterial multilayered polylactic acid/halloysite/gentamicin membranes for bone regeneration application. *Appl. Clay Sci.* **2018**, *160*, 95–105. [CrossRef]
16. Chen, Y.F.; Xu, J.Y.; Tan, Q.G.; Zhang, Z.L.; Zheng, J.; Xu, X.Y.; Li, Y. End-group functionalization of polyethylene glycol-polylactic acid copolymer and its application in the field of pharmaceutical carriers. *J. Biobased Mater. Bioenergy* **2019**, *13*, 690–698.
17. Torres-Giner, S.; Martinez-Abad, A.; Gimeno-Alcañiz, J.V.; Ocio, M.J.; Lagaron, J.M. Controlled delivery of gentamicin antibiotic from bioactive electrospun polylactide-based ultrathin fibers. *Adv. Eng. Mater.* **2012**, *14*, B112–B122. [CrossRef]
18. Agüero, A.; Morcillo, M.d.C.; Quiles-Carrillo, L.; Balart, R.; Boronat, T.; Lascano, D.; Torres-Giner, S.; Fenollar, O. Study of the influence of the reprocessing cycles on the final properties of polylactide pieces obtained by injection molding. *Polymers* **2019**, *11*, 1908. [CrossRef]
19. Quiles-Carrillo, L.; Montanes, N.; Garcia-Garcia, D.; Carbonell-Verdu, A.; Balart, R.; Torres-Giner, S. Effect of different compatibilizers on injection-molded green composite pieces based on polylactide filled with almond shell flour. *Compos. Part B Eng.* **2018**, *147*, 76–85. [CrossRef]
20. Quiles-Carrillo, L.; Montanes, N.; Pineiro, F.; Jorda-Vilaplana, A.; Torres-Giner, S. Ductility and toughness improvement of injection-molded compostable pieces of polylactide by melt blending with poly(ε-caprolactone) and thermoplastic starch. *Materials* **2018**, *11*, 2138. [CrossRef]
21. Bioplastics market data 2018. In *Global Production Capacities of Bioplastics 2018–2023*; European Bioplastics: Berlin, Germany, 2018; p. 4.
22. Valerio, O.; Pin, J.M.; Misra, M.; Mohanty, A.K. Synthesis of glycerol-based biopolyesters as toughness enhancers for polylactic acid bioplastic through reactive extrusion. *ACS Omega* **2016**, *1*, 1284–1295. [CrossRef] [PubMed]
23. Zhang, B.; Bian, X.C.; Xiang, S.; Li, G.; Chen, X.S. Synthesis of PLLA-based block copolymers for improving melt strength and toughness of PLLA by in situ reactive blending. *Polym. Degrad. Stab.* **2017**, *136*, 58–70. [CrossRef]
24. Zou, J.; Qi, Y.Z.; Su, L.L.; Wei, Y.; Li, Z.L.; Xu, H.Q. Synthesis and characterization of poly(ester amide)s consisting of poly(L-lactic acid) and poly(butylene succinate) segments with 2,2′-bis(2-oxazoline) chain extending. *Macromol. Res.* **2018**, *26*, 1212–1218. [CrossRef]

25. Lan, X.R.; Li, X.; Liu, Z.Y.; He, Z.K.; Yang, W.; Yang, M.B. Composition, morphology and properties of poly(lactic acid) and poly(butylene succinate) copolymer system via coupling reaction. *J. Macromol. Sci. Part A Pure Appl. Chem.* **2013**, *50*, 861–870. [CrossRef]
26. Garcia-Campo, M.J.; Quiles-Carrillo, L.; Masia, J.; Reig-Perez, M.J.; Montanes, N.; Balart, R. Environmentally friendly compatibilizers from soybean oil for ternary blends of poly(lactic acid)-PLA, poly(ε-caprolactone)-PCL and poly(3-hydroxybutyrate)-PHB. *Materials* **2017**, *10*, 1339. [CrossRef]
27. Garcia-Campo, M.J.; Quiles-Carrillo, L.; Sanchez-Nacher, L.; Balart, R.; Montanes, N. High toughness poly(lactic acid) (PLA) formulations obtained by ternary blends with poly(3-hydroxybutyrate) (PHB) and flexible polyesters from succinic acid. *Polym. Bull.* **2019**, *76*, 1839–1859. [CrossRef]
28. Sathornluck, S.; Choochottiros, C. Modification of epoxidized natural rubber as a PLA toughening agent. *J. Appl. Polym. Sci.* **2019**, *136*, 48267. [CrossRef]
29. Su, S.; Kopitzky, R.; Tolga, S.; Kabasci, S. Polylactide (PLA) and its blends with poly(butylene succinate) (PBS): A brief review. *Polymers* **2019**, *11*, 1193. [CrossRef]
30. Zhang, B.; Sun, B.; Bian, X.C.; Li, G.; Chen, X.S. High melt strength and high toughness PLLA/PBS blends by copolymerization and in situ reactive compatibilization. *Ind. Eng. Chem. Res.* **2017**, *56*, 52–62. [CrossRef]
31. Fortelny, I.; Ujcic, A.; Fambri, L.; Slouf, M. Phase structure, compatibility, and toughness of PLA/PCL blends: A review. *Front. Mater.* **2019**, *6*, 206. [CrossRef]
32. Wang, Y.; Mei, Y.; Wang, Q.; Wei, W.; Huang, F.; Li, Y.; Li, J.Y.; Zhou, Z.W. Improved fracture toughness and ductility of PLA composites by incorporating a small amount of surface-modified helical carbon nanotubes. *Compos. Part B Eng.* **2019**, *162*, 54–61. [CrossRef]
33. Li, J.J.; Li, J.; Feng, D.J.; Zhao, J.F.; Sun, J.R.; Li, D.G. Excellent rheological performance and impact toughness of cellulose nanofibers/PLA/ionomer composite. *RSC Adv.* **2017**, *7*, 28889–28897. [CrossRef]
34. Gonzalez-Ausejo, J.; Gamez-Perez, J.; Balart, R.; Lagaron, J.M.; Cabedo, L. Effect of the addition of sepiolite on the morphology and properties of melt compounded PHBV/PLA blends. *Polym. Compos.* **2019**, *40*, E156–E168. [CrossRef]
35. Tsou, C.H.; Gao, C.; De Guzman, M.; Wu, D.Y.; Hung, W.S.; Yuan, L.; Suen, M.C.; Yeh, J.T. Preparation and characterization of poly(lactic acid) with adipate ester added as a plasticizer. *Polym. Polym. Compos.* **2018**, *26*, 446–453. [CrossRef]
36. Huang, H.C.; Chen, L.J.; Song, G.L.; Tang, G.Y. An efficient plasticization method for poly(lactic acid) using combination of liquid-state and solid-state plasticizers. *J. Appl. Polym. Sci.* **2018**, *135*, 46669. [CrossRef]
37. Kang, H.L.; Li, Y.S.; Gong, M.; Guo, Y.L.; Guo, Z.; Fang, Q.H.; Li, X. An environmentally sustainable plasticizer toughened polylactide. *RSC Adv.* **2018**, *8*, 11643–11651. [CrossRef]
38. Carbonell-Verdu, A.; Ferri, J.M.; Dominici, F.; Boronat, T.; Sanchez-Nacher, L.; Balart, R.; Torre, L. Manufacturing and compatibilization of PLA/PBAT binary blends by cottonseed oil-based derivatives. *Express Polym. Lett.* **2018**, *12*, 808–823. [CrossRef]
39. Quiles-Carrillo, L.; Blanes-Martínez, M.M.; Montanes, N.; Fenollar, O.; Torres-Giner, S.; Balart, R. Reactive toughening of injection-molded polylactide pieces using maleinized hemp seed oil. *Eur. Polym. J.* **2018**, *98*, 402–410. [CrossRef]
40. Quiles-Carrillo, L.; Duart, S.; Montanes, N.; Torres-Giner, S.; Balart, R. Enhancement of the mechanical and thermal properties of injection-molded polylactide parts by the addition of acrylated epoxidized soybean oil. *Mater. Des.* **2018**, *140*, 54–63. [CrossRef]
41. Ferri, J.M.; Garcia-Garcia, D.; Sanchez-Nacher, L.; Fenollar, O.; Balart, R. The effect of maleinized linseed oil (mlo) on mechanical performance of poly(lactic acid)-thermoplastic starch (PLA-TPS) blends. *Carbohydr. Polym.* **2016**, *147*, 60–68. [CrossRef]
42. Ferri, J.M.; Garcia-Garcia, D.; Montanes, N.; Fenollar, O.; Balart, R. The effect of maleinized linseed oil as biobased plasticizer in poly(lactic acid)-based formulations. *Polym. Int.* **2017**, *66*, 882–891. [CrossRef]
43. Notta-Cuvier, D.; Murariu, M.; Odent, J.; Delille, R.; Bouzouita, A.; Raquez, J.M.; Lauro, F.; Dubois, P. Tailoring polylactide properties for automotive applications: Effects of co-addition of halloysite nanotubes and selected plasticizer. *Macromol. Mater. Eng.* **2015**, *300*, 684–698. [CrossRef]
44. Luzi, F.; Dominici, F.; Armentano, I.; Fortunati, E.; Burgos, N.; Fiori, S.; Jimenez, A.; Kenny, J.M.; Torre, L. Combined effect of cellulose nanocrystals, carvacrol and oligomeric lactic acid in PLA-PHB polymeric films. *Carbohydr. Polym.* **2019**, *223*, 115131. [CrossRef]

45. Burgos, N.; Martino, V.P.; Jimenez, A. Characterization and ageing study of poly(lactic acid) films plasticized with oligomeric lactic acid. *Polym. Degrad. Stab.* **2013**, *98*, 651–658. [CrossRef]
46. Battegazzore, D.; Bocchini, S.; Frache, A. Crystallization kinetics of poly(lactic acid)-talc composites. *Express Polym. Lett.* **2011**, *5*, 849–858. [CrossRef]
47. Kaygusuz, B.; Ozerinc, S. Improving the ductility of polylactic acid parts produced by fused deposition modeling through polyhydroxyalkanoate additions. *J. Appl. Polym. Sci.* **2019**, *136*, 48154. [CrossRef]
48. Lule, Z.; Kim, J. Nonisothermal crystallization of surface-treated alumina and aluminum nitride-filled polylactic acid hybrid composites. *Polymers* **2019**, *11*, 1077. [CrossRef]
49. Quiles-Carrillo, L.; Montanes, N.; Sammon, C.; Balart, R.; Torres-Giner, S. Compatibilization of highly sustainable polylactide/almond shell flour composites by reactive extrusion with maleinized linseed oil. *Ind. Crop. Prod.* **2018**, *111*, 878–888. [CrossRef]
50. Jing, X.; Mi, H.Y.; Peng, X.F.; Turng, L.S. The morphology, properties, and shape memory behavior of polylactic acid/thermoplastic polyurethane blends. *Polym. Eng. Sci.* **2015**, *55*, 70–80. [CrossRef]
51. Shen, T.F.; Liang, L.Y.; Lu, M.G. Novel biodegradable shape memory composites based on PLA and PCL crosslinked by polyisocyanate. In *Advances in Biomedical Engineering*; Hu, J., Ed.; Information Engineering Research Inst, USA: Newark, NJ, USA, 2011; pp. 302–305.
52. Zhang, Z.X.; He, Z.Z.; Yang, J.H.; Huang, T.; Zhang, N.; Wang, Y. Crystallization controlled shape memory behaviors of dynamically vulcanized poly(l-lactide)/poly(ethylene vinyl acetate) blends. *Polym. Test.* **2016**, *51*, 82–92. [CrossRef]
53. Shao, L.N.; Dai, J.; Zhang, Z.X.; Yang, J.H.; Zhang, N.; Huang, T.; Wang, Y. Thermal and electroactive shape memory behaviors of poly(l-lactide)/thermoplastic polyurethane blend induced by carbon nanotubes. *RSC Adv.* **2015**, *5*, 101455–101465. [CrossRef]
54. Ambrosio-Martin, J.; Fabra, M.J.; Lopez-Rubio, A.; Lagaron, J.M. An effect of lactic acid oligomers on the barrier properties of polylactide. *J. Mater. Sci.* **2014**, *49*, 2975–2986. [CrossRef]
55. Courgneau, C.; Domenek, S.; Guinault, A.; Averous, L.; Ducruet, V. Analysis of the structure-properties relationships of different multiphase systems based on plasticized poly(lactic acid). *J. Polym. Environ.* **2011**, *19*, 362–371. [CrossRef]
56. Fortunati, E.; Puglia, D.; Iannoni, A.; Terenzi, A.; Kenny, J.M.; Torre, L. Processing conditions, thermal and mechanical responses of stretchable poly(lactic acid)/poly(butylene succinate) films. *Materials* **2017**, *10*, 809. [CrossRef]
57. Ferri, J.M.; Samper, M.D.; García-Sanoguera, D.; Reig, M.J.; Fenollar, O.; Balart, R. Plasticizing effect of biobased epoxidized fatty acid esters on mechanical and thermal properties of poly(lactic acid). *J. Mater. Sci.* **2016**, *51*, 5356–5366. [CrossRef]
58. Chee, W.K.; Ibrahim, N.A.; Zainuddin, N.; Abd Rahman, M.F.; Chieng, B.W. Impact toughness and ductility enhancement of biodegradable poly(lactic acid)/poly(ε-caprolactone) blends via addition of glycidyl methacrylate. *Adv. Mater. Sci. Eng.* **2013**, *2013*, 976373. [CrossRef]
59. Xue, B.; He, H.Z.; Zhu, Z.W.; Li, J.Q.; Huang, Z.X.; Wang, G.Z.; Chen, M.; Zhan, Z.M. A facile fabrication of high toughness poly(lactic acid) via reactive extrusion with poly(butylene succinate) and ethylene-methyl acrylate-glycidyl methacrylate. *Polymers* **2018**, *10*, 1401. [CrossRef]
60. Wang, X.; Peng, S.X.; Chen, H.; Yu, X.L.; Zhao, X.P. Mechanical properties, rheological behaviors, and phase morphologies of high-toughness PLA/PBAT blends by in-situ reactive compatibilization. *Compos. Part B Eng.* **2019**, *173*, 107028. [CrossRef]
61. Lascano, D.; Quiles-Carrillo, L.; Torres-Giner, S.; Boronat, T.; Montanes, N. Optimization of the curing and post-curing conditions for the manufacturing of partially bio-based epoxy resins with improved toughness. *Polymers* **2019**, *11*, 1354. [CrossRef]
62. Burgos, N.; Tolaguera, D.; Fiori, S.; Jimenez, A. Synthesis and characterization of lactic acid oligomers: Evaluation of performance as poly(lactic acid) plasticizers. *J. Polym. Environ.* **2014**, *22*, 227–235. [CrossRef]
63. Ljungberg, N.; Wesslén, B. The effects of plasticizers on the dynamic mechanical and thermal properties of poly(lactic acid). *J. Appl. Polym. Sci.* **2002**, *86*, 1227–1234. [CrossRef]
64. Xing, Q.; Zhang, X.Q.; Dong, X.; Liu, G.M.; Wang, D.J. Low-molecular weight aliphatic amides as nucleating agents for poly (L-lactic acid): Conformation variation induced crystallization enhancement. *Polymer* **2012**, *53*, 2306–2314. [CrossRef]

65. Maróti, P.; Kocsis, B.; Ferencz, A.; Nyitrai, M.; Lőrinczy, D. Differential thermal analysis of the antibacterial effect of PLA-based materials planned for 3D printing. *J. Therm. Anal. Calorim.* **2019**, 1–8. [CrossRef]
66. Maiza, M.; Benaniba, M.T.; Quintard, G.; Massardier-Nageotte, V. Biobased additive plasticizing polylactic acid (PLA). *Polímeros* **2015**, *25*, 581–590. [CrossRef]
67. Jia, S.K.; Yu, D.M.; Zhu, Y.; Wang, Z.; Chen, L.G.; Fu, L. Morphology, crystallization and thermal behaviors of pla-based composites: Wonderful effects of hybrid GO/PEG via dynamic impregnating. *Polymers* **2017**, *9*, 528. [CrossRef] [PubMed]
68. Shi, X.T.; Zhang, G.C.; Phuong, T.V.; Lazzeri, A. Synergistic effects of nucleating agents and plasticizers on the crystallization behavior of poly(lactic acid). *Molecules* **2015**, *20*, 1579–1593. [CrossRef]
69. Lu, X.L.; Sun, Z.J.; Cai, W.; Gao, Z.Y. Study on the shape memory effects of poly(l-lactide-co-ε-caprolactone) biodegradable polymers. *J. Mater. Sci. Mater. Med.* **2008**, *19*, 395–399. [CrossRef]
70. Leones, A.; Sonseca, A.; Lopez, D.; Fiori, S.; Peponi, L. Shape memory effect on electrospun PLA-based fibers tailoring their thermal response. *Eur. Polym. J.* **2019**, *117*, 217–226. [CrossRef]

© 2019 by the authors. Licensee MDPI, Basel, Switzerland. This article is an open access article distributed under the terms and conditions of the Creative Commons Attribution (CC BY) license (http://creativecommons.org/licenses/by/4.0/).

Article

Effect of Almond Shell Waste on Physicochemical Properties of Polyester-Based Biocomposites

Marina Ramos [1], Franco Dominici [2], Francesca Luzi [2], Alfonso Jiménez [1], Maria Carmen Garrigós [1,*], Luigi Torre [2] and Debora Puglia [2,*]

1. Department of Analytical Chemistry, Nutrition & Food Sciences, University of Alicante, San Vicente del Raspeig, ES-03690 Alicante, Spain; marina.ramos@ua.es (M.R.); alfjimenez@ua.es (A.J.)
2. Department of Civil and Environmental Engineering, University of Perugia, 05100 Terni, Italy; francodominici1@gmail.com (F.D.); francesca.luzi@unipg.it (F.L.); luigi.torre@unipg.it (L.T.)
* Correspondence: mc.garrigos@ua.es (M.C.G); debora.puglia@unipg.it (D.P.)

Received: 15 March 2020; Accepted: 1 April 2020; Published: 6 April 2020

Abstract: Polyester-based biocomposites containing INZEA F2® biopolymer and almond shell powder (ASP) at 10 and 25 wt % contents with and without two different compatibilizers, maleinized linseed oil and Joncryl ADR 4400®, were prepared by melt blending in an extruder, followed by injection molding. The effect of fine (125–250 m) and coarse (500–1000 m) milling sizes of ASP was also evaluated. An improvement in elastic modulus was observed with the addition of< both fine and coarse ASP at 25 wt %. The addition of maleinized linseed oil and Joncryl ADR 4400 produced some compatibilizing effect at low filler contents while biocomposites with a higher amount of ASP still presented some gaps at the interface by field emission scanning electron microscopy. Some decrease in thermal stability was shown which was related to the relatively low thermal stability and disintegration of the lignocellulosic filler. The added modifiers provided some enhanced thermal resistance to the final biocomposites. Thermal analysis by differential scanning calorimetry and thermogravimetric analysis suggested the presence of two different polyesters in the polymer matrix, with one of them showing full disintegration after 28 and 90 days for biocomposites containing 25 and 10 wt %, respectively, under composting conditions. The developed biocomposites have been shown to be potential polyester-based matrices for use as compostable materials at high filler contents.

Keywords: almond shell waste; reinforcing; polyester-based biocomposites; physicochemical properties; disintegration

1. Introduction

Almond is characterized by its high nutritional value, although information reported so far mainly concerns its edible kernel or meat. Other parts also present in the almond fruit are the middle shell, outer green shell cover or almond hull and a thin leathery layer known as brown skin of meat or seed coat [1]. Almonds are used as a fruit in snack foods and as ingredients in a variety of processed foods, especially in bakery and confectionery products. However, almond production generates large amounts of almond by-products since the nutritional and commercial relevance of almonds is restricted to the kernel. In particular, almond shell is the name given to the ligneous material forming the thick endocarp or husk of the almond (*Prunus amygdalus* L.) tree fruit. It is principally composed of cellulose (ranging from 29.8 to 50.7 wt %), hemicellulose (from 19.3 to 29.0 wt %) and lignin (from 20.4 to 50.7 wt %) [1]. This by-product is normally incinerated or dumped without control, which results in the production of large amounts of waste and pollution [2]. Several researchers have focused on different alternatives for using almond shell wastes based on their potential uses as biomass to produce renewable energy [3]; as a source of organic biopesticides [4], heavy metal adsorbents [5], dye adsorbents [6], growing media [7], the preparation of activated carbons [8] and xylo-oligosaccharides [9], antioxidants [10] or as additives in eco-friendly composites [11–13].

The development of eco-friendly composites arises from the need for reducing environmental problems generated by industrial processes. In this scenario, agricultural waste utilization has become a potential option for the development of eco-friendly composites. This powerful area of interest presents several benefits such as biodegradability in combination with bio-based or natural polymers, light weight, low cost and easy processing [14,15]. Among the wide variety of lignocellulosic wastes, almond shell powder has been already considered as filler for commodity plastics, such as polypropylene [16–20], polyethylene [21], poly(methyl methacrylate) [22] and toughened epoxies [23,24].

Looking at a more environmentally friendly use, the role of ASP has been recently studied in enhancing the mechanical performance of some melt compounded biopolymers [25–27]. Nonetheless, due to the lack of miscibility between hydrophobic polymer matrices and highly hydrophilic almond shell fillers, the obtained green composites usually presented poor ductility and low thermal stability. In order to increase the interaction between them, several solutions have been proposed, such as silanization, acetylation and maleic anhydride modification [28]. Plasticizers could also act as internal lubricants, thus allowing chain mobility, which enhances processability and improves thermal stability and ductility. Recently, vegetable oils have been proposed as environmentally friendly compatibilizers as an alternative to conventional petroleum-based ones [29]. Specifically, maleinized linseed oil (MLO) has been used as a compatibilizer in biopolymer/ASP composites [11,30,31]. In these works, authors discussed plasticization and compatibilization effects provided by MLO due to the interaction between succinic anhydride polar groups contained in MLO and hydroxyl groups in ASP (hydroxyl groups in cellulose). The compatibilizing effect was obtained by melt grafting for the formation of new carboxylic ester bonds through the reaction of maleic anhydride functionalities present in MLO with the hydroxyl groups of both the polyester terminal chains and cellulose on the ASP surface. On the other hand, the possibility of improving the stress transfer between the filler and the polymer can be realized by reactive processing with chain extenders [32].

The main aim of the present work is the development and characterization of new biocomposites prepared using a commercial INZEA® biopolymer (mainly composed of a polyester-based matrix) containing almond shell powder at 10 and 25 wt % contents. The effect of adding two different milling sizes (125–250 μm and 500–1000 μm) in the biocomposites preparation was also evaluated. In addition, the potential of maleinized linseed oil as a compatibilizer was studied. The effect of this vegetable-oil-derived compatibilizer was also compared with a conventional epoxy styrene-acrylic oligomer (Joncryl ADR 4400) in terms of mechanical properties, thermal stability and blend morphology. Biocomposites containing 10 and 25 wt % of ASP at two grinding levels were submitted to a disintegration test in order to verify the effectiveness of the developed polyester/ASP composites to be used as compostable materials.

2. Materials and Methods

2.1. Materials

INZEA® biopolyester commercial grade, with a density of 1.23 g cm^{-3} measured at 23 °C, a moisture content <0.5% and a melt flow rate of 19 g/10 min (2.16 kg, 190 °C), was kindly supplied by Nurel (Zaragoza, Spain). Almond shell (AS) waste used as filler was supplied by Fecoam (Murcia, Spain) as an agricultural by-product and pulverized with a high-speed rotor mill (Ultra Centrifugal Mill ZM 200, RETSCH, Haan, Germany). The obtained particles were sieved by selecting the sizes of the ground shells in the ranges of 125–250 μm as fine grain (F) and 500–1000 μm as coarse grain (C) to evaluate the effect of particle size in the composites. Two different compatibilizers were selected to improve the compatibility between the polyester-based matrix and the natural filler: a synthetic polymer chain extender with recognized efficacy as compatibilizer supplied as Joncryl ADR 4400® (J44) (BASF S.A, Barcelona, Spain) and a biodegradable additive obtained by the maleinizing treatment of linseed oil, supplied as Veomer Lin by Vandeputte (Mouscron, Belgium, viscosity of 10 dPa s at 20 °C and an acid value of 105–130 mg KOH g^{-1}).

2.2. Biocomposites Preparation

Biocomposite materials were obtained using the melt blending method by mixing the biopolymer matrix with the almond particles, obtained by grinding and sieving the almond shells as previously described and the additives according to the proportions shown in Table 1. A co-rotating twin-screw extruder, Xplore 5 & 15 Micro Compounder by DSM, was used by mixing at a rotating speed of 90 rpm for 3 min and setting a temperature profile of 190–195–200 °C in the three heating zones from feeding section to die. A Micro Injection Molding Machine 10 cc by DSM, coupled to the extruder and equipped with adequate molds, was used to produce samples for flexural tests according to the standards. An appropriate pressure/time profile was used for the injection of each type of sample, while the temperatures of the injection barrel and the molds were set, respectively, at 210 and 30 °C.

Table 1. Formulations obtained in this work and their codification.

Formulation	Biopolymer (wt %)	Milled ASP (wt %)	Grain Size *	J44 (wt %)	MLO (wt %)
INZEA	100.00				
INZ_10ASF	90.00	10	F		
INZ_10ASC	90.00	10	C		
INZ_25ASF	75.00	25	F		
INZ_25ASC	75.00	25	C		
INZ_10ASF_1J	89.10	10	F	0.90	
INZ_10ASC_1J	89.10	10	C	0.90	
INZ_25ASF_1J	74.25	25	F	0.75	
INZ_25ASC_1J	74.25	25	C	0.75	
INZ_10ASF_5MLO	85.50	10	F		4.50
INZ_10ASC_5MLO	85.50	10	C		4.50
INZ_25ASF_5MLO	71.25	25	F		3.75
INZ_25ASC_5MLO	71.25	25	C		3.75

* Fine grain (F): 125–250 µm; coarse grain (C): 500–1000 µm. J44: Joncryl ADR 4400; MLO: maleinized linseed oil.

2.3. Almond and Biocomposites Characterization

2.3.1. Field Emission Scanning Electron Microscopy

Morphological characterization of ASP was carried out using a field emission scanning electron microscope (FESEM), Supra 25 by Zeiss (Oberkochen, Germany). The surfaces and the fractures of biocomposites were analyzed with a FESEM Merlin VP Compact by ZEISS. In both cases, micrographs were taken using an accelerating voltage of 5 kV at different magnifications. Samples were previously gold-sputtered with an Automatic Sputter Coater, B7341 by Agar Scientific (Stansted, Essex, UK), operating with a vacuum atmosphere (0.1–0.005 mbar) and low current (0–50 mA) to provide electric conductivity.

2.3.2. Thermal Characterization

Thermogravimetric analysis of ASP was performed with thermogravimetric analysis (TGA; (Seiko Exstar 6300, Tokyo, Japan). Approximately 5 mg of samples were heated from 30 to 600 °C at 10 °C min^{-1} under nitrogen atmosphere (flow rate 200 mL min^{-1}).

Differential scanning calorimetry (DSC) tests were conducted for the determination of thermal events by using a DSC (Q1000, TA Instruments, New castle, DE, USA) under a nitrogen atmosphere (50 mL min^{-1}). A 3 mg amount of samples were introduced in aluminum pans (40 µL) and they were submitted to the following thermal program: −30 °C to 250 °C at 10 °C min^{-1}, with two heating and one cooling scans.

The thermal degradation behavior of biocomposites in composting conditions was evaluated by thermogravimetric analysis (TGA/SDTA851e/SF/1100, Mettler Toledo, (Schwarzenbach, Switzerland). Around 5 mg of samples were used to perform dynamic tests in a nitrogen atmosphere (200 mL min^{-1}) from 30 °C to 700 °C at 10 °C min^{-1}.

2.3.3. Mechanical Properties

Flexural tests were carried out by using a universal test machine LR30K (Lloyd Instruments Ltd., Bognor Regis, UK) at room temperature. A minimum of five different samples was tested using a 0.5 kN load cell, setting the crosshead speed to 2 mm min^{-1} for three points bending test, as suggested by ISO 178 Standard.

2.3.4. Disintegrability in Composting Conditions

Disintegration tests in composting conditions were performed, in triplicate, by following the ISO 20200 Standard method using a commercial compost with a certain amount of sawdust, rabbit food, starch, oil and urea [33]. Tested samples were obtained from the previously prepared dog-bone-shaped bars, which were cut in pieces (5 × 10 × 2 mm^3), buried at a 5 cm depth in perforated boxes and incubated at 58 °C. The aerobic conditions were guaranteed by mixing the compost softly and by the periodical addition of water according to the standard requirements.

Different disintegration times were selected to recover samples from burial and further tested: 0, 4, 7, 15, 21, 28, 40, 69 and 90 days. Samples were immediately washed with distilled water to remove traces of compost extracted from the container and further dried at 40 °C for 24 h before gravimetric analysis. The disintegrability value for each material at different times was obtained by normalizing the sample weight with the value obtained at the initial time.

The evolution of disintegration was monitored by taking photographs of recovered samples for visual evaluation of physical alterations with disintegration time. In addition, thermal (DSC, TGA) properties upon disintegrability tests were also studied.

2.4. Statistical Analysis

Statistical analysis of experimental data was performed by one-way analysis of variance (ANOVA) using SPSS 15.0 (IBM, Chicago, IL, USA) and expressed as means ± standard deviation. Differences between average values were assessed based on the Tukey test at a confidence level of 95% ($p < 0.05$).

3. Results

3.1. Characterization of Almond Shell Powder

3.1.1. Morphological Analysis

Low-magnification FESEM micrographs of almond shell waste at the two different studied sizes (125–250 μm, fine, and 500–1000 μm, coarse), reported in Figure 1, showed a general view typical of fillers obtained after grinding and sieving processes. Almond shell was shown to have a sheet-like structure. In addition, a series of 1 μm pores were observed on the almond surface (see inserts) [34]. Most of the particles were characterized by a spherical shape, though some aggregates, as well as flat and long rod-like particles, were also observed. A detail of the particle surface can be seen in the high-magnification FESEM images. The micrographs revealed that the particles were irregular in shape and presented a rough surface, more likely resulting from the crushing process due to the high hardness of this type of filler. Some granular features can also be observed, which resemble the original grainy and wavy structure of almond shell [35].

Figure 1. FESEM images of fine (**a**) and coarse (**b**) almond shell powders.

3.1.2. Thermal Properties

TG and derivative DTG profiles obtained for ASP (coarse size) under the nitrogen atmosphere at a heating rate of 10 °C min^{-1}, reported in Figure 2, showed the typical thermal degradation profile for biomasses with three well-demarked steps for moisture release, devolatilization and char formation. Weight loss in the lower temperature region can be attributed to the loss of moisture, while major weight loss was observed at temperatures ranging from 225 to 365 °C, over which hemicellulose and cellulose decomposition occurs, leading to the formation of pyrolysis products (volatiles, gases and primary biochar) [36]. This phenomenon was followed by a slow weight loss until 600 °C, which was attributable to the continuous devolatilization of biochar caused by a further breakdown of C–C and C–H bonds.

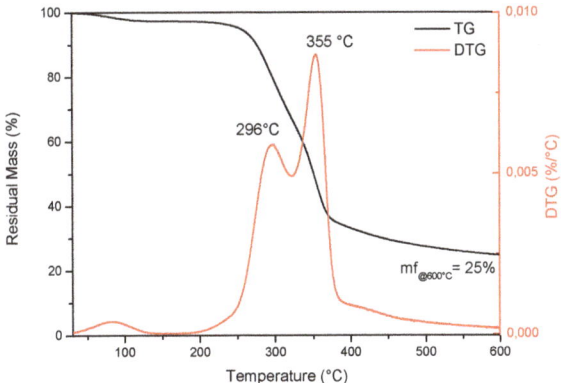

Figure 2. TGA analysis of almond coarse shell powder.

3.2. Characterization of ASP Biocomposites

3.2.1. Flexural Tests

Flexural tests provided information on the effect of the amount and size of ASP incorporated in the composite, as well as on the effect of adding the studied compatibilizing additives. In general, results show that the addition of the filler did not improve the maximum strength or strain at break with respect to the reference polymer matrix (Table 2). The elastic modulus of the biocomposites was improved only in formulations containing 25 wt % of both fine and coarse filler. The comparison of flexural tests for formulations with 10 wt % of ASP (Table 2 and Figure 3) showed that the size of the coarse grain gives greater rigidity than the fine grain by increasing strength and modulus but slightly reducing elongation. In fact, the presence of 10 wt % of fine filler in the INZEA_10ASF biocomposite produced a maximum strength value of 44 MPa with a flexural modulus of 1473 MPa, lower than the biocomposite with the coarse filler INZEA_10ASC, which showed σ_{max} = 47 MPa and E = 1699 MPa values. Nabinejad et al. [37] reported that the surface roughness and high surface area of coarse powder particles could have a positive effect on the mechanical performance of composites, being able to restrict the polymer chain mobility. Porosity and roughness of the hydrophilic surface for coarse almond shell powder would be expected to increase its wettability by the polymer matrix. So, as a result, the INZEA_10ASC composite showed high stiffness values compared to composites containing fine filler (INZEA_10ASF) with low surface roughness and porosity. Additionally, results from Zaini et al [38] confirmed that composites filled with a larger-sized filler showed higher modulus, tensile and impact strengths, particularly at high filler loadings.

Table 2. Flexural parameters for almond shell powder (ASP)-based biocomposites (mean ± SD, n = 5).

Formulation	σ_{max} (MPa)	ε(%) at σ_{max}	E (MPa)
INZEA	66 ± 2	7.9 ± 0.2	1913 ± 13
INZEA_10ASF	44 ± 2	4.4 ± 0.4	1473 ± 16
INZEA_10ASF_5MLO	36 ± 1	4.9 ± 0.3	1170 ± 15
INZEA_10ASF_1J	48 ± 2	5.5 ± 0.6	1414 ± 8
INZEA_25ASF	50 ± 1	3.0 ± 0.2	2537 ± 35
INZEA_25ASF_5MLO	44 ± 1	4.0 ± 0.2	1838 ± 22
INZEA_25ASF_1J	56 ± 2	3.5 ± 0.1	2555 ± 61
INZEA_10ASC	47 ± 2	4.0 ± 0.3	1699 ± 32
INZEA_10ASC_5MLO	38 ± 2	5.4 ± 0.6	1300 ± 33
INZEA_10ASC_1J	47 ± 2	4.2 ± 0.4	1653 ± 24
INZEA_25ASC	47 ± 2	3.0 ± 0.2	2392 ± 77
INZEA_25ASC _5MLO	35 ± 1	3.5 ± 0.1	1732 ± 7
INZEA_25ASC _1J	53 ± 1	3.7 ± 0.2	2394 ± 25

σ_{max}: flexural strength; ε at σ_{max}: strain at maximum stress; E: Young's Modulus.

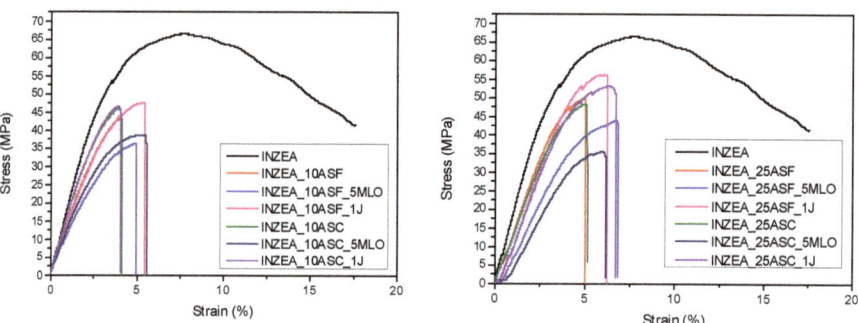

Figure 3. Stress–strain curves of polyester-based biocomposites containing 10 and 25 wt % of ASP at two grinding levels (F, C), with or without compatibilizers.

The addition of 5 wt % of MLO in the biocomposite formulation produced a compatibilizing effect lower than expected. In fact, the improvement in deformability moved from 4.4% to 4.9% of INZEA_10ASF_5MLO and from 4.0% to 5.4% of INZEA_10ASC_5MLO in the formulations with MLO and 10 wt % ASP fine and coarse, respectively, with a consistent reduction in both flexural strength and flexural modulus. If the modest plasticizing effect is excluded, the MLO did not produce improvements of the interface strength between the natural filler and the biopolymer matrix. Effectiveness of maleinized linseed oil has been demonstrated in composites based on lignin fillers and, specifically, from ground almond shells and some biodegradable polymers, such as poly(lactic acid) (PLA) and lignin. In this case, the poor result obtained with the INZEA matrix should be attributed to the particular composition of the commercial biopolymer used (a bio-based blend mainly composed of a polyester matrix) [30,39,40]. Joncryl ADR 4400 added to the formulation with fine particles INZEA_10ASF_1J produced an increase in flexural strength from 44 to 48 MPa. This improvement is due to the compatibilizing effect of the polymer chain extender, which improves the adhesion between the matrix and the particles allowing a greater deflection. This effect is highlighted by the perfect overlap between the σ–ε curve of the unmodified INZEA_10ASF and the INZEA_10ASF_1J curve which, thanks to the J44, extends up to 5.5% increasing the flexural strength [41]. The effect of Joncryl was negligible on biocomposites with coarse grain, since the compatibilization effect at the interface between the matrix and the filler was much lower (about 6%) than biocomposites with fine particles. In fact, when simplifying the particles as spherical and calculating the ratios between coarse and fine surfaces, with the same wt % content, an area ratio of 0.0625 was obtained.

When increasing the quantity of filler to 25 wt %, an increase in the rigidity of the biocomposites was obtained. The result is a general increase in flexural strength and moduli, which corresponds to a reduction in deflection. Even in formulations with a higher content of the natural filler, MLO did not produce any other effects than those already shown in the set with 10 wt % of ASP. The flexural strength of the biocomposite containing 25 wt % of fine particle (INZEA_25ASF) rises to 50 MPa (44 MPa for INZEA_10ASF), while the INZEA_25ASC composite maintains the same value of 47 MPa as 10 wt % of filler, highlighting the achievement of the plateau of the coarse grain reinforcement. In the case of the higher amount of ASP, the compatibilizing effect of Joncryl appears even more evident, since a better interface bonding between the matrix and the filler was achieved. INZEA_25ASF_1J showed a further increase in flexural strength reaching 56 MPa, with an improvement in modulus to 2555 MPa and without excessively reducing the flexural deflection. In addition, INZEA_25ASC_1J with Joncryl showed an improvement in strength going up from 47 to 53 MPa, but in this case, the compatibilizing effect of J44 was less evident, because it occurred on a smaller interface surface, compared to fine-grain-sized biocomposites, due to the larger particle size [13].

3.2.2. Morphological and Thermal Analysis

In Figure 4, FESEM images of fractured surfaces for INZEA/ASP composites (uncompatibilized and compatibilized biocomposites) are reported. As it can be observed, in the case of the unmodified matrix, the polymer-particle adhesion was very poor, both at low and high ASP contents, so important gaps can be found between the particles and the surrounding polyester matrix [11]. This morphological observation correlates with the above-described mechanical performance of the unmodified INZEA composites, in which the presence of ASP did not contribute to an improvement in mechanical performance. The addition of 5 wt % MLO provides some interaction as the gap seems to be reduced, indicating a limited but good compatibilizing effect of MLO modifier, but at the same time, the presence of microsized voids in the matrix compatibilized with 5 wt % of MLO was visible [30]. In the presence of a styrene-acrylic-based compatibilizer and oligomeric agent, an improvement in the interface with ASF and ASC was noted at higher contents (25 wt %), even if biocomposites containing the higher amount presented some gaps at the interface. So, it can be concluded the chain extender was effective at both filler contents [42].

Figure 5a,b shows the TG/DTG thermograms of the INZEA matrix with the addition of 25 wt % ASP at the two different grinding sizes. Only the trends in thermal behavior for the biocomposites containing the higher ASP content have been reported, being the ones at 10 wt % essentially inline (data not shown). The presence of a double degradation peak for the INZEA neat matrix gives us an indication of a material that degrades in two steps around 350 and 400 °C, that could match with the possible degradation temperatures of PLA and poly(butylene succinate) (PBS) polyesters. It is also important to note that at 900 °C, even the unmodified INZEA matrix maintains a residual mass of ca. 5 wt %, which is increased by the presence of the fillers. This is in accordance with the possible presence of an inorganic filler in the formulation of the commercial product.

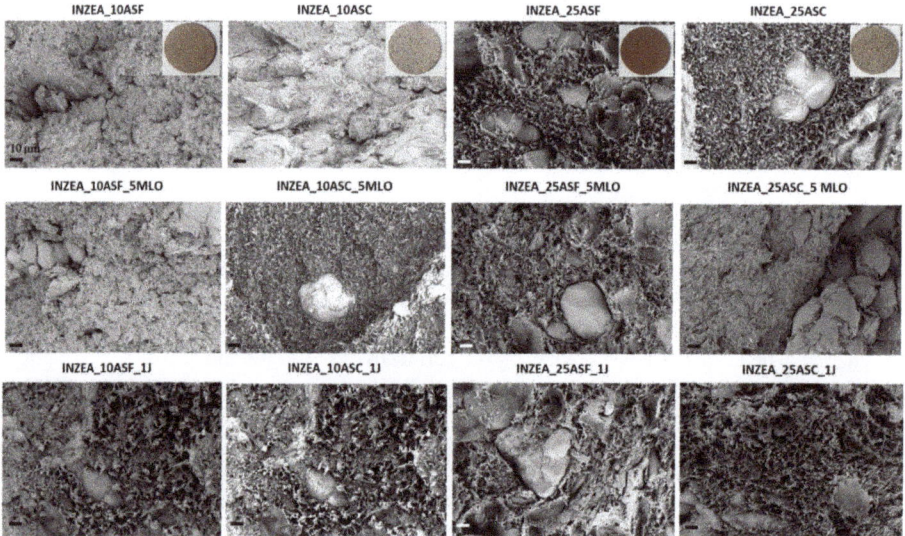

Figure 4. FESEM images of INZEA-based biocomposites with 10 and 25 wt % of ASP at the two grinding levels (fine, coarse), with or without compatibilizers.

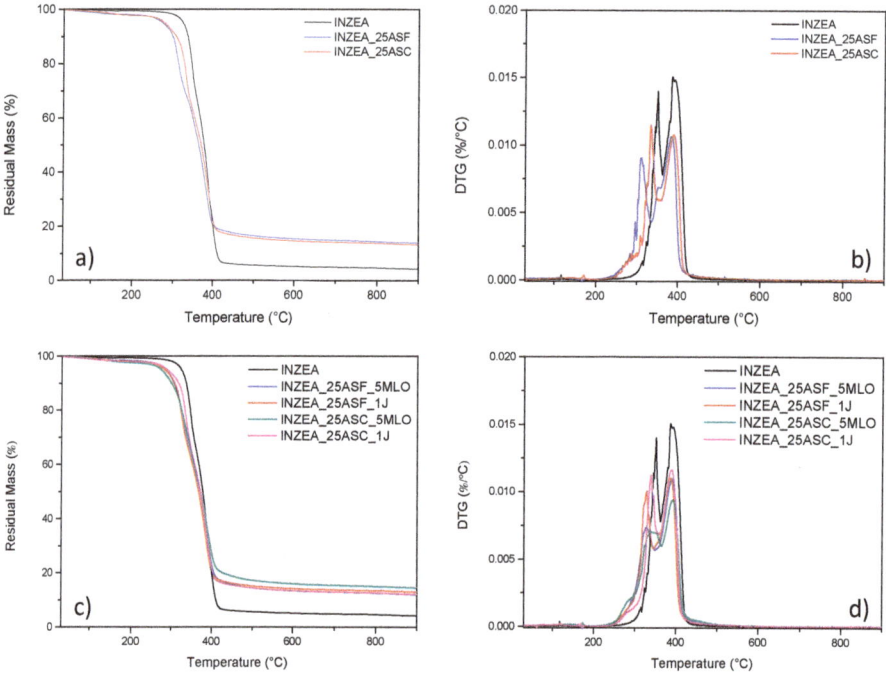

Figure 5. TG/DTG curves of INZEA-based biocomposites with 25 wt % of ASP at the two grinding levels (**a**,**b**) and INZEA biocomposites with 25 wt % of ASP in the presence of MLO or Joncryl compatibilizers (**c**,**d**).

The TG curves corresponding to INZEA/ASP composites showed an evident decrease at the onset degradation temperature from 324 °C for neat INZEA to 275 °C and 280 °C, respectively, for INZEA_25ASC and INZEA_25ASF (Table 3). This behavior is essentially due to the relatively low thermal stability of the lignocellulosic filler, which initiated its degradation at 198 °C (Figure 2), and negatively contributed to the reduction of the global thermal stability of the INZEA-based biocomposites, in agreement with previous studies on the same bio-based reinforcement [17,20]. The introduction of the ASP filler mainly affected the thermal stability of the polyester component with the lower T_{peak} temperature (maximum mass loss rate): in detail, T_{peak1} moved from 351 °C to 311 °C and 334 °C for INZEA_25ASF and INZEA_25ASC, respectively, while the temperature for the second peak (T_{peak2}) remained practically unchanged for all the different composites (Table 3). This behavior was related to the higher amount of ASP incorporated and its degradation over this temperature range. According to other authors, cellulose, hemicelluloses and lignin show a broad temperature range starting at about 250 °C and ending at 450 °C in a progressive weight loss process, and this degradation can reduce the thermal stability of biopolyesters [43–45]. Similar findings were reported by Liminana et al. [31] who observed a decrease of 11.2 °C in thermal stability of PBS with the addition of 30 wt % of almond shells.

Table 3. T_{onset} and T_{peak} values of INZEA-based biocomposites with 25 wt % of ASP at the two grinding levels with and without MLO or Joncryl compatibilizers (mean ± SD, n = 3).

Formulation	T_{onset} (°C)	T_{peak1} (°C)	T_{peak2} (°C)
INZEA	324 ± 1	351 ± 2	385 ± 1
INZEA_25ASF	280 ± 2	311 ± 2	391 ± 2
INZEA_25ASF_5MLO	285 ± 2	328 ± 2	388 ± 3
INZEA_25ASF_1J	285 ± 1	329 ± 1	390 ± 2
INZEA_25ASC	275 ± 2	334 ± 1	390 ± 2
INZEA_25ASC_5MLO	275 ± 3	341 ± 3	392 ± 4
INZEA_25ASC_1J	291 ± 3	339 ± 2	390 ± 3

T_{onset}: initial degradation temperature; T_{peak1} and T_{peak2}: first and second maximum degradation temperatures, respectively.

The addition of MLO (Figure 5c,d) exerted a limited positive effect on the overall thermal stability of the biocomposites (Table 3). Specifically, the onset degradation temperature increased to 285 °C for INZEA_25ASF_5MLO and substantially remained unchanged for INZEA_25ASC_5MLO (if compared to uncompatibilized matrices). On the other hand, the temperature of T_{peak1} was significantly improved, in comparison with the unmodified ASP biocomposites, at the two grinding sizes. In particular, T_{peak1} was delayed up to 328 °C for the INZEA_25ASF_5MLO material and 341 °C for the INZEA_25ASC_5MLO material. This increase in thermal stability could be directly related to the chemical interaction achieved by MLO, due to the establishment of covalent bonds between the lignocellulosic fillers and the polyester matrix. In addition, MLO could also provide a physical barrier that obstructs the removal of volatile products produced during decomposition. A similar effect on thermal stability was recently reported for PLA and epoxidized palm oil blends [46].

It has been also reported that the reactive extrusion of aliphatic polyesters, such as PLA, with styrene-epoxy acrylic oligomers can provide an increase in thermal stability due to the branching effect obtained during the extrusion process. As it has been found in our case (Figure 5c,d), the presence of Joncryl provided enhanced thermal resistance for the low T_{peak} polyester phase, moving this value from 311 °C and 334 °C (INZEA_25ASF and INZEA_25ASC), respectively, to 329 °C and 339 °C for INZEA_25ASF_1J and INZEA_25ASC_1J (Table 3). This effect was already observed by Lascano et al in poly(lactic acid)/poly(butylene succinate-co-adipate) blends containing Joncryl [47].

3.3. Disintegration Tests

According to the characterization results previously obtained, the disintegrability of polyester-based biocomposites with 10 and 25 wt % of ASP at two grinding levels (F, C) was studied to evaluate their degradation in natural environments. Formulations including the two studied compatibilizing agents were not included in this study, since the main goal of this characterization was the analysis of ASP size and content on disintegration behavior of the reference matrix.

The visual evaluation of all samples at different degradation times was carried out and results are shown in Figure 6. Some changes in sample surfaces submitted to composting conditions were clearly appreciable, showing all samples considerable modifications in color and morphology after 15 treatment days. These modifications can be associated to the beginning of the polymer matrix degradation, which can be related to the direct contact of biocomposites with the compost, by a gradually microorganism erosion from the surface to the bulk and to the moisture absorption, as it has been reported in the literature [45]. After 69 days of study, samples with 25 wt % of ASP have completely lost their morphology, and small fragments can be observed (in particular for INZEA_25ASC). In addition, differences in color could be observed between samples with different amounts of ASP. Formulations with 10 wt % of ASP showed disintegration behavior similar to that of INZEA neat during all the study.

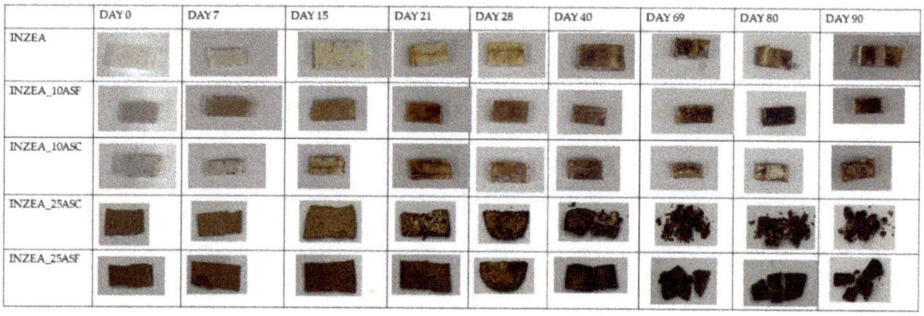

Figure 6. Visual appearance of INZEA-based biocomposites with 10 and 25 wt % of ASP at two grinding levels (F, C) at different testing days at 58 °C.

Figure 7 shows the evolution of disintegrability values (%) as a function of testing time for all biomaterials. According to de Olivera et al. [48], the first stage of the biodegradation mechanism is the release of enzymes that can cause the hydrolysis of the polymer matrix and break of polymer chains, creating functional groups capable of improving hydrophilicity and the microorganism's adhesion on the surface of the polymer matrix. The results obtained at longer times suggested that physical degradation progressed slowly with burial time, indicating that microorganisms required more time to produce suitable enzymes capable to break down polymer chains, resulting in an incomplete loss of the initial morphology and general rupture after 90 days for formulations with 10 wt % of ASP and the INZEA control. Formulations with 25 wt % of ASP significantly increased their disintegrability ratio compared to formulations with 10 wt % of ASP and INZEA control after 28 days of study, probably due to the higher amount of ASP, which enhances the high biodegradability of lignocellulosic residues, and to the poor fiber/matrix adhesion allowing and facilitating microorganisms attack and biodegradation rate by promoting biofouling and the adhesion of microorganisms to the surface [49]. Moreover, this increase in the disintegrability rate of the polymer matrix could be due to the presence of hydroxyl groups in ASP [49], which could play a catalytic role on the hydrolysis of the polymer, inducing an acceleration of polymer weight loss due to the higher filler addition [50]. Similar results were found by Wu [51] and de Oliveira et al. [48] when studying the biodegradation of composites obtained with poly(butylene adipate-co-terephthalate) and different natural fillers. The authors observed that the biodegradation rate of the composites increased with filler content. In this sense, the presence of high amounts of ASP can be related to a greater discontinuity in the polymer matrix which could facilitate water penetration into the biocomposites producing a huge modification of the surface and generating a natural environment conducive to the growth of microorganisms [48]. After 90 days of study, almost 50% of the materials were disintegrated under composting conditions. However, lower weight loss ratios were obtained with lower amounts of ASP. This behavior suggests that the disintegration rate is more influenced and dependent on the polymer matrix being slower at lower ASP contents.

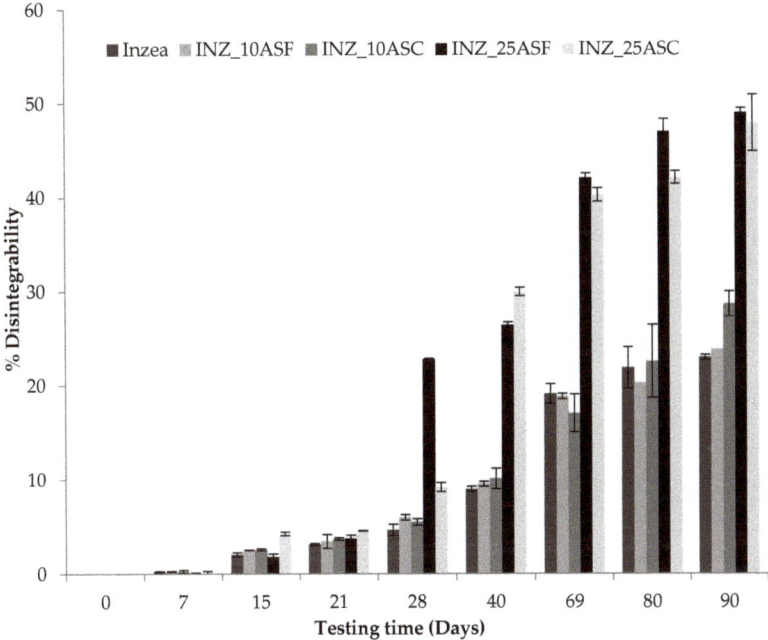

Figure 7. Disintegrability (%) of INZEA-based biocomposites with 10 and 25 wt % of ASP at two grinding levels (F, C) as a function of degradation time under composting conditions at 58 °C (mean ± SD, $n = 3$).

Figure 8 shows the DSC thermograms obtained during the second heating scan for all formulations as a function of composting time (0, 28 and 90 days). Two different peaks were observed at day 0, around 110 and 170 °C, indicating the presence of two main polyesters in the polymer matrix, in agreement with the behavior previously observed by TGA (Section 3.2.2). A similar DSC profile for the first and second peaks was described by Liminana et al. and Quiles-Carrillo et al. for the characterization of PBS-based and PLA-based composites reinforced with similar amounts of almond shells, respectively [12,52]. Both melting peak temperatures remained practically invariable after the addition of ASP at the two studied contents, 10 and 25 wt % (Table 4), showing a slight modification which could be related to the formation of more perfect crystals into the polymer matrix. Similar behavior was observed by Quiles-Carrillo et al. for PLA-based composites with 25 wt % of almond shell [12]. A slight decrease in the melting enthalpy of the first peak was observed with the addition of ASP which was more pronounced at 25 wt % This effect could be related to the nucleating effect exerted by the lignocellulosic filler on semicrystalline polymers acting the cellulose crystals of the almond shells as nucleating points [11].

Regarding the disintegration study, the DSC melting temperature of the second endothermic peak moved from 168.4 ± 3.7 °C for INZEA at day 0 to 142.4 ± 0.8 °C at day 90 (Table 4). Formulations with ASP did not show significant differences at day 0 compared to the values obtained for INZEA control. This decrease in around 26 °C was related to a rapid molecular mass reduction, implying that small and imperfect crystals disappeared with degradation time [53,54]. INZEA and formulations with 10 wt % of ASP did not show significant differences in DSC values at the same disintegration time, maintaining similar values throughout the whole study. In this sense, the addition of 10 wt % of ASP could not be enough to achieve an acceptable weight loss ratio into the disintegration process under composting conditions.

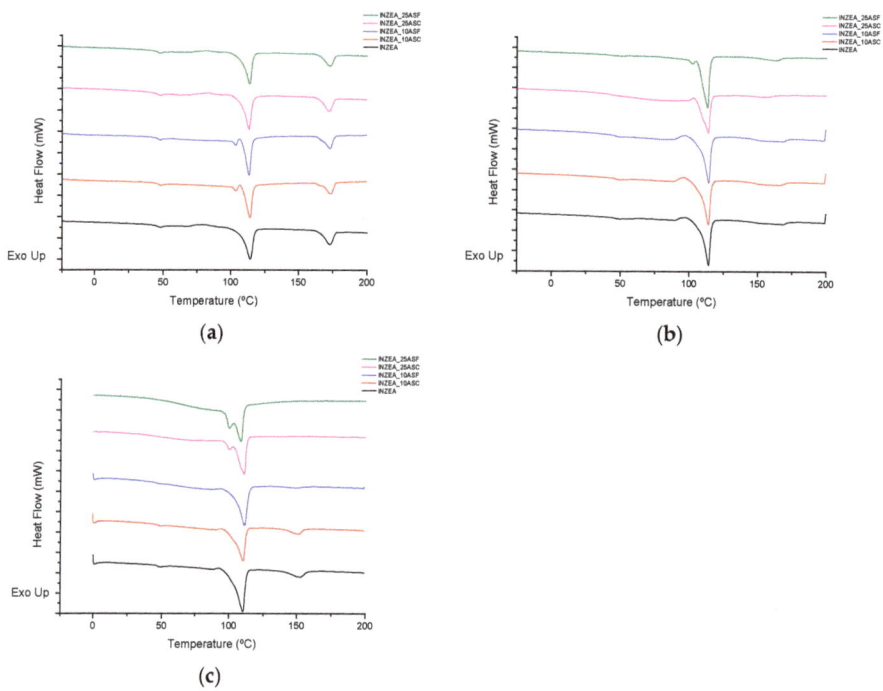

Figure 8. DSC thermograms of INZEA-based biocomposites with 10 and 25 wt % of ASP at two grinding levels (F, C) after different degradation times ((**a**): 0 days, (**b**): 28 days, (**c**): 90 days) at 58 °C during the first heating scan (10 °C min^{-1}).

Under composting conditions, a different behavior was observed for formulations with 25 wt % of ASP versus time, as the second DSC peak initially appearing around 170 °C started to disappear after 28 days of study (Figure 8). After 90 days, the appearance of the thermogram suggested that this polyester-based polymer was totally disintegrated by disappearing the corresponding glass transition temperature (around 50 °C) and melting peak around 170 °C (Table 4). This result was also related to the final appearance and percentage of disintegrability achieved in samples after 90 days. On the other hand, the observed peak around 110 °C remained unchanged after 90 days of study, with slight modifications on its profile, indicating that this polymer was not degraded yet.

This behavior was also confirmed by TGA analysis (Table 4). The obtained results demonstrated that after 28 days of study the maximum degradation temperature of the first polymer peak, T_{peak1}, decreased around 70 °C respect to day 0, disappearing this degradation peak from the TGA curve after 40 days under composting conditions. These results confirm those obtained by DSC where some modification of the thermal profile was observed due to the disintegration of one main polyester component of the polymer matrix.

Table 4. Thermal parameters of INZEA-based biocomposites with 10 and 25 wt % of ASP at two grinding levels (F, C) after different degradation times at 58 °C (mean ± SD, $n = 3$).

Day		TGA			DSC (2nd Heating)				
		T_{peak1} (°C)	T_{peak2} (°C)	T_{m1} (°C) *	ΔH_{m1} (J/g) *	T_{m2} (°C) **	ΔH_{m2} (J/g) **	T_g (°C)	
DAY 0	INZEA	353.7 ± 4.0 ab	390.7 ± 1.5 a	112.3 ± 0.5 a	35.0 ± 0.9 a	168.4 ± 3.7 a	9.3 ± 1.8 a	47.2 ± 0.6 a	
	INZEA_10ASF	355.7 ± 1.2 a	392.7 ± 0.6 a	113.6 ± 0.3 b	30.3 ± 0.5 bc	170.2 ± 0.2 a	12.6 ± 0.2 bc	45.3 ± 0.1 a	
	INZEA_10ASC	339.9 ± 6.0 cd	381.7 ± 11.3 a	113.3 ± 0.1 b	33.6 ± 1.5 ab	170.2 ± 0.1 a	13.5 ± 0.5 c	45.2 ± 0.6 a	
	INZEA_25ASF	331.0 ± 2.0 d	388.7 ± 0.6 a	113.7 ± 0.2 b	25.6 ± 1.7 d	169.8 ± 0.2 a	10.7 ± 0.2 ab	45.8 ± 0.3 a	
	INZEA_25ASC	345.0 ± 1.0 bc	393.3 ± 1.2 a	114.0 ± 0.1 b	28.9 ± 2.0 dc	170.2 ± 0.2 a	11.3 ± 1.0 abc	45.9 ± 0.4 a	
DAY 15	INZEA	290.7 ± 1.5 ab	394.3 ± 2.1 a	113.6 ± 0.3 a	37.5 ± 1.1 a	165.9 ± 0.7 a	10.0 ± 0.3 a	50.0 ± 0.2 a	
	INZEA_10ASF	297.5 ± 5.2 b	393.3 ± 0.7 a	113.7 ± 0.2 a	35.6 ± 0.2 ab	164.3 ± 2.0 a	9.8 ± 1.2 a	45.2 ± 0.2 a	
	INZEA_10ASC	293.3 ± 1.5 ab	394.0 ± 1.0 a	113.7 ± 0.2 a	38.2 ± 1.1 a	166.5 ± 0.2 a	10.3 ± 0.5 a	44.9 ± 0.3 a	
	INZEA_25ASF	306.0 ± 5.6 c	388.7 ± 0.6 b	113.3 ± 0.1 a	29.6 ± 0.3 c	159.9 ± 0.9 ab	5.5 ± 0.7 b	45.4 ± 0.2 a	
	INZEA_25ASC	286.0 ± 1.0 a	391.7 ± 1.5 ab	112.7 ± 0.1 b	31.3 ± 3.6 bc	150.0 ± 0.4 c	4.5 ± 0.5 b	45.4 ± 0.2 a	
DAY 28	INZEA	300.3 ± 0.6 a	393.3 ± 0.6 a	113.6 ± 0.2 a	41.4 ± 0.8 a	166.1 ± 0.6 a	5.2 ± 3.3 a	45.2 ± 0.2 a	
	INZEA_10ASF	287.3 ± 1.2 b	391.8 ± 1.3 ab	113.7 ± 0.2 a	37.6 ± 0.4 a	166.2 ± 0.4 a	4.5 ± 0.6 a	45.5 ± 0.3 ab	
	INZEA_10ASC	299.7 ± 1.2 a	393.3 ± 0.6 a	113.8 ± 0.3 a	39.9 ± 0.6 a	166.3 ± 0.8 a	6.2 ± 2.4 a	45.5 ± 0.3 ab	
	INZEA_25ASF	271.3 ± 5.8 c	394.0 ± 1.0 a	110.9 ± 0.7 b	50.2 ± 2.1 b	n.d.	n.d.	44.9 ± 0.2 a	
	INZEA_25ASC	275.3 ± 2.1 c	389.7 ± 0.6 b	112.0 ± 0.2 c	40.0 ± 4.0 a	n.d.	n.d.	45.9 ± 0.2 b	
DAY 40	INZEA	298.0 ± 3.6 a	394.3 ± 2.1 a	113.6 ± 0.2 a	44.1 ± 0.1 a	166.6 ± 0.2 a	13.4 ± 0.6 a	44.3 ± 0.2 a	
	INZEA_10ASF	287.3 ± 3.2 b	393.3 ± 0.6 a	113.2 ± 0.1 ab	41.1 ± 0.3 a	167.0 ± 2.8 a	4.5 ± 2.9 b	46.6 ± 0.1 b	
	INZEA_10ASC	287.7 ± 3.2 b	392.7 ± 0.6 a	113.3 ± 0.1 ab	40.9 ± 2.9 a	166.9 ± 2.6 a	5.5 ± 1.0 b	45.0 ± 0.1 c	
	INZEA_25ASF	n.d.	382.7 ± 1.2 b	112.9 ± 0.3 b	38.7 ± 1.1 a	n.d.	n.d.	n.d.	
	INZEA_25ASC	n.d.	389.0 ± 1.0 c	112.3 ± 0.3 c	44.6 ± 5.3 a	n.d.	n.d.	n.d.	
DAY 69	INZEA	292.3 ± 6.7 a	392.7 ± 2.1 a	111.5 ± 0.4 ab	47.4 ± 3.0 a	144.8 ± 2.1 a	4.1 ± 1.3 a	44.8 ± 0.3 a	
	INZEA_10ASF	286.3 ± 2.5 a	392.3 ± 1.5 a	110.9 ± 0.3 a	42.0 ± 0.8 a	151.7 ± 4.7 a	3.9 ± 0.6 a	44.5 ± 0.3 a	
	INZEA_10ASC	283.3 ± 3.8 a	391.0 ± 1.7 a	111.7 ± 0.7 a	50.0 ± 7.2 a	144.9 ± 2.6 a	4.0 ± 0.6 a	45.0 ± 0.2 a	
	INZEA_25ASF	n.d.	382.3 ± 0.6 b	111.1 ± 0.1 ab	39.5 ± 0.4 a	n.d.	n.d.	n.d.	
	INZEA_25ASC	n.d.	391.7 ± 1.5 a	112.6 ± 0.1 b	45.6 ± 7.4 a	n.d.	n.d.	n.d.	
DAY 90	INZEA	284.7 ± 4.7 a	392.3 ± 1.2 a	108.9 ± 0.9 a	46.0 ± 0.8 a	142.4 ± 0.8 a	5.4 ± 0.3 a	44.8 ± 0.1 a	
	INZEA_10ASF	281.7 ± 1.2 a	388.7 ± 0.6 a	110.6 ± 0.5 ab	42.8 ± 4.4 ab	n.d.	n.d.	n.d.	
	INZEA_10ASC	276.7 ± 3.2 a	392.7 ± 0.6 a	109.7 ± 1.5 ab	35. ± 3.4 bc	n.d.	n.d.	n.d.	
	INZEA_25ASF	n.d.	374.3 ± 0.6 b	110.6 ± 0.3 ab	32.0 ± 2.5 c	n.d.	n.d.	n.d.	
	INZEA_25ASC	n.d.	384.3 ± 6.4 a	111.8 ± 0.3 b	40.9 ± 5.6 abc	n.d.	n.d.	n.d.	

DSC: * and ** correspond to the first and second melting peaks appearing in Figure 8, respectively. Different superscripts (a, b, c, ab, abc) within the same day of study indicate statistically significant different values ($p < 0.05$). T_{peak1} and T_{peak2}: first and second maximum degradation temperatures, respectively; T_{m1} and T_{m2}: first and second melting temperatures, respectively; ΔH_{m1} and ΔH_{m2}: First and second enthalpies of fusion, respectively; T_g: glass transition temperature.

4. Conclusions

In this work, biocomposite materials were obtained based on a polyester matrix (INZEAF2) and almond shell as a reinforcing agent at 10 and 25 wt % and two different milling sizes (125–250 µm and 500–1000 µm). MLO and Joncryl ADR 4400 were studied as compatibilizers. The reinforced effect of the addition of ASP was shown by increasing the elastic modulus of the biocomposites with both fine and coarse ASP at 25 wt % The addition of MLO and Joncryl produced some compatibilizing effect at 10 wt % while some gaps at the interface, visible by FESEM at 25 wt %, did not substantially affect the overall flexural properties. A double degradation pattern was obtained by TGA for the INZEA neat matrix indicating the degradation of the material in two steps and the presence of two different polyesters in the matrix. Some decrease in thermal stability of biocomposites was shown which was more pronounced at high ASP contents and it was related to the relatively low thermal stability and disintegration of the lignocellulosic filler over the studied temperature range. Some limited positive enhancement in thermal stability was obtained by adding the studied modifiers. The addition of 10 wt % of ASP was not enough to achieve an acceptable weight loss ratio of disintegration under composting conditions, whereas around 50% of disintegration (nearly double of the polymer control) was obtained by adding 25 wt % of ASP after 90 days. At these conditions, one of the polyester-based polymers was totally disintegrated by disappearing the corresponding glass transition temperature (around 50 °C), melting peak (around 170 °C) and degradation temperature (around 350 °C). On the other hand, the second polymer with T_m around 110 °C was not degraded yet.

The developed composites represent an interesting approach to reduce the overall cost of bio-based polyesters increasing also the added-value potential of almond agricultural wastes, obtaining environmentally friendly materials with specific reinforced and aesthetic functionalities and contributing to the circular economy approach. Further work will be needed in order to evaluate a possible improvement in filler–matrix adhesion and disintegration rate of the biocomposites by using different compatibilizers and adding higher amounts of lignocellulosic filler without compromising the final mechanical and thermal properties.

Author Contributions: Conceptualization, A.J., M.C.G., L.T. and D.P.; methodology M.R., F.D., and F.L.; validation, A.J., M.C.G., L.T. and D.P.; formal analysis, M.R., F.D., and F.L.; investigation, M.R., F.D., and F.L.; resources, A.J., M.C.G., L.T. and D.P.; data curation, M.R., F.D., F.L., A.J., M.C.G. and D.P.; writing—original draft preparation, M.R., F.D., and F.L.; writing—review and editing, A.J., M.C.G. and D.P.; supervision, A.J., M.C.G., L.T. and D.P.; funding acquisition, A.J., M.C.G., L.T. and D.P. All authors have read and agreed to the published version of the manuscript.

Funding: The authors express their gratitude to the Bio Based Industries Consortium and European Commission for the financial support to the project BARBARA: Biopolymers with advanced functionalities for building and automotive parts processed through additive manufacturing. This project has received funding from the Bio Based Industries Joint Undertaking under the European Union's Horizon 2020 research and innovation programme under grant agreement No 745578.

Conflicts of Interest: The authors declare no conflicts of interest.

References

1. Prgomet, I.; Gonçalves, B.; Domínguez-Perles, R.; Pascual-Seva, N.; Barros, A.I. Valorization challenges to almond residues: Phytochemical composition and functional application. *Molecules* **2017**, *22*, 1774. [CrossRef] [PubMed]
2. Esfahlan, A.J.; Jamei, R.; Esfahlan, R.J. The importance of almond (*Prunus amygdalus* L.) and its by-products. *Food Chem.* **2010**, *120*, 349–360. [CrossRef]
3. García, R.; Pizarro, C.; Lavín, A.G.; Bueno, J.L. Characterization of Spanish biomass wastes for energy use. *Bioresour. Technol.* **2012**, *103*, 249–258. [CrossRef] [PubMed]
4. Fernandez-Bayo, J.D.; Shea, E.A.; Parr, A.E.; Achmon, Y.; Stapleton, J.J.; VanderGheynst, J.S.; Hodson, A.K.; Simmons, C.W. Almond processing residues as a source of organic acid biopesticides during biosolarization. *Waste Manag.* **2020**, *101*, 74–82. [CrossRef] [PubMed]

5. Khan, A.M.; Ahmad, C.S.; Farooq, U.; Mahmood, K.; Sarfraz, M.; Balkhair, K.S.; Ashraf, M.A. Removal of metallic elements from industrial waste water through biomass and clay. *Front. Life Sci.* **2015**, *8*, 223–230. [CrossRef]
6. Ben Arfi, R.; Karoui, S.; Mougin, K.; Ghorbal, A. Adsorptive removal of cationic and anionic dyes from aqueous solution by utilizing almond shell as bioadsorbent. *Euro-Mediterr. J. Environ. Integr.* **2017**, *2*, 20. [CrossRef]
7. Oliveira, I.; Meyer, A.; Silva, R.; Afonso, S.; Gonçalves, B. Effect of almond shell addition to substrates in Phaseolus vulgaris L. (cv. Saxa) growth, and physiological and biochemical characteristics. *Int. J. Recycl. Org. Waste Agric.* **2019**, *8*, 179–186. [CrossRef]
8. Nazem, M.A.; Zare, M.H.; Shirazian, S. Preparation and optimization of activated nano-carbon production using physical activation by water steam from agricultural wastes. *RSC Adv.* **2020**, *10*, 1463–1475. [CrossRef]
9. Singh, R.D.; Nadar, C.G.; Muir, J.; Arora, A. Green and clean process to obtain low degree of polymerisation xylooligosaccharides from almond shell. *J. Clean. Prod.* **2019**, *241*, 118237. [CrossRef]
10. Prgomet, I.; Gonçalves, B.; Domínguez-Perles, R.; Pascual-Seva, N.; Barros, A.I.R.N.A. A Box-Behnken Design for Optimal Extraction of Phenolics from Almond By-products. *Food Anal. Methods* **2019**, *12*, 2009–2024. [CrossRef]
11. Liminana, P.; Garcia-Sanoguera, D.; Quiles-Carrillo, L.; Balart, R.; Montanes, N. Optimization of maleinized linseed oil loading as a biobased compatibilizer in poly(butylene succinate) composites with almond shell flour. *Materials* **2019**, *12*, 685. [CrossRef] [PubMed]
12. Quiles-Carrillo, L.; Montanes, N.; Garcia-Garcia, D.; Carbonell-Verdu, A.; Balart, R.; Torres-Giner, S. Effect of different compatibilizers on injection-molded green composite pieces based on polylactide filled with almond shell flour. *Compos. Part B Eng.* **2018**, *147*, 76–85. [CrossRef]
13. McCaffrey, Z.; Torres, L.; Flynn, S.; Cao, T.; Chiou, B.S.; Klamczynski, A.; Glenn, G.; Orts, W. Recycled polypropylene-polyethylene torrefied almond shell biocomposites. *Ind. Crop. Prod.* **2018**, *125*, 425–432. [CrossRef]
14. Dungani, R.; Karina, M.; Subyakto, A.S.; Hermawan, D.; Hadiyane, A. Agricultural waste fibers towards sustainability and advanced utilization: A review. *Asian J. Plant Sci.* **2016**, *15*, 42–55. [CrossRef]
15. Kellersztein, I.; Shani, U.; Zilber, I.; Dotan, A. Sustainable composites from agricultural waste: The use of steam explosion and surface modification to potentialize the use of wheat straw fibers for wood plastic composite industry. *Polym. Compos.* **2019**, *40*, E53–E61. [CrossRef]
16. El Mechtali, F.Z.; Essabir, H.; Nekhlaoui, S.; Bensalah, M.O.; Jawaid, M.; Bouhfid, R.; Qaiss, A. Mechanical and Thermal Properties of Polypropylene Reinforced with Almond Shells Particles: Impact of Chemical Treatments. *J. Bionic Eng.* **2015**, *12*, 483–494. [CrossRef]
17. Essabir, H.; Nekhlaoui, S.; Malha, M.; Bensalah, M.O.; Arrakhiz, F.Z.; Qaiss, A.; Bouhfid, R. Bio-composites based on polypropylene reinforced with Almond Shells particles: Mechanical and thermal properties. *Mater. Des.* **2013**, *51*, 225–230. [CrossRef]
18. Tasdemir, M. Polypropylene/olive pit & almond shell polymer composites: Wear and friction. *IOP Conf. Ser. Mater. Sci. Eng.* **2017**, *204*, 012015.
19. Lashgari, A.; Eshghi, A.; Farsi, M. A Study on Some Properties of Polypropylene Based Nanocomposites Made Using Almond Shell Flour and Organoclay. *Asian J. Chem.* **2013**, *25*, 1043–1049. [CrossRef]
20. Hosseinihashemi, S.K.; Eshghi, A.; Ayrilmis, N.; Khademieslam, H. Thermal Analysis and Morphological Characterization of Thermoplastic Composites Filled with Almond Shell Flour/Montmorillonite. *BioResources* **2016**, *11*, 6768–6779. [CrossRef]
21. Altay, L.; Guven, A.; Atagur, M.; Uysalman, T.; Tantug, G.S.; Ozkaya, M.; Sever, K.; Sarikanat, M.; Seki, Y. Linear Low Density Polyethylene Filled with Almond Shells Particles: Mechanical and Thermal Properties. *Acta Physica Polonica A.* **2019**, *135*. [CrossRef]
22. Sabbatini, A.; Lanari, S.; Santulli, C.; Pettinari, C. Use of almond shells and rice husk as fillers of poly (methyl methacrylate)(PMMA) composites. *Materials* **2017**, *10*, 872. [CrossRef] [PubMed]
23. Chaudhary, A.K.; Gope, P.C.; Singh, V.K. Water absorption and thickness swelling behavior of almond (*Prunus amygdalus* L.) shell particles and coconut (*Cocos nucifera*) fiber hybrid epoxy-based biocomposite. *Sci. Eng. Compos. Mater.* **2015**, *22*, 375–382. [CrossRef]

24. Singh, V.K.; Bansal, G.; Agarwal, M.; Negi, P. Experimental determination of mechanical and physical properties of almond shell particles filled biocomposite in modified epoxy resin. *J. Mater. Sci. Eng.* **2016**, *5*, 246.
25. García, A.M.; García, A.I.; Cabezas, M.Á.L.; Reche, A.S. Study of the Influence of the Almond Variety in the Properties of Injected Parts with Biodegradable Almond Shell Based Masterbatches. *Waste Biomass Valorization* **2015**, *6*, 363–370. [CrossRef]
26. Singh, R.; Kumar, R.; Pawanpreet; Singh, M.; Singh, J. On mechanical, thermal and morphological investigations of almond skin powder-reinforced polylactic acid feedstock filament. *J. Thermoplast. Compos. Mater.* 2019. [CrossRef]
27. Sutivisedsak, N.; Cheng, H.N.; Burks, C.S.; Johnson, J.A.; Siegel, J.P.; Civerolo, E.L.; Biswas, A. Use of Nutshells as Fillers in Polymer Composites. *J. Polym. Environ.* **2012**, *20*, 305–314. [CrossRef]
28. Pashaei, S.; Hosseinzadeh, S. Investigation on Physicomechanical, Thermal, and Morphological of Dipodal Silane-Modified Walnut Shell Powder-Filled Polyurethane Green Composites and Their Application for Removal of Heavy Metal Ions from Water. *Polym. Plast. Technol. Eng.* **2018**, *57*, 1197–1208. [CrossRef]
29. Balart, J.F.; Fombuena, V.; Fenollar, O.; Boronat, T.; Sánchez-Nacher, L. Processing and characterization of high environmental efficiency composites based on PLA and hazelnut shell flour (HSF) with biobased plasticizers derived from epoxidized linseed oil (ELO). *Compos. Part B Eng.* **2016**, *86*, 168–177. [CrossRef]
30. Quiles-Carrillo, L.; Montanes, N.; Sammon, C.; Balart, R.; Torres-Giner, S. Compatibilization of highly sustainable polylactide/almond shell flour composites by reactive extrusion with maleinized linseed oil. *Ind. Crop. Prod.* **2018**, *111*, 878–888. [CrossRef]
31. Liminana, P.; Quiles-Carrillo, L.; Boronat, T.; Balart, R.; Montanes, N. The Effect of Varying Almond Shell Flour (ASF) Loading in Composites with Poly(Butylene Succinate) (PBS) Matrix Compatibilized with Maleinized Linseed Oil (MLO). *Materials* **2018**, *11*, 2179. [CrossRef]
32. Frenz, V.; Scherzer, D.; Villalobos, M.; Awojulu, A.; Edison, M.; Van Der Meer, R. Multifunctional polymers as chain extenders and compatibilizers for polycondensates and biopolymers. *ANTEC* **2008**, *3*, 1682–1686.
33. ISO-20200:2015. *Determination of the Degree of Disintegration of Plastic Materials under Simulated Composting Conditions in a Laboratory-Scale Test*; ISO: Geneva, Switzerland, 2015.
34. Bordbar, M. Biosynthesis of Ag/almond shell nanocomposite as a cost-effective and efficient catalyst for degradation of 4-nitrophenol and organic dyes. *RSC Adv.* **2017**, *7*, 180–189. [CrossRef]
35. Montava-Jordà, S.; Quiles-Carrillo, L.; Richart, N.; Torres-Giner, S.; Montanes, N. Enhanced interfacial adhesion of polylactide/poly(ε-caprolactone)/walnut shell flour composites by reactive extrusion with maleinized linseed oil. *Polymers* **2019**, *11*, 758. [CrossRef] [PubMed]
36. Li, X.; Liu, Y.; Hao, J.; Wang, W. Study of almond shell characteristics. *Materials* **2018**, *11*, 1782. [CrossRef]
37. Nabinejad, O.; Sujan, D.; Rahman, M.E.; Davies, I.J. Effect of oil palm shell powder on the mechanical performance and thermal stability of polyester composites. *Mater. Des. (1980–2015)* **2015**, *65*, 823–830. [CrossRef]
38. Zaini, M.J.; Fuad, M.Y.A.; Ismail, Z.; Mansor, M.S.; Mustafah, J. The effect of filler content and size on the mechanical properties of polypropylene/oil palm wood flour composites. *Polym. Int.* **1996**, *40*, 51–55. [CrossRef]
39. Dominici, F.; Samper, M.D.; Carbonell-Verdu, A.; Luzi, F.; López-Martínez, J.; Torre, L.; Puglia, D. Improved Toughness in Lignin/Natural Fiber Composites Plasticized with Epoxidized and Maleinized Linseed Oils. *Materials* **2020**, *13*, 600. [CrossRef]
40. Ferri, J.M.; Garcia-Garcia, D.; Montanes, N.; Fenollar, O.; Balart, R. The effect of maleinized linseed oil as biobased plasticizer in poly(lactic acid)-based formulations. *Polym. Int.* **2017**, *66*, 882–891. [CrossRef]
41. Carbonell-Verdu, A.; Ferri, J.M.; Dominici, F.; Boronat, T.; Sanchez-Nacher, L.; Balart, R.; Torre, L. Manufacturing and compatibilization of PLA/PBAT binary blends by cottonseed oil-based derivatives. *Express Polym. Lett.* **2018**, *12*, 808–823. [CrossRef]
42. Zhang, Y.; Yuan, X.; Liu, Q.; Hrymak, A. The Effect of Polymeric Chain Extenders on Physical Properties of Thermoplastic Starch and Polylactic Acid Blends. *J. Polym. Environ.* **2012**, *20*, 315–325. [CrossRef]
43. Nabinejad, O.; Sujan, D.; Rahman, M.E.; Davies, I.J. Determination of filler content for natural filler polymer composite by thermogravimetric analysis. *J. Therm. Anal. Calorim.* **2015**, *122*, 227–233. [CrossRef]

44. Seggiani, M.; Cinelli, P.; Mallegni, N.; Balestri, E.; Puccini, M.; Vitolo, S.; Lardicci, C.; Lazzeri, A. New bio-composites based on polyhydroxyalkanoates and posidonia oceanica fibres for applications in a marine environment. *Materials* **2017**, *10*, 326. [CrossRef] [PubMed]
45. Sánchez-Safont, E.L.; Aldureid, A.; Lagarón, J.M.; Gámez-Pérez, J.; Cabedo, L. Biocomposites of different lignocellulosic wastes for sustainable food packaging applications. *Compos. Part B Eng.* **2018**, *145*, 215–225. [CrossRef]
46. Chieng, B.; Ibrahim, N.; Then, Y.; Loo, Y. Epoxidized vegetable oils plasticized poly(lactic acid) biocomposites: Mechanical, thermal and morphology properties. *Molecules* **2014**, *19*, 16024–16038. [CrossRef] [PubMed]
47. Lascano, D.; Quiles-Carrillo, L.; Balart, R.; Boronat, T.; Montanes, N. Toughened poly(lactic acid)—PLA formulations by binary blends with poly(butylene succinate-co-adipate)—PBSA and their shape memory behaviour. *Materials* **2019**, *12*, 622. [CrossRef] [PubMed]
48. De Oliveira, T.A.; de Oliveira Mota, I.; Mousinho, F.E.P.; Barbosa, R.; de Carvalho, L.H.; Alves, T.S. Biodegradation of mulch films from poly(butylene adipate co-terephthalate), carnauba wax, and sugarcane residue. *J. Appl. Polym. Sci.* **2019**, *136*, 48240. [CrossRef]
49. Valdés García, A.; Ramos Santonja, M.; Sanahuja, A.B.; Selva, M.D.C.G. Characterization and degradation characteristics of poly(ε-caprolactone)-based composites reinforced with almond skin residues. *Polym. Degrad. Stab.* **2014**, *108*, 269–279. [CrossRef]
50. Fortunati, E.; Puglia, D.; Monti, M.; Santulli, C.; Maniruzzaman, M.; Foresti, M.L.; Vazquez, A.; Kenny, J.M. Okra (*Abelmoschus esculentus*) Fibre Based PLA Composites: Mechanical Behaviour and Biodegradation. *J. Polym. Environ.* **2013**, *21*, 726–737. [CrossRef]
51. Wu, C.-S. Characterization and biodegradability of polyester bioplastic-based green renewable composites from agricultural residues. *Polym. Degrad. Stab.* **2012**, *97*, 64–71. [CrossRef]
52. Liminana, P.; Garcia-Sanoguera, D.; Quiles-Carrillo, L.; Balart, R.; Montanes, N. Development and characterization of environmentally friendly composites from poly(butylene succinate) (PBS) and almond shell flour with different compatibilizers. *Compos. Part B Eng.* **2018**, *144*, 153–162. [CrossRef]
53. Luo, Y.; Lin, Z.; Guo, G. Biodegradation Assessment of Poly(Lactic Acid) Filled with Functionalized Titania Nanoparticles (PLA/TiO$_2$) under Compost Conditions. *Nanoscale Res. Lett.* **2019**, *14*, 56. [CrossRef] [PubMed]
54. Ramos, M.; Fortunati, E.; Peltzer, M.; Dominici, F.; Jiménez, A.; del Carmen Garrigós, M.; Kenny, J.M. Influence of thymol and silver nanoparticles on the degradation of poly(lactic acid) based nanocomposites: Thermal and morphological properties. *Polym. Degrad. Stab.* **2014**, *108*, 158–165. [CrossRef]

© 2020 by the authors. Licensee MDPI, Basel, Switzerland. This article is an open access article distributed under the terms and conditions of the Creative Commons Attribution (CC BY) license (http://creativecommons.org/licenses/by/4.0/).

Article

Mechanical Recycling of Partially Bio-Based and Recycled Polyethylene Terephthalate Blends by Reactive Extrusion with Poly(styrene-*co*-glycidyl methacrylate)

Sergi Montava-Jorda [1], Diego Lascano [2,3], Luis Quiles-Carrillo [2], Nestor Montanes [2], Teodomiro Boronat [2], Antonio Vicente Martinez-Sanz [4], Santiago Ferrandiz-Bou [2] and Sergio Torres-Giner [5,*]

1. Department of Mechanical and Materials Engineering (DIMM), Universitat Politècnica de València (UPV), Plaza Ferrándiz y Carbonell 1, 03801 Alcoy, Spain; sermonjo@mcm.upv.es
2. Technological Institute of Materials (ITM), Universitat Politècnica de València (UPV), Plaza Ferrándiz y Carbonell 1, 03801 Alcoy, Spain; dielas@epsa.upv.es (D.L.); luiquic1@epsa.upv.es (L.Q.-C.); nesmonmu@upvnet.upv.es (N.M.); tboronat@dimm.upv.es (T.B.); sferrand@mcm.upv.es (S.F.-B.)
3. Escuela Politécnica Nacional, Ladrón de Guevara E11-253, Quito 17-01-2759, Ecuador
4. Institute of Design and Manufacturing (IDF), Universitat Politècnica de València (UPV), Plaza Ferrándiz y Carbonell 1, 03801 Alcoy, Spain; anmarsan@mcm.upv.es
5. Novel Materials and Nanotechnology Group, Institute of Agrochemistry and Food Technology (IATA), Spanish National Research Council (CSIC), Calle Catedrático Agustín Escardino Benlloch 7, 46980 Paterna, Spain
* Correspondence: storresginer@iata.csic.es; Tel.: +34-963-900-022

Received: 5 December 2019; Accepted: 3 January 2020; Published: 9 January 2020

Abstract: In the present study, partially bio-based polyethylene terephthalate (bio-PET) was melt-mixed at 15–45 wt% with recycled polyethylene terephthalate (r-PET) obtained from remnants of the injection blowing process of contaminant-free food-use bottles. The resultant compounded materials were thereafter shaped into pieces by injection molding for characterization. Poly(styrene-*co*-glycidyl methacrylate) (PS-*co*-GMA) was added at 1–5 parts per hundred resin (phr) of polyester blend during the extrusion process to counteract the ductility and toughness reduction that occurred in the bio-PET pieces after the incorporation of r-PET. This random copolymer effectively acted as a chain extender in the polyester blend, resulting in injection-molded pieces with slightly higher mechanical resistance properties and nearly the same ductility and toughness than those of neat bio-PET. In particular, for the polyester blend containing 45 wt% of r-PET, elongation at break (ε_b) increased from 10.8% to 378.8% after the addition of 5 phr of PS-*co*-GMA, while impact strength also improved from 1.84 kJ·m^{-2} to 2.52 kJ·m^{-2}. The mechanical enhancement attained was related to the formation of branched and larger macromolecules by a mechanism of chain extension based on the reaction of the multiple glycidyl methacrylate (GMA) groups present in PS-*co*-GMA with the hydroxyl (–OH) and carboxyl (–COOH) terminal groups of both bio-PET and r-PET. Furthermore, all the polyester blend pieces showed thermal and dimensional stabilities similar to those of neat bio-PET, remaining stable up to more than 400 °C. Therefore, the use low contents of the tested multi-functional copolymer can successfully restore the properties of bio-based but non-biodegradable polyesters during melt reprocessing with their recycled petrochemical counterparts and an effective mechanical recycling is achieved.

Keywords: bio-PET; r-PET; chain extenders; reactive extrusion; secondary recycling; food packaging

1. Introduction

The transition of the plastic industry from its traditional Linear Economy to a Circular Economy, a more valuable and sustainable model, is being spearheaded by the European Union (EU), where legislative measures are being introduced to eliminate excessive waste [1]. To this end, it is first necessary to promote sustainable polymer technologies that decouple plastics from fossil feedstocks [2]. Furthermore, it is also important to increase the quality and uptake of plastic recycling [3]. In this context, the food packaging sector it is among the most heavily scrutinized, given the large production and the short life cycle of their products [4]. Polyethylene terephthalate (PET) is a thermoplastic polyester that is widely used in the manufacture of bottles for water or beverages and also food trays due to its good mechanical properties, chemical resistance, clarity, thermal stability, barrier properties, and low production cost [5]. PET is fully recyclable and it is, indeed, one of the most recycled plastics in the world. Since 2012, PET monomaterial packaging has showed recycling rates of approximately 52% in the EU and 31% in the United States (US) [6]. Moreover, the recycling of PET articles is expected to increase, as exemplified by companies like Coca-Cola, which have already announced a full switch to recycled plastics for their beverage bottles in some EU countries, starting with at least 50% rates by the end of 2023 [7].

In recent years, the replacement of conventional or petrochemical polymer materials with those obtained from natural and renewable sources is of great interest for research due to the depletion of fossil resources and the cost of extracting them. The main interest and advantage of using bio-based polymers reflects the concept of the so-called "biorefinery system design" due to its ability to improve the environmental impact of a product by reducing greenhouse gas emissions, economizing fossil resources, exploring the possibility of using a local resource, and the use of by-products or even wastes [8]. Bio-based polymers, which can be either biodegradable or non-biodegradable, are certified according to international standards such as EN 16640:2015, ISO 16620-4:2016, ASTM 6866-18, and EN 16785-1:2015. In particular, EN 16640:2015, ISO 16620-4:2016, and ASTM 6866-18 measure the bio-based carbon content in a material through 14C measurements, while EN 16785-1:2015 measures the bio-based content of a material using radiocarbon and elemental analyses [9]. Indeed, the pattern of production of biopolymers is shifting from biodegradable to bio-based. Bio-based but non-biodegradable polymers represented 57.1% of the total biopolymer production in 2018, whereas biodegradable polymers accounted for 42.9% [10]. This is in contradiction to the public perception that most biopolymers are biodegradable. Among them, partially bio-based polyethylene terephthalate (bio-PET) is currently the most produced bioplastic, reaching ~27% of the total production in 2018, that is, 0.54 million tons per year. [11]. This is based on the fact that this "green polyester" offers an almost identical chemical structure and properties to its petrochemical counterpart, that is, PET.

Several studies on PET recycling methods have indicated that mechanical recycling appears to be the most desirable method for the management of PET waste, as compared to chemical recycling and incineration [12–14]. The chemical or tertiary recycling method involves depolymerization of the PET polymer by chemical agents (chemolysis) or temperature (pyrolysis) to obtain its constituent monomers, that is, monoethylene glycol (MEG) and terephthalic acid (TA), or their derivatives, which can be used for new polymerization processes [15,16]. In addition, the char from pyrolysis of washed PET wastes can successfully replace up to 50 wt% the epoxy resin used in the production of thermosets [17]. In mechanical or secondary recycling, PET waste is subjected to mechanical processes including shredding, grinding, melting, and, when necessary, as in the case of contaminated articles, is combined with washing and/or drying [18]. In this process, the polymers stay intact, which permits multiple reuses in the same or similar products. However, mechanical recycling is currently the preferred option for monomaterial packaging since it has the advantages of simplicity and low cost, requires little investment, uses established equipment, and has little adverse environmental impact. Despite these advantages, mechanical recycling of PET is difficult due to complexity and waste contamination [12,19,20]. Another issue observed during the mechanical recycling of PET by "fusion reprocessing" is that the polymer is habitually subjected to chemical, mechanical, thermal, and oxidative degradation, which decreases the molar mass and, finally, causes the deterioration of the performance and transparency of PET articles [21,22].

Different studies have conventionally supported the notion that the use of recycled polymer streams in a mixture with virgin polymer of the same nature is a good solution for improving the properties of recycled polymer materials and achieving mechanical recycling. Elamri et al. [23] investigated fibers from blends of recycled polyethylene terephthalate (r-PET) and virgin PET. An improvement in the melt processing of r-PET and its fibers, with similar mechanical characteristics to those obtained from virgin PET, was reported. In another study, Scarfato and La Mantia [24] studied recycled and virgin mixtures of polyamide 6 (PA6), showing mechanical and rheological properties similar to those of the virgin polyamide. Moreover, Elamri et al. [25] also investigated blends of recycled and virgin high-density polyethylene (HDPE), reporting a predictable linear behavior in the mechanical properties as a function of the recycled content. A novel, plausible solution to increase the properties of r-PET during melt reprocessing is the use of additives such as chain extenders [26–29]. This option represents a more economical and attractive strategy than chemical recycling processes for monomaterial packaging since these additives can be used during reprocessing cycles by extrusion or injection molding [26,27]. Chain extenders are additives containing at least two functional groups that are capable of reacting with the end groups of different macromolecular fragments, thereby creating new covalent bonds. During melt reprocessing, these additives are able to "reconnect" the previously broken polymer chains, leading to a polymer with higher molecular weight (M_W) and restored properties [27,28].

The use of chain extenders during PET reprocessing has been widely investigated. Some studies have reported the use of pyromellitic dianhydride (PMDA) [30,31], organic phosphites [32,33], bis-oxazolines [34–36], bis-anhydrides [30,37], diisocyanates [29,34,38], diepoxides [34], epoxy-based oligomers [34,39], or oligomeric polyisocyanates [40]. However, new commercial chain extenders based on random copolymers of poly(styrene-co-glycidyl methacrylate) (PS-co-GMA) have recently arisen. In particular, the glycidyl methacrylate (GMA) multi-functionality has excellent affinity with condensation polymers, not only PET, but also polybutylene terephthalate (PBT) or polylactide (PLA). It performs by connecting the polymer chains through a chemical reaction of the GMA groups with the hydroxyl (–OH), carboxyl (–COOH), and amine (–NH_2) terminal groups of the polycondensation polymers. This process can result in a branched structure that also offers higher melt strength for optimal processing conditions. Although some other block copolymers based on GMA groups have been previously used as a chain extenders in other types of polymer blends, such as poly(trimethylene terephthalate) (PTT)/polystyrene (PS) [41] or PET/polypropylene (PP) [42], their use in PET systems remains almost unexplored. Only Benvenuta et al. [43] recently synthesized reactive tri-block copolymers of styrene glycidyl methacrylate (SGMA) and butyl acrylate (BA) (SGMA-co-BA-co-SGMA) by reversible addition-fragmentation chain transfer (RAFT) and used them as chain extenders of r-PET. These additives turned out to be very effective at increasing the molar mass and intrinsic viscosity, diminishing the melt flow, and improving the melt elasticity and processability of r-PET.

Despite the large amount of work that has been undertaken on the recycling of PET and r-PET blends, those based on bio-PET are barely beginning to be studied. Furthermore, while the price of virgin PET remains stable [44], the use of r-PET is significantly increasing, and thus, novel and more sustainable technologies for PET recycling could encourage more competitive prices in the industry. In the present work, the properties of bio-PET and pre-consumer (uncontaminated) r-PET blends were analyzed to ascertain the potential for mechanical recycling of the next generation of PET materials. To this end, different amounts of r-PET were melt-mixed with bio-PET in a twin-screw industrial extruder and then shaped into pieces by injection molding. During melt compounding, PS-co-GMA was added and the effect of the random copolymer on bio-PET/r-PET blends was assessed through a detailed characterization, including measurements of the mechanical, thermal, and thermomechanical properties as well as the rheological behavior.

2. Materials and Methods

2.1. Materials

Bio-PET, commercial grade BioPET 001, was obtained from NaturePlast (Ifs, France). According to the manufacturer, this grade is produced from up to 30 wt% renewable materials. It has a melting temperature (T_m) between 240–260 °C, a true density of 1.3–1.4 g·cm^{-3}, and an intrinsic viscosity between 75–79 mL·g^{-1} while it is supplied with a water content of less than 0.4 wt%. Further details can be found elsewhere [45]. r-PET was obtained from remnants of the injection blowing process of contaminant-free food-use bottles supplied by the local company Plásticos Guadalaviar S.A. (Beniparell, Spain). The pre-consumer PET waste was shredded and supplied in the form of flakes. PS-co-GMA was provided by Polyscope Polymers B.V. (Geleen, The Netherlands) as Xibond™ 920 in the form of pellets. It has a mass average M_W of 50,000 g·mol^{-1} and a glass transition temperature (T_g) of 95 °C. The manufacturer recommends a dosage level of 0.1–5 wt% while not exceeding 330 °C to avoid thermal degradation.

2.2. Preparation of the Bio-PET/r-PET Blends

Both bio-PET and r-PET were dried at 60 °C for 72 h, while PS-co-GMA was dried at 90 °C for 3 h. All materials were dried separately to avoid premature cross-linking and then premixed in a zipper bag. Table 1 summarizes the coding and compositions of the prepared samples.

Table 1. Code and composition of each sample according to the weight content of partially bio-based (bio-PET) and recycled polyethylene terephthalate (r-PET) in which poly(styrene-co-glycidyl methacrylate) (PS-co-GMA) was added as parts per hundred resin (phr) of blend.

Sample	bio-PET (wt%)	r-PET (wt%)	PS-co-GMA (phr)
B100	100	-	-
R100	-	100	-
B85-R15	85	15	-
B70-R30	70	30	-
B55-R45	55	45	-
B55-R45-X1	55	45	1
B55-R45-X3	55	45	3
B55-R45-X5	55	45	5

The compounding process was carried out in a twin-screw extruder from Construcciones Mecánicas Dupra, S.L. (Alicante, Spain) equipped with a screw diameter of 25 mm and a length-to-diameter (L/D) ratio of 24. The rotating speed was set to 25 rpm and the temperature profile from the feeding to the die was: 240 °C (hopper)–245 °C–250 °C–255 °C (nozzle). The strands were cooled in air and granulated in pellets in an air-knife unit. The resultant pellets were dried at 60 °C for 72 h to remove moisture since PET has a high sensitivity to hydrolysis.

Standard pieces for characterization were obtained from the compounded pellets by injection molding in a Meteor 270/75 from Mateu&Solé (Barcelona, Spain). The temperature profile in the injection machine was programmed to 250 °C (hopper)–255 °C–255 °C–260 °C (injection nozzle). The resultant pieces were finally annealed at 60 °C for 72 h to further develop crystallinity, improve their dimensional stability, and remove any residual moisture.

2.3. Mechanical Characterization

Tensile properties were obtained in a universal test machine ELIB-50 from S.A.E. Ibertest (Madrid, Spain) following ISO 527-1:2012. The selected cross-head speed was 5 mm·min^{-1} and the load cell was 5 kN. As recommended by the standard, the tensile modulus (E_t), maximum tensile strength at break (σ_{max}), and elongation at break (ε_b) values were determined. Shore D hardness was obtained in a 676-D durometer from Instruments J. Bot S.A. (Barcelona, Spain) as indicated in ISO

868:2003. The impact strength was obtained in a Charpy's pendulum of 1 J from Metrotec (San Sebastián, Spain) on V-notched samples with a radius notch of 0.25 mm according to ISO 179-1:2010. All the mechanical tests were performed under controlled conditions of 25 °C and 40% relative humidity (RH). At least 6 samples of each material were evaluated.

2.4. Color Measurements

Changes in color were measured in a colorimetric spectrophotometer ColorFlex from Hunterlab (Reston, VA, USA). The selected color space was the chromatic model L* a* b* or CIELab (spherical color space). L* stands for the luminance, where L* = 0 represents dark and L* = 100 indicates clarity or lightness. The a*b* pair represents the chromaticity coordinate, where a* > 0 is red, a* < 0 is green, b* > 0 is yellow, and b* < 0 is blue. The L*a*b* coordinate values were obtained on five different samples and the color difference (ΔE*) was calculated using the following Equation (1):

$$\Delta E^* = \left[(\Delta L^*)^2 + (\Delta a^*)^2 + (\Delta b^*)^2\right]^{0.5} \quad (1)$$

where ΔL*, Δa*, and Δb* correspond to the differences between the color parameters of the tested pieces and the values of the neat bio-PET (B100) piece. Color change was evaluated as follows: Unnoticeable (ΔE* < 1), only an experienced observer can notice the difference (ΔE* ≥ 1 and < 2), an unexperienced observer notices the difference (ΔE* ≥ 2 and < 3.5), clear noticeable difference (ΔE* ≥ 3.5 and < 5), and the observer notices different colors (ΔE* ≥ 5) [46].

2.5. Microscopy

The fracture surfaces of the injection-molded samples after the impact test were covered with a thin metal layer to provide the conducting properties for analysis by field emission scanning electron microscopy (FESEM). The sputtering process was carried out in a cathodic sputter-coater Emitech SC7620 from Quorum Technologies LTD (East Sussex, UK). Subsequently, the morphologies were observed in a CARL ZEISS Ultra-55 FESEM microscope from Oxford Instruments (Abingdon, UK). The acceleration voltage was set to 2.0 kV.

2.6. Thermal Characterization

Thermal characterization was carried out on a differential scanning calorimeter DSC-821 from Mettler-Toledo Inc (Schwarzenbach, Switzerland). Samples with a mean weight of 6.1 ± 1.2 mg were placed in aluminum pans and subjected to a dynamic thermal program in three stages: initial heating from 30 °C to 280 °C, cooling to 0 °C, and a second heating to 350 °C. The heating/cooling rates were set at 10 °C·min^{-1} for all three stages, and the atmosphere was air at a flow-rate of 66 mL·min^{-1}. The DSC runs were obtained in triplicate to obtain reliable results. The values of T_g, cold crystallization temperature (T_{cc}), T_m, as well as the enthalpies corresponding to the melting process (ΔH_m) and the cold crystallization process (ΔH_{cc}) were obtained from the second heating step, whereas the temperature of crystallization (T_c) and enthalpy of the crystallization process (ΔH_c) were obtained from the cooling step. The degree of crystallinity (χ_c) was calculated using Equation (2):

$$\%\chi_c = \frac{(\Delta H_m - \Delta H_{cc})}{\Delta H_{100\%} \cdot W_p} \cdot 100 \quad (2)$$

where W_p stands for the weight fraction of PET (%) in the sample and $\Delta H_{100\%}$ is the melting enthalpy of a theoretic 100% crystalline PET, that is, 140 J·g^{-1} [47,48].

Thermal degradation was evaluated by thermogravimetry (TGA) in a TGA/SDTA851 thermobalance from Mettler-Toledo Inc (Schwarzenbach, Switzerland). Samples with an average weight of 5–7 mg were subjected to a temperature sweep from 30 °C to 700 °C at a constant heating rate of 20 °C·min^{-1} in air atmosphere with a flow-rate of 50 mL·min^{-1}. The onset degradation temperature, which was assumed at a weight loss of 5 wt% ($T_{5\%}$), and the maximum degradation rate temperature (T_{deg}) were obtained from the TGA curves. All thermal tests were run in triplicate.

2.7. Thermomechanical Characterization

Dynamical mechanical thermal analysis (DMTA) was conducted in an oscillatory rheometer AR-G2 from TA Instruments (New Castle, DE, USA), equipped with a special clamp system for solid samples working in a combination of shear-torsion stresses. The maximum shear deformation (%γ) was set to 0.1% at a frequency of 1 Hz. The thermal sweep was from 20 °C to 160 °C at a heating rate of 5 °C min^{-1}. Thermomechanical analysis (TMA) was carried out in a Q-400 thermoanalyzer, also from TA Instruments. A dynamic thermal sweep from 0 °C to 140 °C at a constant heating rate of 2 °C·min^{-1} with an applied load of 20 mN was performed. All thermomechanical tests were also done in triplicate.

3. Results and Discussion

3.1. Mechanical Properties of the Bio-PET/r-PET Blends

Table 2 shows the mechanical properties in terms of E_t, σ_{max}, ε_b, Shore D hardness, and impact strength for bio-PET, r-PET, and their blends. One can observe that neat bio-PET resulted in elastic and flexible pieces, having E_t and σ_{max} values of approximately 600 MPa and 57 MPa, respectively, whereas ε_b nearly reached a value 495%. Shore D hardness and impact strength values were 92.3 kJ·m^{-2} and 2.87 kJ·m^{-2}, respectively. These mechanical properties were similar to those reported in our earlier research work [45], though the present pieces were slightly less rigid and more flexible. The performance differences could be related to variations in the moisture content since the PET polyester can easily absorb water due to the high hygroscopicity that plasticizes it [49]. In contrast, neat r-PET produced pieces with a similar mechanical strength but a significantly lower mechanical ductility and toughness. In particular, ε_b was 23.1%, while the impact strength was 0.82 kJ·m^{-2}, that is, a respective 21- and 3.5-fold reduction in comparison with virgin bio-PET. This ductility impairment has been associated with the decrease in the PET's M_W, mainly caused by a phenomenon of chain scission [50,51]. Therefore, as expected, increasing the amount of r-PET in the blend induced a progressive reduction in the ductile performance of the pieces, reaching ε_b values in the 10–11% range for the highest contents. The incorporation of r-PET in contents of 30 wt% and 45 wt%, however, increased the E value by approximately 40%. This mechanical behavior has been previously ascribed by Mancini and Zanin to an increase in crystallinity [50] but it can also be associated with the attained reduction in ductility.

Table 2. Summary of the mechanical properties of the partially bio-based and recycled polyethylene terephthalate (bio-PET/r-PET) blends in terms of tensile modulus (E_t), maximum tensile strength (σ_{max}), elongation at break (ε_b), Shore D hardness, and impact strength.

Piece	E_t (MPa)	σ_{max} (MPa)	ε_b (%)	Shore D Hardness	Impact Strength (kJ·m^{-2})
B100	600.3 ± 19.1	56.8 ± 0.8	494.6 ± 9.7	92.3 ± 0.9	2.87 ± 0.42
R100	558.8 ± 11.7	56.1 ± 0.7	23.1 ± 4.7	79.8 ± 0.8	0.82 ± 0.13
B85-R15	665.4 ± 19.2	56.6 ± 2.7	20.4 ± 4.9	77.8 ± 1.2	1.73 ± 0.06
B70-R30	822.8 ± 25.6	57.5 ± 1.0	10.5 ± 1.1	77.3 ± 1.3	1.84 ± 0.29
B55-R45	820.5 ± 30.2	57.7 ± 2.8	10.8 ± 1.4	76.0 ± 1.4	1.84 ± 0.38
B55-R45-X1	742.9 ± 10.9	58.1 ± 1.1	11.2 ± 0.9	79.8 ± 1.5	1.89 ± 0.24
B55-R45-X3	852.9 ± 21.8	58.3 ± 0.3	312.9 ± 7.3	81.0 ± 1.2	2.43 ± 0.29
B55-R45-X5	847.8 ± 11.7	57.4 ± 0.1	378.8 ± 8.4	82.8 ±1.8	2.52 ± 0.25

The incorporation of 1 phr of PS-*co*-GMA into the blend system had a slight influence on the mechanical performance of the bio-PET/r-PET pieces. However, the use of contents of 3 phr and 5 phr successfully yielded a significant increase in the ductility and toughness in the r-PET/bio-PET pieces. In particular, for the blend piece containing 45 wt% r-PET, ε_b increased from 10.8% to 312.9% and

378.8% for the same pieces that were melt-processed with 3 phr and 5 phr of PS-*co*-GMA, respectively. The same trend was observed for the impact strength values, increasing from 1.84 kJ·m^{-2} to 2.43 kJ·m^{-2} and 2.52 kJ·m^{-2}, respectively. This improved ductility can be ascribed to the chain extension mechanism of PET achieved during extrusion by the presence of the PS-*co*-GMA. This reactive copolymer contains multiple GMA groups that can react with the –OH and –COOH terminal groups of both bio-PET and r-PET. As a result of the so-called reactive extrusion process, a macromolecule of higher M_W based on a linear chain-extended, branched, or even cross-linked structure is formed. The proposed chain extension mechanism is shown in Figure 1. Thus, the resultant material shows a macromolecular structure with a higher degree of entanglements to resist mechanical deformation. A similar mechanical improvement was observed, for instance, by the addition of diisocyanates, in which the ε_b value notably increased from approximately 5% to 300% [34]. Similarly, maleinized hemp seed oil (MHO) and acrylated epoxidized soybean oil (AESO) have been used to increase the mechanical performance of PLA materials [52].

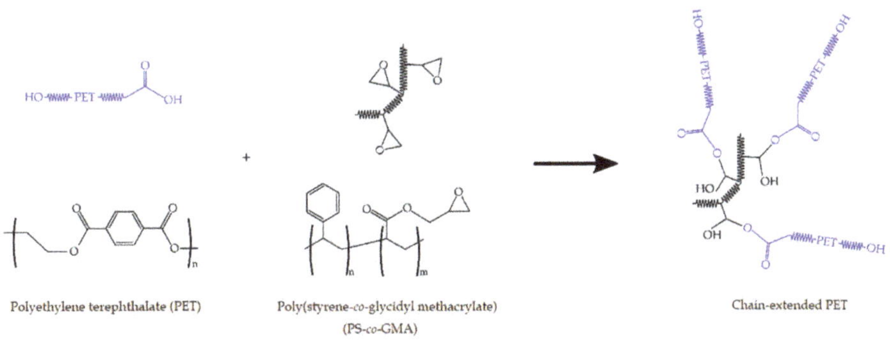

Figure 1. Proposed mechanism of chain extension of polyethylene terephthalate (PET) with poly(styrene -*co*-glycidyl methacrylate) (PS-*co*-GMA).

3.2. Morphologgy and Appareance of the Bio-PET/r-PET Blends

Figure 2 shows the appearance of the injection-molded neat bio-PET, r-PET, and their blends. Color changes of the bio-PET/r-PET pieces were quantified by the L*a*b* coordinates and are reported in Table 3. PET is a semi-crystalline polyester that can be manufactured during injection molding into articles of different transparencies by selecting the appropriate cooling conditions, being very transparent when full amorphous or opaque when it is highly crystallized [53]. The neat bio-PET piece presented a natural bright color, but it was opaque, indicating that the biopolymer developed a certain crystallinity during cooling in the injection mold and in the subsequent annealing. The r-PET pieces were also opaque, but slightly grey and less bright. One can observe that the bio-PET pieces developed a yellowish color (b* > 0) as the weight percentage of r-PET increased, being also opaque. This effect has been described by Torres et al. [29], who attributed it to the presence of spherulitic crystallization as a result of the addition of r-PET. One can also observe that the introduction of PS-*co*-GMA reduced the yellow tonality in the pieces, though the effect of the random copolymer on their visual appearance was relatively low. Therefore, although a different color was noticed (ΔE* ≥ 5) in all the pieces, the color differences were significantly less noticeable after the incorporation of PS-*co*-GMA.

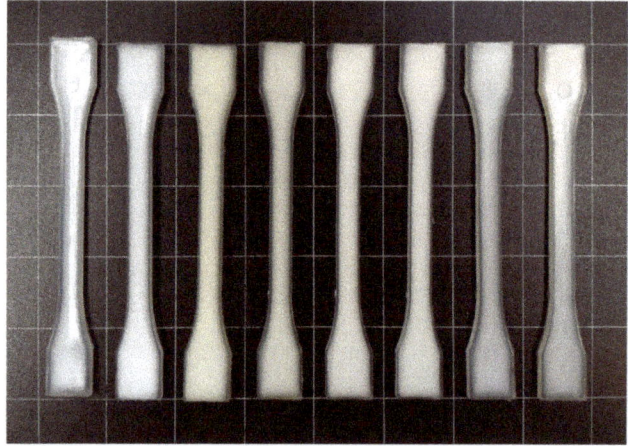

Figure 2. Visual aspect of the injection-molded pieces of the partially bio-based and recycled polyethylene terephthalate (bio-PET/r-PET) blends corresponding, from left to right, to: B100, R100, B85-R15, B70-R30, B55-R45, B55-R45-X1, B55-R45-X3, and B55-R45-X5.

Table 3. Color coordinates by CIElab color space (L*a*b*) of the partially bio-based and recycled polyethylene terephthalate (bio-PET/r-PET) blend pieces.

Piece	L*	a*	b*	ΔE*
B100	75.67 ± 0.58	−2.72 ± 0.16	−5.27 ± 0.32	-
R100	42.86 ± 0.37	−0.54 ± 0.11	−2.62 ± 0.10	32.99 ± 1.31
B85-R15	68.03 ± 0.44	−3.63 ± 0.07	5.27 ± 0.25	10.84 ± 0.57
B70-R30	64.85 ± 0.48	−2.95 ± 0.14	5.62 ± 0.18	15.35 ± 1.18
B55-R45	63.44 ± 0.96	−2.61 ± 0.14	5.41 ± 0.50	16.24 ± 0.90
B55-R45-X1	69.88 ± 0.52	−1.97 ± 0.17	0.57 ± 0.23	8.26 ± 1.13
B55-R45-X3	72.71 ± 0.86	−2.56 ± 0.13	0.33 ± 0.20	6.34 ± 0.77
B55-R45-X5	73.73 ± 0.98	−2.34 ± 0.17	1.35 ± 0.31	6.91 ± 0.84

Figure 3 presents the FESEM images corresponding to fracture surfaces of the injection-molded pieces after the impact tests. All morphologies showed the formation of microcracks that were responsible for the final fracture of the material. Figure 3a, which corresponds to the FESEM micrograph of the neat bio-PET piece, shows that the size of the cracking steps was larger with somewhat more rounded edges compared with those of the r-PET piece shown in Figure 3b. This morphological change is due to more energy being required during the breakage process so that the fracture produced a rougher surface. In both cases, the cracking steps were uniform. In contrast, the FESEM micrographs corresponding to the bio-PET/r-PET blends, shown in Figure 3c–e, yielded very heterogeneous fracture surfaces. This type of fracture can be mainly ascribed to the hardness increase of the injection-molded piece [54]. Furthermore, although the morphologies revealed the presence of a single phase, this type of surface can also be ascribed to the lack of chemical interaction between the matrix, that is, bio-PET, and the dispersed domains of r-PET. One can observe in Figure 3f that a similar morphology was obtained when the polyester blend was melt-processed with 1 phr of PS-*co*-GMA. However, as seen in Figure 3g,h, the addition of the 3 phr and 5 phr of PS-*co*-GMA produced pieces with fracture surfaces very similar to those observed for neat bio-PET. In particular, one can observe that larger cracks with rounded edges were produced due to the aforementioned increase in toughness. Therefore, the fracture surfaces correlated well with the mechanical strength and impact strength performance attained during the mechanical analysis. A similar morphology was reported by Badia et al. [55], who showed a smoother surface loss in the surface fractures of virgin PET after the addition of r-PET. This behavior was also noted in other recycled polymers [56,57].

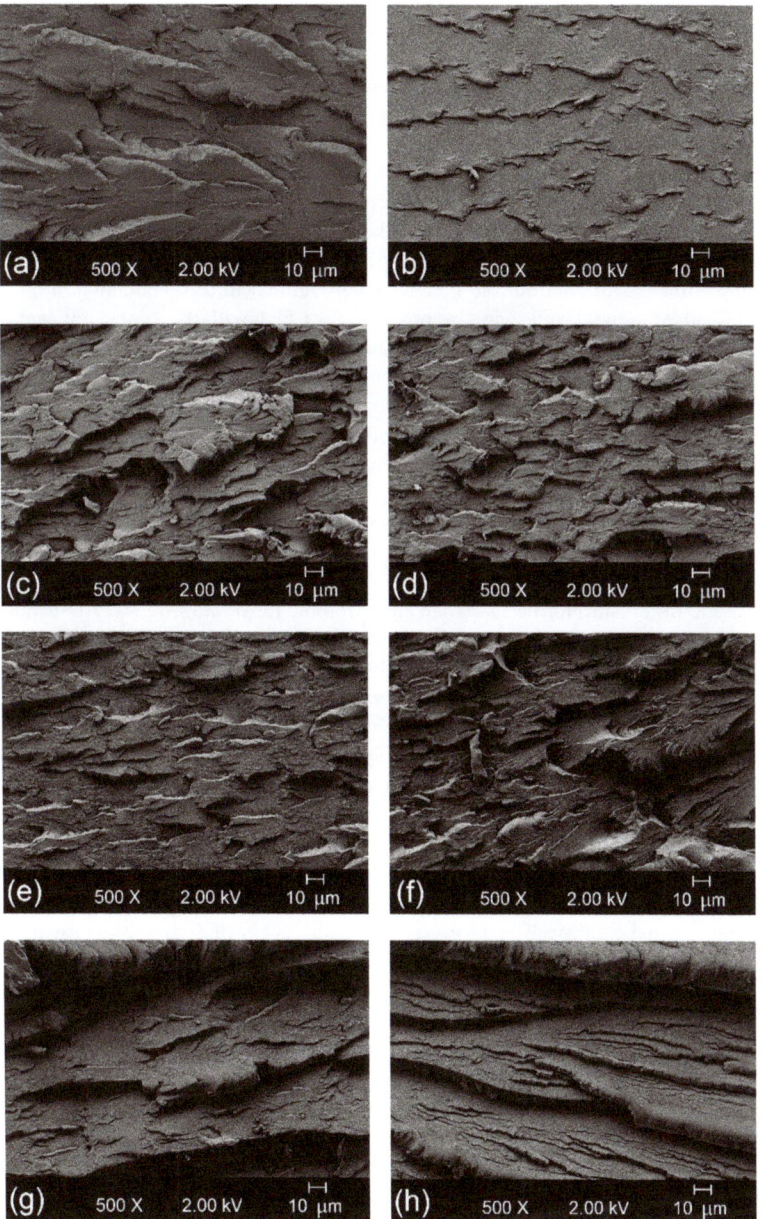

Figure 3. Field emission scanning electron microscopy (FESEM) images, taken at 500×, corresponding to fracture surfaces of the injection-molded pieces of the partially bio-based and recycled polyethylene terephthalate (bio-PET/r-PET) blends: (**a**) B100; (**b**) R100; (**c**) B85-R15; (**d**) B70-R30; (**e**) B55-R45; (**f**) B55-R45-X1; (**g**) B55-R45-X3; (**h**) B55-R45-X5. Scale markers of 10 μm.

3.3. Thermal Characterization of the Bio-PET/r-PET Blends

The thermal transitions and degree of crystallinity are parameters of great technological importance that reflect the chemical structure of the polymer and they can be correlated with its mechanical properties [58]. Figure 4 shows a comparative plot of the DSC thermograms of the bio-PET/r-PET blends during the cooling step (Figure 4a) and second heating step (Figure 4b). The main thermal properties obtained from the DSC analysis are summarized in Table 4. In relation to the bio-PET sample, the glass transition region was seen as a step in the baseline located around 82 °C in the heating thermogram. Then, one can observe in the cooling thermogram that the biopolyester did not crystallize from the melt but developed a cold crystallization process during heating in the thermal range between 140 °C and 180 °C, showing a maximum exothermic peak, the so-called T_{cc}, at 161 °C. Finally, shown in the heating thermogram, the endothermic peak between 220 °C and 260 °C corresponds to the melting process of the total crystalline fraction in the biopolyester, showing a T_m value of nearly 245 °C. The thermal parameters attained for bio-PET were very similar to those obtained in our previous work [45]. A similar thermal profile was also observed for r-PET, with the most significant difference being the lower T_{cc} value observed, which was located at approximately 150 °C. This observed value can be ascribed to the lower M_W of r-PET, which compromises shorter chains that can more easily cold crystallize due to the fact that the polyester was subjected to a second round of processing [59].

Interestingly, the bio-PET/r-PET blends showed a different thermal profile during DSC analysis. One can observe that the polyester blends crystallized in all cases during cooling, showing values of T_c in the 180–190 °C range. This observation suggests that crystallization was thermodynamically favored in the PET blends. This phenomenon can be ascribed to the potential role of r-PET as a nucleating agent in bio-PET. In this regard, the nucleating effect of r-PET has been previously related to the presence of impurities [29]. Therefore, the spherulitic crystallization of r-PET occurs at lower temperatures and the crystals formed can thereafter promote the homogenous crystallization of PET chains at higher temperatures. This effect was further confirmed by the increase in the T_c values and the higher intensity of the exothermic peaks observed during crystallization from the melt with higher r-PET contents. Thus, χ_c value increased from 10.6% in the neat bio-PET sample to 29% for the bio-PET blend containing 15 wt% of r-PET, whereas these values reached 32.3% and 37% for the blends containing 30 wt% and 45 wt% of r-PET, respectively. It is also worth mentioning that while the melting process of the neat bio-PET sample occurred in a single melt peak, all the bio-PET/r-PET blends showed two overlapping peaks during melting. This double-melting peak phenomenon can be explained by the presence of dominant and subsidiary crystals [60]. The two melting peaks (T_{m1} and T_{m2}) were observed at approximately 238 and 248 °C, with the latter temperature being assigned to primary folded chain crystals with a higher melting point than those observed for the neat bio-PET. This observation further confirms that the presence of r-PET in bio-PET favors the formation of more perfect crystalline structures, which were obtained from the melt of imperfect structures or crystals with lower lamellae thicknesses [61]. Similar results were observed by Spinacé and De Paoli [62], who indicated that the chain scission, occurring during the processing of PET, improved chain packing, increasing the crystallite size and consequently shifting T_c and T_m to higher values.

The incorporation of PS-co-GMA barely modified the T_g, T_{m1}, and T_{m2} values, though the use of contents of 3 phr and 5 phr decreased the crystallinity to 17.7% and 25.9%, respectively. The reactive random copolymer also caused a slight reduction in the T_c values, showing values around 185 °C, indicating that the structural disorder introduced by the chain extender potentially hindered both nucleation and crystallization. These observations agree with the aforementioned assumption that the induced M_W increase could decrease the molecular mobility, resulting in a low crystallization rate and lower crystallinities [63]. Moreover, Kiliaris et al. [64] also concluded that the ability of the PET chains to crystallize in folded lamellae was limited after chain extension, which led to the formation of smaller and less perfect crystallites. Furthermore, if one considers branching as the prevailing chain-extension reaction, the presence of the resultant branching points could effectively hinder chain symmetry, restricting the segment movements and yielding crystal defects.

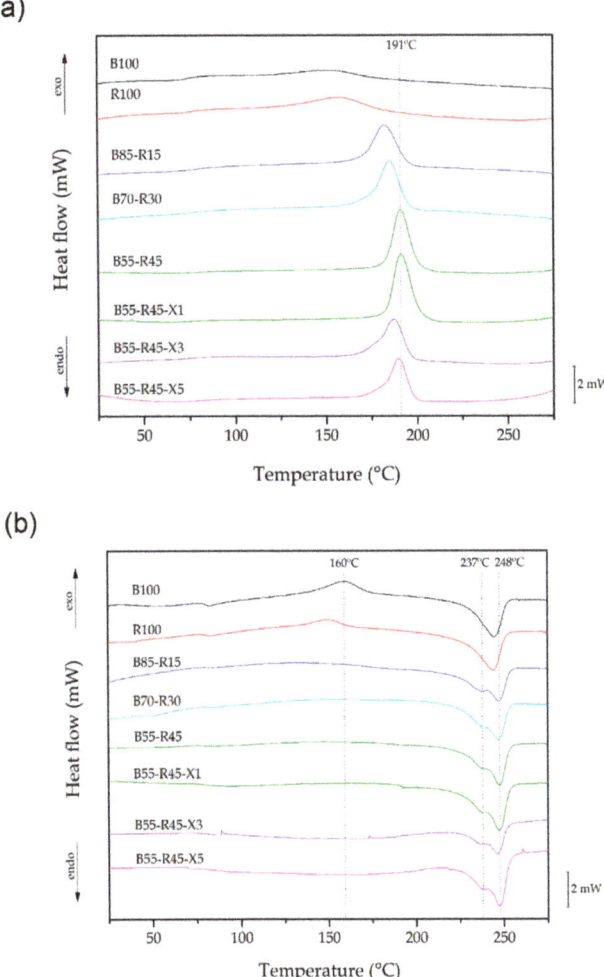

Figure 4. Differential scanning calorimetry (DSC) curves during cooling (a) and second heating (b) of the injection-molded pieces of the partially bio-based and recycled polyethylene terephthalate (bio-PET/r-PET) blends.

Table 4. Summary of the main thermal properties of the injection-molded pieces of the partially bio-based and recycled polyethylene terephthalate (bio-PET/r-PET) blends in terms of glass transition temperature (T_g), cold crystallization temperature (T_{cc}), crystallization temperature (T_c), melting temperature (T_m), and degree of crystallinity (χ_c).

Piece	T_g (°C)	T_{cc} (°C)	T_c (°C)	T_{m1} (°C)	T_{m2} (°C)	χ_c (%)
B100	82.0 ± 0.7	150.8 ± 0.4	-	244.9 ± 0.9	-	10.6 ± 0.5
R100	80. 4± 0.8	157.6 ± 0.2	-	244.5 ± 1.1	-	24.2 ± 0.5
B85-R15	81.2 ± 0.6	-	182.7 ± 0.3	238.4 ± 0.8	247.4 ± 1.0	29.0 ± 0.8
B70-R30	81.5 ± 0.8	-	185.1 ± 0.2	238.9 ± 0.7	247.2 ± 0.8	32.3 ± 0.9
B55-R45	82.6 ± 0.9	-	191.2 ± 0.3	237.9 ± 1.0	247.8 ± 1.1	37.0 ± 0.7
B55-R45-X1	81.0 ± 0.7	-	191.4 ± 0.1	238.0 ± 1.1	247.5 ± 0.9	38.1 ± 0.8
B55-R45-X3	80.9 ± 0.8	-	187.4 ± 0.2	236.8 ± 0.8	247.9 ± 0.8	17.7 ± 0.5
B55-R45-X5	80.6 ± 0.7	-	189.8 ± 0.2	237.1 ± 0.7	247.9 ± 1.2	25.9 ± 0.9

One of the main problems related to the mechanical recycling of thermoplastic materials is the loss of thermal stability. Figure 5 gathers the TGA curves (Figure 5a) and the first derivatives of the curves (DTG) (Figure 5b) of the bio-PET/r-PET blends, while Table 5 presents the thermal values obtained from the TGA curves. It can be observed that bio-PET degradation occurred in two stages, as previously reported by Vannier et al. [65]. The first stage, occurring from 350 °C to 460 °C, corresponds to the main degradation of the biopolymer backbone and the formation of char with a mass loss of ~83%. The second step, which was associated with a loss of nearly the totality of the remaining biopolyester, was related to the thermo-oxidative degradation of the char. This occurred from 490 °C to 560 °C. It has also been reported that the decomposition mechanism of PET consists of an hererolytic scission via a six-membered ring intermediate, where the hydrogen from a β-carbon to the ester group is transferred to the ester carbonyl, followed by scission at the ester links [66]. Dzięcioł and Trzeszczyński [67,68] also reported that the degradation of PET leads to the formation of acetaldehyde, oligomers with terminal carboxyl groups, CO, and CO_2, among other compounds. In all cases, the bio-PET/r-PET samples resulted in a residual mass of 1–2 wt% at 700 °C.

One can also observe that the thermal degradation profile of r-PET and the bio-PET/r-PET blends was similar to that of the neat bio-PET. The addition of PS-*co*-GMA yielded an increase in the onset temperature of degradation that was measured as the degradation temperature at 5% of mass loss, that is, $T_{5\%}$. In particular, the beginning of the thermal degradation was delayed from approximately 393 °C for the blends containing 30 wt% and 45 wt% of r-PET to nearly 403 °C. The thermal degradation peaks (T_{deg1} and T_{deg2}), determined when the maximal degradation rates were produced, remained nearly constant during the first and main mass loss, while they slightly increased during the second one in the PET blends processed without PS-*co*-GMA. In particular, the T_{deg2} values were increased by up to nearly 19 °C, which suggests a delay in the thermo-oxidative degradation of the char. In any case, the influence of the reactive copolymer on the thermal degradation of the bio-PET/r-PET blends was relatively low, since the thermal stability of the polyester blends was already high.

Table 5. Summary of the main thermal properties of the injection-molded pieces of the partially bio-based and recycled polyethylene terephthalate (bio-PET/r-PET) blends in terms of the degradation temperature at 5% of mass loss ($T_{5\%}$), degradation temperature (T_{deg}), and residual mass at 700 °C.

Piece	$T_{5\%}$ (°C)	T_{deg1} (°C)	T_{deg2} (°C)	Residual Mass (%)
B100	405.4 ± 0.2	446.8 ± 0.1	546.0 ± 0.2	1.26 ± 0.04
R100	400.8 ± 0.4	452.3 ± 0.2	552.7 ± 0.1	1.43 ± 0.07
B85-R15	400.2 ± 0.2	452.2 ± 0.3	549.4 ± 0.2	1.22 ± 0.05
B70-R30	393.2 ± 0.3	452.3 ± 0.2	545.7 ± 0.1	1.64 ± 0.18
B55-R45	393.0 ± 0.2	452.9 ± 0.3	559.5 ± 0.2	2.18 ± 0.05
B55-R45-X1	403.2 ± 0.2	452.9 ± 0.2	578.3 ± 0.2	1.27 ± 0.08
B55-R45-X3	403.1 ± 0.1	452.4 ± 0.1	576.6 ± 0.4	1.73 ± 0.15
B55-R45-X5	403.0 ± 0.1	451.8 ± 0.2	576.1 ± 0.2	2.29 ± 0.06

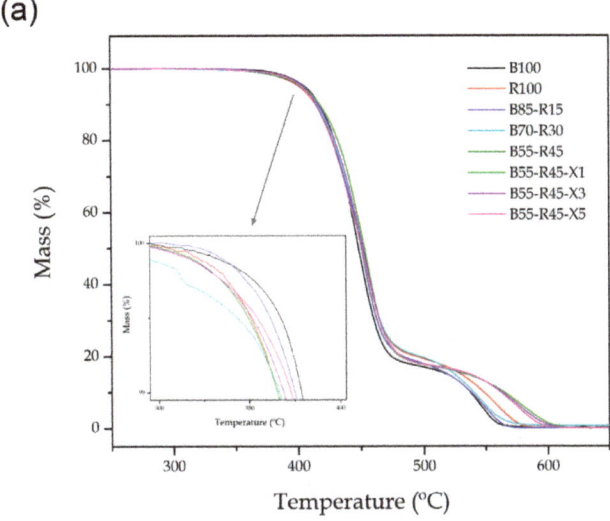

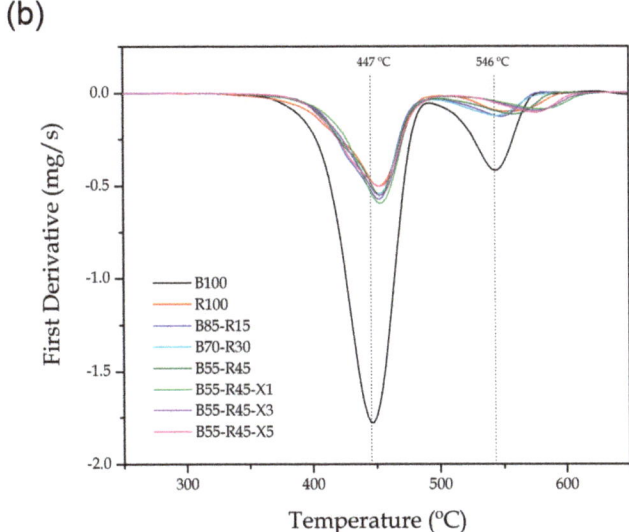

Figure 5. (a) Thermogravimetric analysis (TGA) and (b) first derivative thermogravimetric (DTG) curves of the injection-molded pieces of the partially bio-based and recycled polyethylene terephthalate (bio-PET/r-PET) blends.

3.4. Thermomechanical Properties of the Bio-PET/r-PET Blends

The prepared bio-PET/r-PET blends and the effect of the reactive random copolymer on these polyester blends were analyzed in terms of their thermomechanical properties by means of DMTA and TMA. Figure 6 shows the DMTA curves of the bio-PET/r-PET blend pieces. Figure 6a shows the storage modulus as a function of temperature. The evolution of the storage modulus of the bio-PET was characterized by a single thermal transition with a temperature increase in the 50–130 °C range. Up to approximately 70 °C, the variation in the storage modulus values was nearly negligible since the biopolyester was in a glassy state. At room temperature, all the pieces showed similar storage modulus values, being around 1 GPa and slightly lower for the neat bio-PET due to its reduced crystallinity. Thereafter, one can observe a remarkable decrease in the storage modulus values in the temperature range from 70 °C to 85 °C. This process notably involved a decrease of more than two orders of magnitude in the storage modulus, which is representative for the glass-to-rubber transition of bio-PET. At higher temperatures, the variation in the storage modulus was nearly negligible, indicating that the biopolyester reached its rubber state and that further temperature increase did not affect the mechanical properties. Table 6 presents the values of the storage modulus obtained at 60 °C and 100 °C since these temperatures are representative of the mechanical rigidity of the bio-PET pieces before and after the glass transition region of bio-PET. While at 60 °C, the value of the storage modulus of the neat bio-PET piece was nearly 900 MPa, this value was reduced to approximately 3 MPa at 100 °C. It can also be observed that the r-PET piece and all the bio-PET/r-PET pieces showed similar thermochemical performance with slightly higher storage modulus values. The incorporation of PS-co-GMA led to an increase in the storage modulus values, in both the glassy and the rubber states. The highest enhancement was obtained for the bio-PET/r-PET blend piece processed with 5 phr of PS-co-GMA, in which it reached values of close to 1.2 GPa and 30 MPa at 60 °C and 100 °C, respectively. This thermomechanical increase can be ascribed to the aforementioned formation of a chain-extended macromolecular structure with higher M_W that showed increased resistance to flow during the application of external forces as a result of the large number of entanglements associated with the long chains or branches [69].

Figure 6b shows the evolution of the damping factor (*tan δ*) as a function of temperature. The maximum peak of the *tan δ* curves corresponds to alpha (α)-transition of bio-PET, which is related to its T_g. These values are also given in Table 6, where it can be seen that for neat bio-PET, a T_g value of approximately 81 °C was observed, being nearly the same as that obtained above by DSC. A slight lower value of T_g, that is, ~80 °C, was observed for r-PET, with this potentially being related to its lower chain lengths and/or the presence of plasticizers. The bio-PET/r-PET blends showed intermediate values of T_g, ranging between 80 °C and 81 °C, being also similar to those obtained by DSC. The biopolyester blends melt-processed with PS-co-GMA showed nearly the same T_g values but with a remarkable reduction in the α-peak intensities. This observation supports the higher crystallinity and the large number of entanglements achieved in the polyester blends based on the fact that the amorphous phase content was reduced due to the presence of r-PET, which partially suppressed the relaxation of the bio-PET chains and lowered the number of molecules undergoing α-transition [70].

Table 6. Storage modulus measured at 60 °C and 100 °C, glass transition temperature (T_g), and coefficient of linear thermal expansion (CLTE) of the injection-molded pieces of the partially bio-based and recycled polyethylene terephthalate (bio-PET/r-PET) blends.

Piece	Storage Modulus (MPa)		T_g (°C)	CLTE (μm·m^{-1}·°C^{-1})	
	60 °C	100 °C		Below T_g	Above T_g
B100	927.4 ± 20.5	2.8 ± 0.1	81.0 ± 0.1	78.9 ± 1.9	80.5 ± 1.9
R100	1018.5 ± 20.2	1.7 ± 0.2	80.3 ± 0.1	90.9 ± 1.7	94.9 ± 0.9
B85-R15	943.4 ± 23.2	2.5 ± 0.1	81.2 ± 0.1	80.7 ± 1.1	90.1 ± 1.0
B70-R30	931.6 ± 32.3	2.3 ± 0.1	81.3 ± 0.1	84.1 ± 1.1	103.2 ± 0.2
B55-R45	924.3 ± 13.2	2.6 ± 0.1	81.5 ± 0.1	93.1 ± 1.2	111.8 ± 0.9
B55-R45-X1	977.7 ± 17.3	3.4 ± 0.2	81.1 ± 0.1	90.1 ± 0.5	104.0 ± 1.4
B55-R45-X3	984.4 ± 27.0	4.7 ± 0.2	80.9 ± 0.2	74.9 ± 0.1	101.3 ± 1.1
B55-R45-X5	1019.7 ± 18.4	4.9 ± 0.2	80.8 ± 0.2	70.6 ± 0.3	99.9 ± 1.8

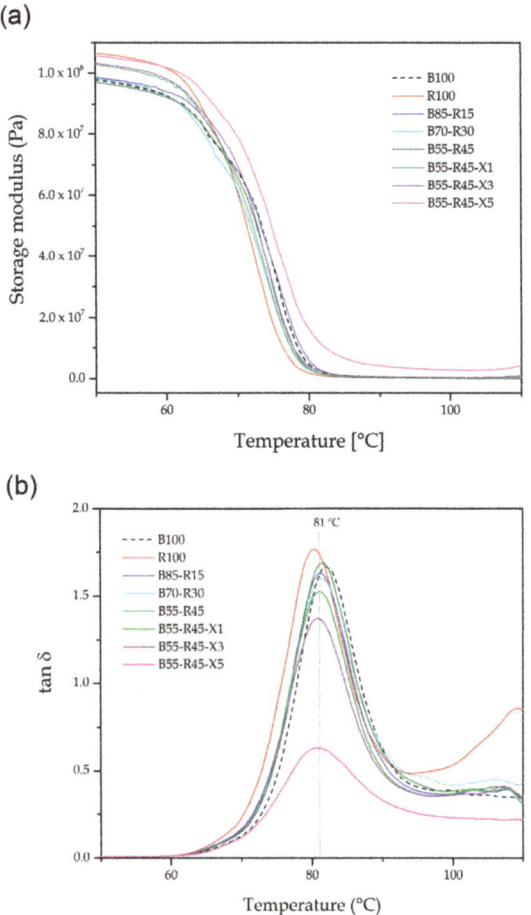

Figure 6. (a) Storage modulus and (b) damping factor (*tan δ*) of the injection-molded pieces of the partially bio-based and recycled polyethylene terephthalate (bio-PET/r-PET) blends.

Finally, the dimensional stability the bio-PET/r-PET blend pieces was determined by measuring the CLTE values below and above the glass transition region. As also shown in Table 6, below T_g, bio-PET showed a lower value than r-PET, that is, 78.9 µm·m^{-1}·°C^{-1} versus 90.9 µm·m^{-1}·°C^{-1}. This thermomechanical difference can be ascribed to the lower M_W attained in the recycled polyester due to the aforementioned phenomenon of chain scission during reprocessing. Then, the bio-PET/r-PET blends presented intermediate values according to their r-PET content. The incorporation of PS-*co*-GMA induced a reduction in the CLTE values, increasing with the reactive random copolyester content. In particular, a value of 70.6 µm·m^{-1}·°C^{-1} was reached for the bio-PET/r-PET blend containing 5 phr PS-*co*-GMA. This observation further supports the formation of a branched and larger macromolecule that reduced the effect of temperature on the dimensional stability, which is also in agreement with the DMTA results. A similar trend with more significant differences was observed for the CLTE values measured above T_g.

4. Conclusions

PS-*co*-GMA, a multi-functional random copolymer, was melt-mixed with bio-PET/r-PET blends at contents of 1–5 phr of polyester blend and shaped into pieces by injection molding. The resultant pieces were characterized to ascertain the potential use of PS-*co*-GMA as a chain extender for the mechanical recycling of these polyester blends. The results showed that while the incorporation of 1 phr of PS-*co*-GMA had a slight influence on the mechanical and thermal performance of the bio-PET/r-PET blend pieces, the contents of 3 phr and 5 phr successfully yielded a significant increase in their ductility and toughness. In particular, ε_b increased from 10.8%, for the blend piece containing 45 wt% of r-PET, to 312.9% and 378.8%, for the same pieces that were melt-processed with 3 phr and 5 phr of PS-*co*-GMA, respectively. In addition, the impact strength values increased from 1.84 kJ·m^{-2} to 2.43 kJ·m^{-2} and 2.52 kJ·m^{-2}, respectively. This mechanical improvement was ascribed to the chain extension mechanism of PET by reactive extrusion due to the reaction of the multiple GMA groups that are present in PS-*co*-GMA with the –OH and –COOH terminal groups of both bio-PET and r-PET. Furthermore, PS-*co*-GMA reduced the crystallinity of the PET blends, suggesting that the structural disorder introduced by the chain extender potentially hindered both the nucleation and crystallization of r-PET on the bio-PET chains, contributing to increase material flexibility. In addition, the influence of the reactive copolymer on the thermal stability of the bio-PET/r-PET blends was low, since the blends were already thermally stable up to nearly 400 °C. The thermomechanical properties of the bio-PET/r-PET blends also improved after the addition of PS-*co*-GMA due to the formation of a branched and larger macromolecule.

Bio-PET can be effectively mixed with its recycled petrochemical counterpart, that is, r-PET, and then mechanically recycled in existing recycling facilities by means of reactive extrusion with PS-*co*-GMA. Therefore, the original biopolymer properties can be successfully restored and the ultimate performance of bio-PET articles would be retained for a given number of reprocessing cycles. This will permit the recovery of upcoming bio-PET streams with current r-PET waste to manufacture the same or similar products. According to this scenario, mechanical recycling for bio-based but non-biodegradable polymers will be appropriate from both an economic and environmental point of view. This will potentially contribute to accelerating the transition of the plastic packaging industry from its traditional linear model to a more valuable and sustainable circular model.

Author Contributions: Conceptualization was devised by N.M., S.F.-B., and S.T.-G.; data curation and writing-original draft by S.M.-J., L.Q.-C., and T.B.; methodology, validation, and formal analysis by S.M.-J., D.L., and A.V.M.-S.; Investigation, resources, Writing—Review and editing by S.T.-G.; project administration, S.F.-B. and S.T.-G. All authors have read and agreed to the published version of the manuscript.

Funding: This research work was funded by the Spanish Ministry of Science, Innovation, and Universities (MICIU) project numbers RTI2018-097249-B-C21 and MAT2017-84909-C2-2-R.

Acknowledgments: L.Q.-C. wants to thank Generalitat Valenciana (GVA) for his FPI grant (ACIF/2016/182) and the Spanish Ministry of Education, Culture, and Sports (MECD) for his FPU grant (FPU15/03812). D.L. thanks Universitat Politècnica de València (UPV) for the grant received through the PAID-01-18 program. S.T.-G. is recipient of a Juan de la Cierva contract (IJCI-2016-29675) from MICIU. S.R.-L. is recipient of a Santiago Grisolía contract (GRISOLIAP/2019/132) from GVA. J.I.-M. wants to thank UPV for an FPI grant PAID-01-19 (SP2019001). Microscopy services at UPV are acknowledged for their help in collecting and analyzing FESEM images. Authors also thank Polyscope Polymers B.V. for kindly supplying Xibond^{TM} 920.

Conflicts of Interest: The authors declare no conflict of interest.

References

1. European Commission. *A European Strategy for Plastics in a Circular Economy*; European Comission: Brussels, Belgium, 2018.
2. Payne, J.; McKeown, P.; Jones, M.D. A circular economy approach to plastic waste. *Polym. Degrad. Stab.* **2019**, *165*, 170–181. [CrossRef]
3. Niaounakis, M. Recycling of biopolymers—The patent perspective. *Eur. Polym. J.* **2019**, *114*, 464–475. [CrossRef]

4. Torres-Giner, S.; Gil, L.; Pascual-Ramírez, L.; Garde-Belza, J.A. Packaging: Food waste reduction. *Encycl. Polym. Appl.* **2018**, *3*, 1990–2009.
5. Tavares, A.A.; Silva, D.F.; Lima, P.S.; Andrade, D.L.; Silva, S.M.; Canedo, E.L. Chain extension of virgin and recycled polyethylene terephthalate. *Polym. Test.* **2016**, *50*, 26–32. [CrossRef]
6. Leblanc, R. Recycling Polyethylene Terephthalate. Available online: https://www.thebalancesmb.com/recycling-polyethylene-terephthalate-pet-2877869 (accessed on 5 December 2019).
7. Reuters. Coca-Cola Switches to Recycled Plastic for PET Bottles in Sweden. Available online: https://www.reuters.com/article/us-coca-cola-plastic/coca-cola-switches-to-recycled-plastic-for-pet-bottles-in-sweden-idUSKBN1XT1QS (accessed on 5 December 2019).
8. European Bioplastics. *Bio-Based Plastics—Fostering a Resource Efficient Circular Economy*; European Bioplastics: Berlin, Germany, 2016.
9. Van Den Oever, M.; Molenveld, K.; Van Der Zee, M.; Bos, H. Bio-based and biodegradable plastics: Facts and figures: Focus on food packaging in the Netherlands—Rapport nr. 1722. In *Bio-Based and Biodegradable Plastics: Facts and Figures: Focus on Food Packaging in the Netherlands (No. 1722)*; Wageningen Food & Biobased Research: Wageningen, The Netherlands, 2017.
10. European Bioplastics. *Global Production Capacities of Bioplastics 2017–2022—Bioplastics Market Data 2017*; European Bioplastics: Berlin, Germany, 2017.
11. European Bioplastics. Available online: https://www.european-bioplastics.org/ (accessed on 28 October 2019).
12. Park, S.H.; Kim, S.H. Poly(ethylene terephthalate) recycling for high value added textiles. *Fash. Text.* **2014**, *1*, 1. [CrossRef]
13. Okuwaki, A. Feedstock recycling of plastics in Japan. *Polym. Degrad. Stab.* **2004**, *85*, 981–988. [CrossRef]
14. Patel, M.; von Thienen, N.; Jochem, E.; Worrell, E. Recycling of plastics in Germany. *Resour. Conserv. Recycl.* **2000**, *29*, 65–90. [CrossRef]
15. Dullius, J.; Ruecker, C.; Oliveira, V.; Ligabue, R.; Einloft, S. Chemical recycling of post-consumer PET: Alkyd resins synthesis. *Prog. Org. Coat.* **2006**, *57*, 123–127. [CrossRef]
16. Shukla, S.; Harad, A.M.; Jawale, L.S. Recycling of waste PET into useful textile auxiliaries. *Waste Manag.* **2008**, *28*, 51–56. [CrossRef]
17. Sogancioglu, M.; Yucel, A.; Yel, E.; Ahmetli, G. Production of Epoxy Composite from the Pyrolysis Char of Washed PET Wastes. *Energy Procedia* **2017**, *118*, 216–220. [CrossRef]
18. Lee, J.H.; Lim, K.S.; Hahm, W.G.; Kim, S.H. Properties of recycled and virgin poly(ethylene terephthalate) blend fibers. *J. Appl. Polym. Sci.* **2013**, *128*, 1250–1256. [CrossRef]
19. Srithep, Y.; Javadi, A.; Pilla, S.; Turng, L.-S.; Gong, S.; Clemons, C.; Peng, J. Processing and characterization of recycled poly(ethylene terephthalate) blends with chain extenders, thermoplastic elastomer, and/or poly(butylene adipate-co-terephthalate). *Polym. Eng. Sci.* **2011**, *51*, 1023–1032. [CrossRef]
20. Garforth, A.A.; Ali, S.; Hernández-Martínez, J.; Akah, A. Feedstock recycling of polymer wastes. *Curr. Opin. Solid State Mater. Sci.* **2004**, *8*, 419–425. [CrossRef]
21. Pawlak, A.; Pluta, M.; Morawiec, J.; Galeski, A.; Pracella, M. Characterization of scrap poly(ethylene terephthalate). *Eur. Polym. J.* **2000**, *36*, 1875–1884. [CrossRef]
22. Pirzadeh, E.; Zadhoush, A.; Haghighat, M. Hydrolytic and thermal degradation of PET fibers and PET granule: The effects of crystallization, temperature, and humidity. *J. Appl. Polym. Sci.* **2007**, *106*, 1544–1549. [CrossRef]
23. Elamri, A.; Lallam, A.; Harzallah, O.; Bencheikh, L. Mechanical characterization of melt spun fibers from recycled and virgin PET blends. *J. Mater. Sci.* **2007**, *42*, 8271–8278. [CrossRef]
24. Scaffaro, R.; La Mantia, F. Characterization of monopolymer blend of virgin and recycled polyamide 6. *Polym. Eng. Sci.* **2002**, *42*, 2412–2417. [CrossRef]
25. Elamri, A.; Abid, K.; Harzallah, O.; Lallam, A. Characterization of recycled/virgin PET polymers and their composites. *Nano* **2015**, *3*, 11–16.
26. Coltelli, M.B.; Savi, S.; Della Maggiore, I.; Liuzzo, V.; Aglietto, M.; Ciardelli, F. A model study of Ti (OBu) 4 catalyzed reactions during reactive blending of polyethylene (PE) and poly(ethylene terephthalate) (PET). *Macromol. Mater. Eng.* **2004**, *289*, 400–412. [CrossRef]
27. Duarte, I.S.; Tavares, A.A.; Lima, P.S.; Andrade, D.L.; Carvalho, L.H.; Canedo, E.L.; Silva, S.M. Chain extension of virgin and recycled poly(ethylene terephthalate): Effect of processing conditions and reprocessing. *Polym. Degrad. Stab.* **2016**, *124*, 26–34. [CrossRef]

28. Li, J.; Tang, S.; Wu, Z.; Zheng, A.; Guan, Y.; Wei, D. Branching and cross-linking of poly(ethylene terephthalate) and its foaming properties. *Polym. Sci. Ser. B* **2017**, *59*, 164–172. [CrossRef]
29. Torres, N.; Robin, J.; Boutevin, B. Study of thermal and mechanical properties of virgin and recycled poly(ethylene terephthalate) before and after injection molding. *Eur. Polym. J.* **2000**, *36*, 2075–2080. [CrossRef]
30. Awaja, F.; Daver, F.; Kosior, E. Recycled poly(ethylene terephthalate) chain extension by a reactive extrusion process. *Polym. Eng. Sci.* **2004**, *44*, 1579–1587. [CrossRef]
31. Incarnato, L.; Scarfato, P.; Di Maio, L.; Acierno, D. Structure and rheology of recycled PET modified by reactive extrusion. *Polymer* **2000**, *41*, 6825–6831. [CrossRef]
32. Jacques, B.; Devaux, J.; Legras, R.; Nield, E. Reactions induced by triphenyl phosphite addition during melt mixing of poly(ethylene terephthalate)/poly(butylene terephthalate) blends: Influence on polyester molecular structure and thermal behaviour. *Polymer* **1996**, *37*, 1189–1200. [CrossRef]
33. Cavalcanti, F.; Teofilo, E.; Rabello, M.; Silva, S. Chain extension and degradation during reactive processing of PET in the presence of triphenyl phosphite. *Polym. Eng. Sci.* **2007**, *47*, 2155–2163. [CrossRef]
34. Torres, N.; Robin, J.; Boutevin, B. Chemical modification of virgin and recycled poly(ethylene terephthalate) by adding of chain extenders during processing. *J. Appl. Polym. Sci.* **2001**, *79*, 1816–1824. [CrossRef]
35. Nascimento, C.R.; Azuma, C.; Bretas, R.; Farah, M.; Dias, M.L. Chain extension reaction in solid-state polymerization of recycled PET: The influence of 2, 2′-bis-2-oxazoline and pyromellitic anhydride. *J. Appl. Polym. Sci.* **2010**, *115*, 3177–3188. [CrossRef]
36. Karayannidis, G.P.; Psalida, E.A. Chain extension of recycled poly(ethylene terephthalate) with 2, 2′-(1, 4-phenylene) bis (2-oxazoline). *J. Appl. Polym. Sci.* **2000**, *77*, 2206–2211. [CrossRef]
37. Xu, X.; Ding, Y.; Qian, Z.; Wang, F.; Wen, B.; Zhou, H.; Zhang, S.; Yang, M. Degradation of poly(ethylene terephthalate)/clay nanocomposites during melt extrusion: Effect of clay catalysis and chain extension. *Polym. Degrad. Stab.* **2009**, *94*, 113–123. [CrossRef]
38. Haralabakopoulos, A.; Tsiourvas, D.; Paleos, C. Chain extension of poly(ethylene terephthalate) by reactive blending using diepoxides. *J. Appl. Polym. Sci.* **1999**, *71*, 2121–2127. [CrossRef]
39. Makkam, S.; Harnnarongchai, W. Rheological and mechanical properties of recycled PET modified by reactive extrusion. *Energy Procedia* **2014**, *56*, 547–553. [CrossRef]
40. Zhang, Y.; Guo, W.; Zhang, H.; Wu, C. Influence of chain extension on the compatibilization and properties of recycled poly(ethylene terephthalate)/linear low density polyethylene blends. *Polym. Degrad. Stab.* **2009**, *94*, 1135–1141. [CrossRef]
41. Huang, J.M. Polymer blends of poly(trimethylene terephthalate) and polystyrene compatibilized by styrene-glycidyl methacrylate copolymers. *J. Appl. Polym. Sci.* **2003**, *88*, 2247–2252. [CrossRef]
42. Pracella, M.; Chionna, D.; Pawlak, A.; Galeski, A. Reactive mixing of PET and PET/PP blends with glycidyl methacrylate–modified styrene-*b*-(ethylene-*co*-olefin) block copolymers. *J. Appl. Polym. Sci.* **2005**, *98*, 2201–2211. [CrossRef]
43. Tapia, J.J.B.; Tenorio-López, J.A.; Martínez-Estrada, A.; Guerrero-Sánchez, C. Application of RAFT-synthesized reactive tri-block copolymers for the recycling of post-consumer r-PET by melt processing. *Mater. Chem. Phys.* **2019**, *229*, 474–481. [CrossRef]
44. Al-Sabagh, A.; Yehia, F.; Eshaq, G.; Rabie, A.; ElMetwally, A. Greener routes for recycling of polyethylene terephthalate. *Egypt. J. Pet.* **2016**, *25*, 53–64. [CrossRef]
45. Montava-Jordà, S.; Torres-Giner, S.; Ferrandiz-Bou, S.; Quiles-Carrillo, L.; Montanes, N. Development of sustainable and cost-competitive injection-molded pieces of partially bio-based polyethylene terephthalate through the valorization of cotton textile waste. *Int. J. Mol. Sci.* **2019**, *20*, 1378. [CrossRef]
46. Agüero, A.; Morcillo, M.d.C.; Quiles-Carrillo, L.; Balart, R.; Boronat, T.; Lascano, D.; Torres-Giner, S.; Fenollar, O. Study of the Influence of the Reprocessing Cycles on the Final Properties of Polylactide Pieces Obtained by Injection Molding. *Polymers* **2019**, *11*, 1908. [CrossRef]
47. Negoro, T.; Thodsaratpreeyakul, W.; Takada, Y.; Thumsorn, S.; Inoya, H.; Hamada, H. Role of crystallinity on moisture absorption and mechanical performance of recycled PET compounds. *Energy Procedia* **2016**, *89*, 323–327. [CrossRef]
48. Badia, J.; Vilaplana, F.; Karlsson, S.; Ribes-Greus, A. Thermal analysis as a quality tool for assessing the influence of thermo-mechanical degradation on recycled poly(ethylene terephthalate). *Polym. Test.* **2009**, *28*, 169–175. [CrossRef]

49. Jabarin, S.A.; Lofgren, E.A. Effects of water absorption on physical properties and degree of molecular orientation of poly(ethylene terephthalate). *Polym. Eng. Sci.* **1986**, *26*, 620–625. [CrossRef]
50. Mancini, S.D.; Zanin, M. Consecutive steps of PET recycling by injection: Evaluation of the procedure and of the mechanical properties. *J. Appl. Polym. Sci.* **2000**, *76*, 266–275. [CrossRef]
51. Oromiehie, A.; Mamizadeh, A. Recycling PET beverage bottles and improving properties. *Polym. Int.* **2004**, *53*, 728–732. [CrossRef]
52. Quiles-Carrillo, L.; Blanes-Martínez, M.M.; Montanes, N.; Fenollar, O.; Torres-Giner, S.; Balart, R. Reactive toughening of injection-molded polylactide pieces using maleinized hemp seed oil. *Eur. Polym. J.* **2018**, *98*, 402–410. [CrossRef]
53. Piergiovanni, L.; Limbo, S. Plastic packaging materials. In *Food Packaging Materials*; Springer: Berlin, Germany, 2016; pp. 33–49.
54. Quiles-Carrillo, L.; Montanes, N.; Garcia-Garcia, D.; Carbonell-Verdu, A.; Balart, R.; Torres-Giner, S. Effect of different compatibilizers on injection-molded green composite pieces based on polylactide filled with almond shell flour. *Compos. Part B Eng.* **2018**, *147*, 76–85. [CrossRef]
55. Badia, J.; Strömberg, E.; Karlsson, S.; Ribes-Greus, A. The role of crystalline, mobile amorphous and rigid amorphous fractions in the performance of recycled poly (ethylene terephthalate)(PET). *Polym. Degrad. Stab.* **2012**, *97*, 98–107. [CrossRef]
56. Vilaplana, F.; Karlsson, S.; Ribes-Greus, A. Changes in the microstructure and morphology of high-impact polystyrene subjected to multiple processing and thermo-oxidative degradation. *Eur. Polym. J.* **2007**, *43*, 4371–4381. [CrossRef]
57. Strömberg, E.; Karlsson, S. The design of a test protocol to model the degradation of polyolefins during recycling and service life. *J. Appl. Polym. Sci.* **2009**, *112*, 1835–1844. [CrossRef]
58. Quiles-Carrillo, L.; Duart, S.; Montanes, N.; Torres-Giner, S.; Balart, R. Enhancement of the mechanical and thermal properties of injection-molded polylactide parts by the addition of acrylated epoxidized soybean oil. *Mater. Des.* **2018**, *140*, 54–63. [CrossRef]
59. Frounchi, M. Studies on degradation of PET in mechanical recycling. In Proceedings of the Macromolecular Symposia; Wiley-VCH: Weinheim, Germany, October 1999; pp. 465–469.
60. Medellin–Rodriguez, F.; Phillips, P.; Lin, J.; Campos, R. The triple melting behavior of poly(ethylene terephthalate): Molecular weight effects. *J. Polym. Sci. Part B Polym. Phys.* **1997**, *35*, 1757–1774. [CrossRef]
61. Awaja, F.; Daver, F.; Kosior, E.; Cser, F. The effect of chain extension on the thermal behaviour and crystallinity of reactive extruded recycled PET. *J. Therm. Anal. Calorim.* **2004**, *78*, 865–884. [CrossRef]
62. Spinacé, M.A.S.; De Paoli, M.A. Characterization of poly(ethylene terephtalate) after multiple processing cycles. *J. Appl. Polym. Sci.* **2001**, *80*, 20–25. [CrossRef]
63. Inata, H.; Matsumura, S. Chain extenders for polyesters. IV. Properties of the polyesters chain-extended by 2, 2′-bis (2-oxazoline). *J. Appl. Polym. Sci.* **1987**, *33*, 3069–3079. [CrossRef]
64. Kiliaris, P.; Papaspyrides, C.D.; Pfaendner, R. Reactive-extrusion route for the closed-loop recycling of poly(ethylene terephthalate). *J. Appl. Polym. Sci.* **2007**, *104*, 1671–1678. [CrossRef]
65. Vannier, A.; Duquesne, S.; Bourbigot, S.; Castrovinci, A.; Camino, G.; Delobel, R. The use of POSS as synergist in intumescent recycled poly(ethylene terephthalate). *Polym. Degrad. Stab.* **2008**, *93*, 818–826. [CrossRef]
66. Levchik, S.V.; Weil, E.D. A review on thermal decomposition and combustion of thermoplastic polyesters. *Polym. Adv. Technol.* **2004**, *15*, 691–700. [CrossRef]
67. Dzięcioł, M.; Trzeszczyński, J. Temperature and atmosphere influences on smoke composition during thermal degradation of poly(ethylene terephthalate). *J. Appl. Polym. Sci.* **2001**, *81*, 3064–3068. [CrossRef]
68. Dzie cioł, M.; Trzeszczyn' ski, J. Studies of temperature influence on volatile thermal degradation products of poly(ethylene terephthalate). *J. Appl. Polym. Sci.* **1998**, *69*, 2377–2381. [CrossRef]

69. Wang, B.; Sun, S. An investigation on the influence of molecular structure upon dynamic mechanical properties of polymer materials. *Dev. Appl. Mater.* **2005**, *3*, 109–118.
70. Torres-Giner, S.; Montanes, N.; Fenollar, O.; García-Sanoguera, D.; Balart, R. Development and optimization of renewable vinyl plastisol/wood flour composites exposed to ultraviolet radiation. *Mater. Des.* **2016**, *108*, 648–658. [CrossRef]

© 2020 by the authors. Licensee MDPI, Basel, Switzerland. This article is an open access article distributed under the terms and conditions of the Creative Commons Attribution (CC BY) license (http://creativecommons.org/licenses/by/4.0/).

Article

Functionalization of Partially Bio-Based Poly(Ethylene Terephthalate) by Blending with Fully Bio-Based Poly(Amide) 10,10 and a Glycidyl Methacrylate-Based Compatibilizer

Maria Jorda [1], Sergi Montava-Jorda [2], Rafael Balart [1], Diego Lascano [1,3,*], Nestor Montanes [1] and Luis Quiles-Carrillo [1]

1. Technological Institute of Materials (ITM), Universitat Politècnica de València (UPV), Plaza Ferrándiz y Carbonell 1, 03801 Alcoy, Spain
2. Department of Mechanical Engineering and Materials (DIMM), Universitat Politècnica de València (UPV), Plaza Ferrándiz y Carbonell 1, 03801 Alcoy, Spain
3. Escuela Politécnica Nacional, Quito 17-01-2759, Ecuador
* Correspondence: dielas@epsa.upv.es; Tel.: +34-966-528-433

Received: 19 July 2019; Accepted: 9 August 2019; Published: 10 August 2019

Abstract: This work shows the potential of binary blends composed of partially bio-based poly(ethyelene terephthalate) (bioPET) and fully bio-based poly(amide) 10,10 (bioPA1010). These blends are manufactured by extrusion and subsequent injection moulding and characterized in terms of mechanical, thermal and thermomechanical properties. To overcome or minimize the immiscibility, a glycidyl methacrylate copolymer, namely poly(styrene-ran-glycidyl methacrylate) (PS-GMA; Xibond™ 920) was used. The addition of 30 wt % bioPA provides increased renewable content up to 50 wt %, but the most interesting aspect is that bioPA contributes to improved toughness and other ductile properties such as elongation at yield. The morphology study revealed a typical immiscible droplet-like structure and the effectiveness of the PS-GMA copolymer was assessed by field emission scanning electron microcopy (FESEM) with a clear decrease in the droplet size due to compatibilization. It is possible to conclude that bioPA1010 can positively contribute to reduce the intrinsic stiffness of bioPET and, in addition, it increases the renewable content of the developed materials.

Keywords: bio-based; poly(ethyelene terephthalate)—PET; poly(amide) 1010—PA1010; mechanical properties; morphology; compatibilization; Xibond™ 920

1. Introduction

In the last decade, there has been a noticeable increase in the sensitiveness and concern about environment. Topics such as sustainable development, circular economy, carbon footprint, petroleum depletion, among others are gaining relevance [1–3]. Therefore, many research works are focused on the development of environmentally friendly materials to positively contribute to a sustainable development. This situation is particularly aggravated in the polymer industry which accounts for the use of large amounts of petroleum-derived plastics with the subsequent environmental impact both at the origin (petroleum) and at the end of the life cycle or disposal (most of the petroleum-based polymers are not biodegradable). For these reasons, the polymer industry is demanding continuously environmentally friendly polymers It is worthy to note the important role that some petroleum-based polymers have acquired in the last decade. In particular, aliphatic polyesters such as poly(ε-caprolactone) (PCL) [4] poly(butylene succinate) (PBS) [5], poly(glycolic acid) (PGA) [6], poly(butylene succinate-*co*-adipate) (PBSA) and their blends/composites with other polymers and lignocellulosic fillers have gained interest in several industrial sectors, despite being petroleum-based, as they can undergo degradation under controlled compost soil [7–9]. Another promising group of environmentally friendly polymers includes polysaccharides (and derivatives), protein-based polymers and bacterial polymers. Poly(lactic acid) (PLA), together with thermoplastic starch (TPS), are perhaps the most studied polymers in this group that can be derived from polysaccharides [10], in particular, from starch-rich materials, i.e., potato, corn, bagasse, and so on. PLA is commercially available at a competitive price. Protein-based polymers include some interesting materials as gluten, soy protein, collagen, rape-seed protein, among others, that find applications in the form of film, parts, fibers, and so on [11–14]. Finally, bacterial polymers include all poly(hydroxyalkanoates) (PHAs) which are expected to invade the market soon [15,16]. Some of the most interesting PHAs include poly(3-hydroxybutyrate) (PHB), poly(3-hydroxybutyrate-co-valerate) (PHBV), poly(3-hydroxybutyrate-co-hexanoate) (PHBH) [17,18].

Despite all of these materials representing a clear environmental efficiency, in general, their properties are far from those of petroleum-derived commodity and engineering plastics. For this reason, many studies have been focused on obtaining commodity and engineering plastics from renewable resources. These show identical properties to their petroleum-based counterparts, but they offer interesting environmental efficiency as they can be totally or partially derived from renewable materials, usually bio-products coming from the food industry and agroforestry. Bio-based poly(ethylene) (bioPE), is a commodity that is synthesised from bioethanol from sugarcane and can reach almost 95% bio-based content. This shows a clear positive environmental efficiency compared to poly(ethylene) from crude oil [19–21]. Currently, bioPE is available worldwide at a relatively cost competitive price and it has been recently used as a base material for 3D printing [22]. In regard to engineering plastics, it is worthy to note the increasing consumption of bio-based poly(ethylene terephthalate) (bioPET) and bio-based poly(amides) (PAs) [23–26]. In recent years, poly(ethylene furanoate) (PEF) has generated great expectations as it can be fully bioderived and could potentially substitute poly(ethylene terephthalate) (PET) polymers [27]. Although in the future bioPET can reach 100% renewable source since there is a bio-route to synthesise terephthalic acid (TA) [28,29], currently its bio-based content is related to the ethylene glycol which can give approximately 30% bio-based content. Regarding bioPAs, castor oil plays a key role as a starting material for PA synthesis [30]. It is worthy to note bioPAs are engineering plastics with different bio-sourced content. Thus, bioPA610 typically offers 60–63% renewable content [31]; bioPA1012 usually offers a renewable content of 45% and bioPA1010 can be 100% derived from renewable resources from sebacic acid and 1,10-decamethylene diamine (DMDA), both derived from ricinoleic acid [32]. Some of these bioPAs have alternative eco-routes and could be fully bioderived. PET and, recently bioPET are widely used in the packaging industry for bottles. Despite this, some beverages (especially oxygen-sensitive beverages) require the use of scavengers that usually are derived from poly(amides), so that poly(amides) are increasingly present in the PET bottle-to-bottle cycle [33].

This work explores the potential of high bio-based content blends of partially bio-based poly(ethylene terephthalate) (bioPET) and fully bio-based poly(amide) 10,10 (bioPA1010) up to 30 wt %. Although bioPET and bioPA1010 show similar properties to their corresponding petroleum-derived counterparts, currently bioPET only contains approximately 30 wt % of biobased content while bioPA10120 can be fully bioderived from castor oil. The production of these partially or totally biobased polymers is increasing in a remarkable way and new biobased routes are being developed for partially biobased polymers to achieve 100% biosourced materials. This can have a positive effect on sustainable development and circular economies as most of the biobased building blocks could be obtained from by-products of the food or agroforestry industries. For these reasons, blending these two polymers is attractive from an environmental standpoint as these blends could reach high biobased contents without compromising other properties, thus leading to engineering blends with potential in the packaging industry. Due to their immiscibility, a glycidyl compatibilizer, namely a poly(styrene-ran-glycidyl methacrylate) copolymer (PS-GMA) Xibond™ 920 was used. The effect of both bioPA1010 and the PS-GMA compatibilizer are evaluated on their mechanical, thermal and thermomechanical properties. The novelty of this work is the high renewable content that can be obtained by these blends together with improved toughness.

2. Experimental

2.1. Materials

The partially bio-based poly(ethylene terephthalate), bioPET and fully bio-based poly(amide) 1010 were supplied by NaturePlast (Ifs, France). Table 1 summarizes the main properties of these commercial grades.

Table 1. Commercial grades and main properties of partially bio-based poly(ethylene terephthalate)—bioPET and fully bio-based poly(amide) 1010, supplied by NaturePlast.

Property	bioPET	bioPA1010
Grade	BioPET 001	NP BioPA1010-201
wt % bio-based	30	100
Melt temperature (°C)	240–260	190–210
Density (g cm^{-3})	1.3–1.4	1.05
Intrinsic viscosity (mL·g^{-1})	75–79	84–90 *

* measured between 230–240 °C.

The selected compatibilizer was a poly(styrene-glycidyl methacrylate) random copolymer (PS-GMA) Xibond™ 920 and was kindly provided by Polyscope (Geleen, The Netherlands). The GMA functionality has excellent affinity with polycondensates which can result in compatibilization, chain extension and/or branching. Figure 1 shows a scheme of the different materials used in this research.

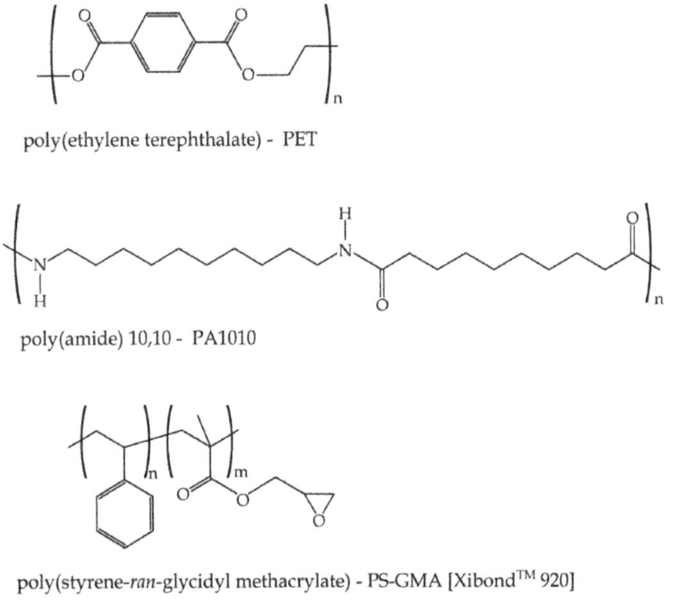

poly(ethylene terephthalate) - PET

poly(amide) 10,10 - PA1010

poly(styrene-ran-glycidyl methacrylate) - PS-GMA [Xibond™ 920]

Figure 1. Schematic representation of the chemical structure of bio(polyethylene terephthalate), bio-based poly(amide) 10,10 and glycidyl copolymer compatibilizer Xibond™ 920.

2.2. Manufacturing of Binary BioPET/BioPA Blends

Initially, all materials (see Table 2 for code and composition) were dried at 60 °C for 24 h to remove residual moisture. After this, the corresponding amounts of each material were mechanically mixed in a zipper bag, and then were fed into the hopper of a twin-screw co-rotating extruder from DUPRA S.L. (Castalla, Spain). The screw diameter was 30 mm and the temperature profile was set to four different barrels as follows (from the hopper to the die): 250 °C, 260 °C, 260 °C and 260 °C. The rotating speed was set to 20 rpm. After this initial compounding stage, the obtained blends were cooled down to room temperature and subsequently pelletized for further processing by injection moulding. The injection moulding machine used was a Mateu & Solé mod. Meteor 270/75 (Barcelona, Spain). The temperature profile was 240 °C (feeding Hopper), 245 °C, 250 °C and 255 °C (injection nozzle). The filling time was set at 1 s and the cooling time was 5 s.

Table 2. The compositions and labeling of binary bioPET/bioPA1010 blends. The bio-based content is calculated considering that bioPET contains an average bio-based content of 30 wt %, bioPA1010 is 100% bio-based and Xibond™ 920 is petroleum-derived (0 wt % bio-based).

Label	bioPET (wt %)	bioPA (wt %)	Xibond™ (phr)*	Bio-based content (wt %)
PET100	100	-	-	30.0
PET90	90	10	-	37.0
PET80	80	20	-	44.0
PET70	70	30	-	51.0
PET70Xibond1	70	30	1	50.5
PET70Xibond3	70	30	3	49.5
PET70Xibond5	70	30	5	48.6

* phr: weight grams of Xibond™ 920 per one hundred grams bioPET/bioPA blend.

2.3. Mechanical Characterization

The tensile properties were obtained through ISO 527-2:2012 standard on injection moulded dog-bone samples using an electromechanical machine ELIB-50 from S.A.E Ibertest (Madrid, Spain). All tests were run at a cross-head speed of 10 mm·min^{-1}, using a 5 kN loadcell. Regarding the impact strength, it was estimated through a Charpy test using a 6-J pendulum from Metrotec S.A. (San Sebastián, Spain) on the unnotched rectangular samples, following indications of ISO 179-1:2010. Finally, the hardness was obtained by using the Shore method in a 673-D durometer from J. Bot Instruments (Barcelona, Spain) as suggested by ISO 868:2003. All mechanical tests were run at room temperature and at least five different samples were tested to obtain the average characteristic parameters.

2.4. Thermal Characterization

Differential scanning calorimetry (DSC) was used to study the main thermal transitions of the manufactured materials. DSC tests were carried out on a Q200 calorimeter from TA Instruments (New Castle, DE, USA). A dynamic thermal program was scheduled in three different stages using standard aluminium crucibles. The first heating from 30 °C up to 280 °C was followed by a cooling down to 0 °C and a second heating up to 350 °C. The heating/cooling rate was set to 10 °C·min^{-1} with a constant nitrogen flow rate of 50 mL min^{-1}. The maximum degree of crystallinity was calculated for both bioPET and bioPA (see Equation 1) by comparing the melt enthalpy (ΔH_m) with the corresponding melt enthalpy of a theoretical 100% crystalline polymer (H_m^0 for PET = 140.1 J·g^{-1} [34], and 244.0 J·g^{-1} for PA1010 [35]), and considering the weight fraction of each polymer in the blend (w).

$$\%\chi_c = \frac{\Delta H_m}{\Delta H_m^0 \cdot w} \qquad (1)$$

Additional thermal characterization was carried out by thermogravimetry (TGA) in a TGA/SDTA 851 thermobalance from Mettler-Toledo (Schwerzenbach, Switzerland). A dynamic heating program from 20 °C to 700 °C at a heating rate of 20 °C·min^{-1} was applied to an average sample weight of 8 mg in an air atmosphere in alumina crucibles.

2.5. Morphology Characterization

Field emission scanning electron microscopy (FESEM) was used to reveal the morphology of the fractured surfaces blends after the impact tests. A FESEM microscope from Oxford Instruments (Abingdon, UK) was used working at an acceleration voltage of 1.5 kV. As the polymer blends were not electrically conducting materials, a metal sputtering process was carried out to provide conducting properties to the samples and to avoid sample charge. All fractured samples were coated with an ultrathin gold-palladium alloy in a Quorum Technologies Ltd. EMITECH model SC7620 sputter coater (East Sussex, UK).

2.6. Thermo-Mechanical Characterization

The effect of the temperature on the mechanical properties and dimensional stability was studied by dynamic-mechanical thermal analysis (DMTA) and thermomechanical analysis (TMA) respectively. Thermomechanical analysis was carried out in a Q400 thermoanalizer from TA Instruments (New Castle, DE, USA). The particular conditions for this test were a dynamic thermal sweep from 0 °C up to 140 °C at a constant heating rate of 2 °C·min^{-1} with an applied load of 20 mN. Regarding dynamic-mechanical thermal characterization, an oscillatory rheometer AR-G2 from TA Instruments (Delaware, USA), equipped with a special clamp system for solid samples (working in a combination of shear-torsion stresses) was used. The maximum shear deformation (%) was set to 0.1% at a frequency of 1 Hz. The thermal sweep was scheduled from 20 °C up to 160 °C at a heating rate of 5 °C·min^{-1}.

3. Results and Discussion

3.1. Mechanical Properties and Morphology of Binary BioPET/BioPA Blends

It has been reported that mechanical properties of PET are highly dependent on the processing conditions [36,37]. In addition, the mechanical properties of PET polymers are also dependent on the thermal treatment, quenching, annealing, etc. As it can be seen in Figure 2, the mechanical and thermal properties of bioPET are highly dependent on the annealing time. To assess this, bioPET has been subjected to different annealing times and studied by dynamic-mechanical thermal analysis (DMTA). Figure 2a shows the evolution of the storage modulus (G') as a function of the increasing temperature. DMTA is based on the use of a dynamical time-dependent stress function, $\sigma = \sigma_0 \sin(\omega t)$ [σ_0 is the maximum stress and ω represents the frequency] which produces a sinusoidal strain (ε) given by $\varepsilon = \varepsilon_0 \sin(\omega t-\delta)$ where ε_0 is the maximum strain and δ is the phase angle which represents the delay (viscous) properties of the material. As the modulus represents the ratio between the maximum stress to the maximum strain, then it is possible to define the complex modulus (G^*) as $\sigma_0 = \varepsilon_0 \, G^* \sin(\omega t + \delta)$. This expression can be expanded to give $\sigma_0 = \varepsilon_0 \, G^* \sin(\omega t) \cos(\delta) + \varepsilon_0 \, G^* \cos(\omega t) \sin(\delta)$, which can be expressed as $\sigma_0 = \varepsilon_0 \, G' \sin(\omega t) + \varepsilon_0 \, G'' \cos(\omega t)$ with $G' = G^* \cos(\delta)$ and $G'' = G^* \sin(\delta)$, thus leading to an elastic response related to G' (storage modulus) and a viscous response related to G'' (loss modulus). As it can be derived, the ratio between the loss modulus (G'') to the storage modulus (G') represents the damping factor or tanδ, which is directly related to the lost energy due to viscoelastic behaviour.

It can be seen that, neat bioPET shows a characteristic DMTA curve characterized by different zones. Below 60 °C, the storage modulus remains almost constant at a temperature range comprising between 60 °C and 80 °C, and a remarkable decrease in G' occurs. This decrease of near three orders of magnitude is representative of the glass transition process. In addition, information is provided about the high amorphous structure due to this three-fold change. Then, at the temperature range comprising between 106 °C and 125 °C, it is possible to observe an increase in G' which is directly related to the cold crystallization process which involves packing of polymer chains in an ordered way which gives increased stiffness. After 15 min annealing time at 110 °C, the morphology of the DMTA curve has changed in a remarkable way. As it can be seen, the decrease in G' is remarkably lower. The glass transition process has shifted to higher temperatures (probably due to the restriction of chain mobility in the crystalized structure) but it is still possible to find a slight increase in G' in the temperature range comprising between 106–120 °C. Nevertheless, with an annealing of 30 min, it is possible to conclude that the maximum crystallinity is achieved. The curves for 30 and 45 min are shifted to the right (higher temperatures) and the cold crystallization process has completely disappeared. Similar findings have been reported by A. Bartolotta et al. [38] who showed a remarkable change in the glass transition onset from 40 °C (cold-drawn PET) up to 90 °C in highly crystalline PET. A. Bartolotta et al. attribute this phenomenon to an increase in density on glassy domains related to the presence of more crystal-packed domains and conclude that there is a link between the chain stiffness since there is a connection between the bulk glass to the ordered structures. Z. Chen et al. [39] have also reported different crystallization mechanisms depending on the annealing temperature with remarkable changes, not only in the glass transition process but also on the melt temperature. In addition, the G' values are higher with increasing annealing time.

Figure 2b shows the evolution of the dynamic damping factor, tan δ. Notably, the maximum damping factor value decreases with the increasing annealing time. This is consistent with the damping factor definition which shows the ratio between the loss modulus (G'') and the storage modulus (G'). As the material becomes stiffer by the cold crystallization process, the denominator is higher, and this leads to lower damping factor values. It is worthy to note that the damping factor is related to the lost energy to stored energy ratio. As the annealing time increases, the material becomes stiffer and this is responsible for less lost energy. There are several methods to assess the glass transition temperature (T_g) by using several methods from DMTA, i.e., the onset of the G' decrease, the peak maximum of G'' or the peak maximum of the damping factor. The peak temperature of tan δ is widely used to give accurate values of T_g. The un-annealed bioPET shows a T_g of 79.9 °C which increases progressively with increasing annealing time at 110 °C resulting in values of 84.8 °C (15 min annealing), 93.7 °C (30 min annealing) and 96.3–96.4 °C for 45 and 60 min annealing. These results are in total accordance with those reported by A. Bartolotta et al. [38], who showed a change in the onset of T_g (using the G' method) from 40 °C up to 90 °C.

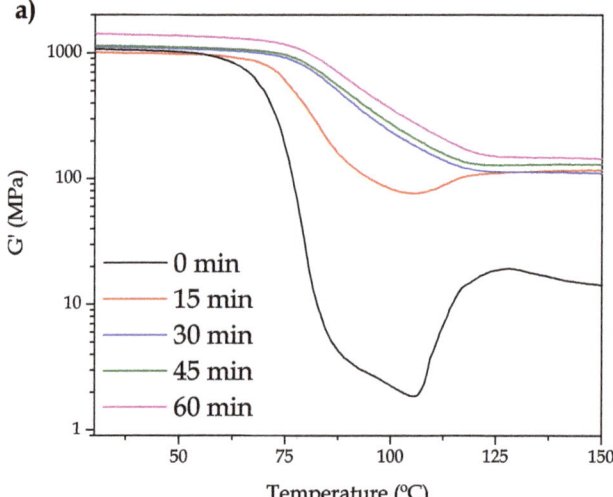

Figure 2. Cont.

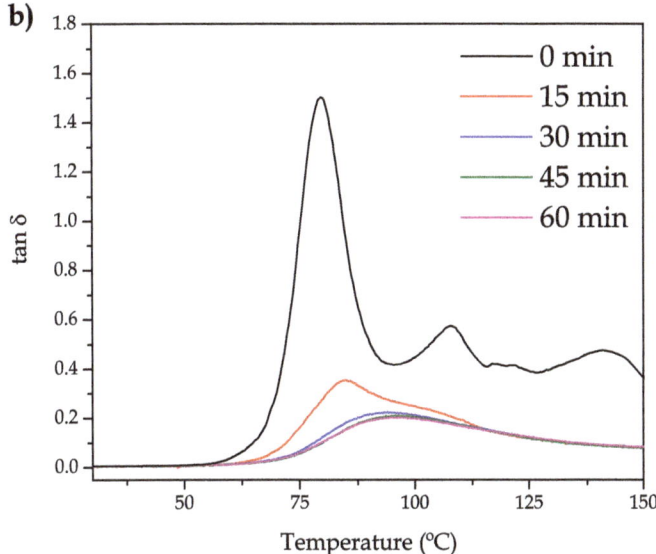

Figure 2. The effect of annealing on the dynamic-mechanical properties of neat bioPET subjected to different annealing times, (**a**) storage modulus, G' and (**b**) dynamic damping factor (tan δ).

In this work, bioPET and its blends have been characterized without any annealing process, just as obtained by the injection moulding. Neat bioPET showed a tensile strength (σ_t) of 46.7 MPa and a relatively low elongation at yield (ε_y) of 3.87% which leads to a stiff material. The effect of the addition of bioPA to bioPET produces two important effects. On the one hand, the tensile strength decreases as expected in an immiscible blend as reported by K.C. Chiou et al. [40], in PA6/PBT binary blends, but the decrease is not so pronounced. In fact, the maximum percentage decrease is close to 11% for the uncompatibilized blend containing 30 wt % bioPA. It is important to remark that the bio-based content of the blend containing 30 wt % bioPA is above 50% which is a positive property from an environmental point of view. On the other hand, the addition of a flexible polymer such as bioPA, provides improved elongation at yield up to values of 4.8% which represents a percentage increase of 24%, thus leading to improved ductile behaviour. The effect of the poly(styrene-*ran*-glycidyl methacrylate) copolymer (PS-GMA) Xibond™ 920 gives interesting results. It is worthy to note that the addition of 3 phr Xibond™ 920 leads to higher tensile strength regarding neat bioPET, reaching values of 47.1 MPa with a parallel increase in elongation up to values of 6.09% (57% increase compared to neat bioPET and 27% the same composition without compatibilizer). This indicates that Xibond™ 920 is providing somewhat compatibilization properties to this binary system. Similar findings have been obtained using a Zn^{2+} ionomer on PET/PA6 blends with improved elongation at break and toughness compared to an uncompatibilized blend [41]. Y. Huang et al. [42], reported the exceptional compatibilizing effect of the glycidyl group by using an epoxy resin (E-44) as a compatibilizer in PET/PA6 blends. C.T. Ferreira et al. [43] reported the potential of reactive extrusion of recycled PET and recycled PA by a reaction of the carboxyl end-chain groups in PET and the amine end-chain groups in PA with a noticeable improvement in mechanical properties using tin(II) 2-ethylhexanoate as a trans-reaction catalyst. Regarding the hardness, as both polymers show similar Shore D values, it is not possible to observe a tendency with varying composition as shown in Table 3.

Table 3. The mechanical properties of binary bioPET/bioPA blends obtained from tensile, hardness and Charpy tests.

Code	σ_b (MPa)	ε_y (%)	Shore D	Impact Strength (kJ·m^{-2})
PET100	46.7 ± 2.3	3.87 ± 0.30	75 ± 2.5	23.1 ± 4.4
PET90	41.5 ± 4.6	4.30 ± 0.39	75 ± 2.5	27.0 ± 3.8
PET80	42.8 ± 2.5	4.71 ± 1.04	75 ± 2.8	30.3 ± 3.6
PET70	41.4 ± 0.6	4.80 ± 0.40	75 ± 2.9	40.5 ± 9.9
PET70Xibond1	41.3 ± 0.8	5.01 ± 0.69	73 ± 2.9	42.9 ± 2.7
PET70Xibond3	47.1 ± 1.1	6.09 ± 0.86	75 ± 2.3	43.4 ± 1.6
PET70Xibond5	40.6 ± 4.5	6.63 ± 1.94	74 ± 2.5	44.6 ± 3.9

Regarding the impact strength which is measured through the Charpy impact test, it is worthy to note that all the developed materials have increased impact strength in comparison to neat bioPET. Neat bioPET offers a quite brittle behaviour with an impact-absorbed energy of 23.1 kJ·m^{-2}. The uncompatibilized binary blend with 10 wt % bioPA1010 offers an increased impact strength of 27.0 kJ·m^{-2} (which represents a percentage increase of approximately 16.9%). This result is in total agreement with other ductile properties such as elongation at yield (ε_y). Y. Yan et al. [44] reported the lubricant effect of PA56 on PET blends at a molecular level with the subsequent effect on mechanical properties. This phenomenon has been observed in other binary blends composed of a brittle polymer matrix in which a rubber-like material is finely dispersed, even with poor miscibility between them. J. Urquijo et al. [45] reported a remarkable increase in both elongation at the break and impact strength in binary blends of poly(lactic acid) (PLA) and different loadings of poly(ε-caprolactone) (PCL). J. Urquijo et al. demonstrated the relevance of the elongation rate on the final elongation and, regarding the impact strength, they attributed the improvement to the small particle size of PCL-rich domains embedded in the brittle PLA-rich matrix which positively contributed to absorb energy in impact conditions. Similar findings have been reported with PLA/PCL, PHB/PCL, PLA/PBS binary blends [46–49], ternary PLA/PHB/PCL blends [50], and some poly(ester) copolymers [51]. This improved toughness is more evident in an uncompatibilzed blend containing 30 wt % bioPA1010 reaching an impact strength of 40.5 kJ·m^{-2} (75.3% increase). The effect of the compatibilizer has a positive effect on improved toughness as it can be seen in Table 2. Xibond™ 920 is a random copolymer of poly(styrene-glycidyl methacrylate) (PS-GMA) and gives excellent results in compatibilizing condensation polymers. This is because of the glycidyl methacrylate group which can interact with both hydroxyl end-groups in bioPET and amine (primary or secondary) in bioPA1010, thus leading to somewhat compatibilization with a marked effect on impact strength. The addition of 5 phr Xibond™ 920 gives an impact strength of 44.6 kJ·m^{-2} (93.1% increase compared to neat bioPET and an additional 10.1% compared to the uncompatibilized blend containing 30 wt % bioPA1010). It is evident the positive effect of the compatibilizer in improved toughness. With regard to the bio-based content, the blends with 30 wt % bioPA with compatibilizer show a bio-based content of approximately 50%. The GMA-based copolymers have been reported as good compatibilizers in different blends containing PAs or polyesters due to their reactivity with both polymers, such as PA6/PP [52,53], PA6/PVF [54], PET/PP [55], HDPE/PET [56].

The improved toughness is directly related to the morphology of the obtained materials. Figure 3 gathers FESEM images of uncompatibilized bioPET/bioPA blends. Figure 3a shows the fracture surface of neat PET with a typical rigid and brittle material, that is, very smooth surfaces resulting from microcrack appearances and their growth without plastic deformation. This brittle behaviour for neat PET has been reported by A.R. McLauchlin et al. [57] in PET/PLA blends. Figure 3b shows the morphology of the uncompatibilized binary blend containing 10 wt % bioPA. This morphology is remarkably different to neat bioPET. In particular, it is possible to observe a brittle fracture surface with a noticeable increase in roughness due to presence of small-sized bioPA immiscible droplets embedded into the bioPET matrix. As the bioPA wt % increases, the roughness is more evident, and the characteristic brittle fracture disappears. In Figure 3d, corresponding to the binary blend with 30 wt %

bioPA, it is possible to observe the characteristic droplet-like structure of an immiscible polymer blend in which bioPA appears in the form of spherical droplets with an average size of approximately 3.9 ± 1.1 µm. This size is higher than the average size observed for lower bioPA content. It is evident that by increasing the bioPA content, the average particle size increases due to the particle coalescence as reported by A.M. Torres-Huerta et al. [58] on PET/PLA and PET/chitosan blends. Y. Yan et al. [44] reported the high immiscibility of PET blends with PA56 (up to 30 wt %) and used dissipative particle dynamics (DPD) to assess the immiscibility of both polymers and how the PA56 domains increase with increasing PA56 content. The poor compatibility between these polymers can be observed at a 30 wt % bioPA in the blends as the morphology shows the typical spherical bioPA droplets embedded in the matrix (sea-island morphology) as well as some voids with the same average diameter consisting of some bioPA droplets on the holes which have been produced after being pulled out during the impact test. This pulling out occurs because of the poor polymer-polymer interactions.

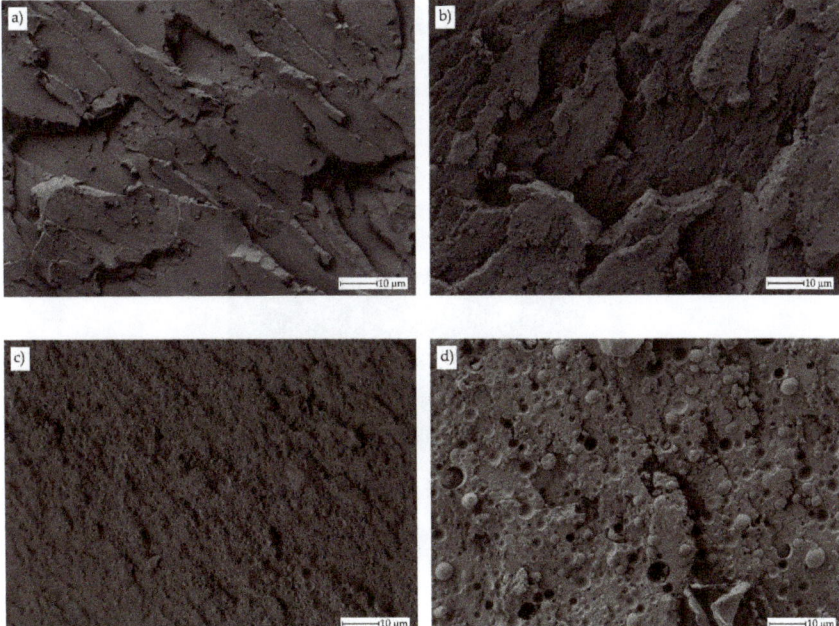

Figure 3. Field emission scanning electron microscope (FESEM) images of the fractured surface from an impact test at 1000x corresponding to uncompatibilized bioPET/bioPA blends with different bioPA content, (**a**) 0 wt % (PET100), (**b**) 10 wt % (PET90), (**c**) 20 wt % (PET80) and (**d**) 30 wt % (PET70).

These results suggest that the poly(styrene-*ran*-glycidyl methacrylate) copolymer (PS-GMA) can positively contribute to a partial compatibilization effect due to the reaction of the glycidyl groups with both hydroxyl terminal groups in bioPET and amine groups in bioPA [54,55]. Y. Huang et al. [42] reported the reactivity of the glycidyl group of an epoxy resin (E-44) in a PET/PA6 binary blend. As the nature of both bioPET and bioPA is the same as petroleum-derived PET and other semicrystalline polyamides, it is possible to assume similar reactions as described in by Y. Huang et al. That study reported the greater tendency of the glycidyl group to react with polyamide due to the presence of many hydrogen bonding (together with carboxylic and amine end-chains) while the reaction of the glycidyl group with PET is restricted to hydroxyl and carboxyl end-chains. Moreover, Y. Huang et al. further reported evidences of these reactions by extracting the polyamide fraction by formic acid which was subjected to FTIR analysis. This analysis showed a shift of the N–H bending from 1560 to 1544

cm^{-1} and shift of the C=O stretching from 1662 to 1642 cm^{-1}, both changes indicating the reaction of PA6 with the glycidyl group in E-44 epoxy resin. In addition, the characteristic peaks of PET at 1730, 1104 and 730 cm^{-1} were also detectable by FTIR thus giving consistency to the grafting process.

The indirect effects of these reactions can also be detectable by a remarkable change in surface morphology as it can be seen in Figure 4. The addition of PS-GMA Xibond™ 920 gives a noticeable decrease in the droplet size of bioPA-rich domains. There is not a great difference between the images corresponding to the compatibilized bioPET/bioPA blends containing 1, 3 or 5 phr PS-GMA. At this magnification (1000×), it can be realized that the droplet diameter has been reduced down to values under 1 μm. This situation can be clearly observed in Figure 5 which shows a comparative FESEM image of the uncompatibilized blend with 30 wt % bioPA and the same blend with 3 phr PS-GMA Xibond™ 920.

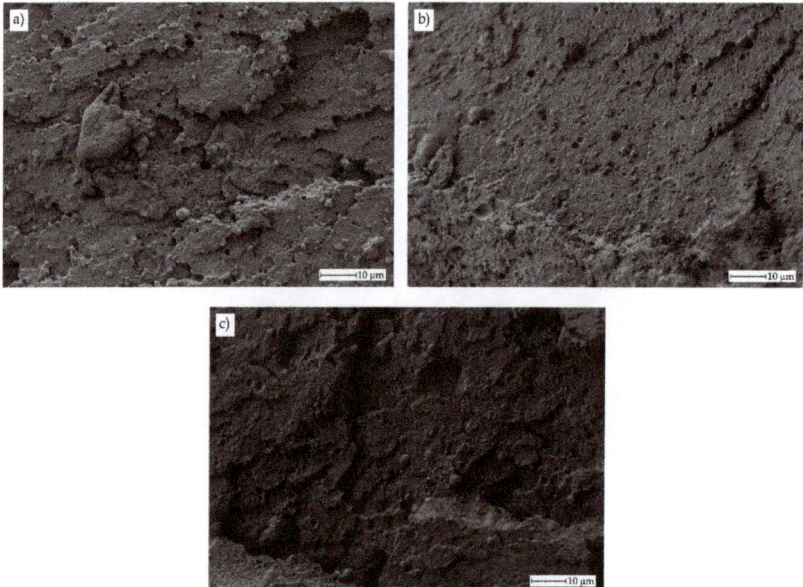

Figure 4. Field emission scanning electron microscope (FESEM) images of the fractured surface from an impact test at 1000x corresponding to compatibilized bioPET/bioPA blends with different loadings of Xibond™ 920 (in phr), (**a**) 1 phr, (**b**) 3 phr and (**c**) 5 phr.

As can be seen in Figure 5, the uncompatibilized blend (Figure 5a) shows a characteristic morphology, typical of poor polymer-polymer interactions. The particle droplet average diameter is 3.9 ± 1.1 μm as indicated previously. It is possible to observe this lack (or poor) interaction between both polymers. In fact, it is evident that some bioPA droplets have been removed (white rectangles) and there is a small gap between the bioPA spheres and the surrounding bioPET matrix (white arrows). Nevertheless, this gap is relatively low compared to other immiscible systems (it is in the nanoscale range), and this contributes to improved tensile properties and toughness as indicated previously. Furthermore, when observing Figure 5b is that the droplet size has been reduced in a remarkable way in the same blend with 3 phr PS-GMA compatibilizer. The new droplet size for the compatibilized blend is 0.62 ± 0.27 μm which is remarkably lower than the average size of bioPA-rich domains in the uncompatibilized blend. In addition, the surface morphology of the polymer matrix is different. As can be seen in Figure 5a for the uncompatibilized blend, the bioPET matrix is quite smooth while this surface is completely different in the compatibilized blend (Figure 5b) since it is remarkably rougher as reported by Y. Pietrasanta et al. [56] on HDPE/PET blends compatibilized with glycidyl polymers. Regarding the gap between the embedded bioPA droplets, the morphology is also different since these embedded bioPA domains seem to be more embedded in the compatibilized blend. Similar findings have been reported for other PET-based immiscible blends such as those developed by C. Carrot et al. [59] (PET/PC) with a clear change in morphology in the presence of compatibilizers, O.M. Jazani et al. [60] (PET/PP), A.M. Torres-Huerta et al. [58] (PET/PLA) and (PET/chitosan), among others.

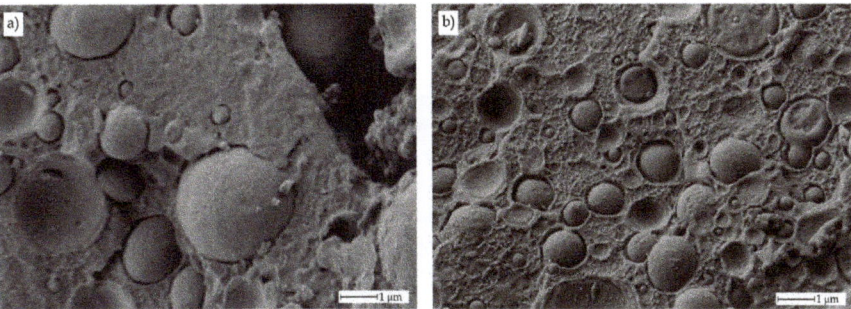

Figure 5. Detailed FESEM images corresponding to (**a**) uncompatibilized bioPET/bioPA blend with 30 wt % bioPA (PET70) and (**b**) compatibilized bioPET/bioPA blend with 30 wt % bioPA and 3 phr Xibond™ 920 (PET70Xibond3).

3.2. Thermal and Thermo-Mechanical Properties of Binary BioPET/BioPA Blends

The main thermal transitions of the developed materials are gathered in Figure 6. The neat bioPET shows in a clear way the main transitions. The step change in the 70–80 °C range corresponds to the glass transition phenomenon (T_g) and it is 75.2 as shown in Table 4. Then, a peak located in the 120–140 °C range corresponds to the cold crystallization process which involves crystallization of the fraction that has not been able to crystallize because of the cooling rate. This process shows a characteristic peak temperature (T_{cc_PET}) of 133.2 °C. Finally, the melt process can be observed at higher temperatures of 225–260 °C with a peak temperature of 248.2 °C. The addition of bioPA up to 30 wt % on uncompatibilized blends does not provide any remarkable change in the T_g with values of approximately 75–76 °C, very similar to neat bioPET. Although these T_g values cannot be clearly seen in Figure 6a,b, the T_g values were obtained from the zoomed DSC thermograms in the 65–85 °C leading to the values shown in Table 4. Regarding the cold crystallization process, bioPA plays a key role in this process. By the addition of 10 wt % bioPA, the peak temperature moved down to values of 121.9 °C, thus indicating bioPA enables crystallization of bioPET. Above 10 wt % bioPA, the cold crystallization process disappears and a slight decrease in the maximum crystallinity of bioPET (calculated with the obtained melt enthalpy values, ΔH_{m_PET}) can be detected as seen in Table 3. In fact, neat bioPET shows a maximum degree of crystallinity of 22.7% and it is slightly reduced to the values of 19.9% for the uncompatibilized blend containing 30 wt % bioPA. The melt peak temperature does not change in a remarkable way for all the developed materials and moves between the 247–248 °C narrow range. The effect of the PS-GMA compatibilizer is interesting. As can be seen in Table 4, a clear decrease in the crystallinity is detected from 19.9% (uncompatibilized blend with 30 wt % bioPA) to 17.3% in same composition with 3 phr Xibond™ 920. These results are in total agreement with those reported by Y. Huang et al. [42] who indicated a key role of the interface on crystallization as the interface is directly related to two relevant phenomena: Crystal nucleation and crystal growth. Y. Huang et al. report the use of an epoxy resin (E-44) as a compatibilizer in PET/PA6 blends and they conclude that although the epoxy resin can positively contribute to improve mechanical properties, a decrease in crystallinity is observed with increasing E-44 content due to the formation of less perfect crystals as a consequence of the increased interactions. Moreover, this study confirmed independent crystallization of PET and PA6 as suggested by the wide angle of the x-ray diffraction spectroscopy

(WAXD). In fact, Y. Huang et al. also report a different effect of epoxy compatibilization on hindering crystallization on both PET and PA6. The glycidyl group has more reactive points with PA6 due to the high number of hydrogen bonding in the structure while the reaction of the glycidyl group with PET is restricted to hydroxyl and carboxyl groups located at the end-chains. Y. Huang et al. reported a percentage decrease in the melt enthalpy of PET of approximately 25.9% while the decrease for PA6 is close to 40%. Due to the nature of both bioPET and bioPA, the same behaviour with the glycidyl compatibilizer is expected as can be seen in Table 4 with a clear large decrease in the melt enthalpy of bioPA compared to bioPET with increasing Xibond™ 920. On the other hand, Quiles-Carrillo et al. [61] reported a clear decrease in crystallinity by reactive extrusion of PLA with maleinized hemp seed oil (MHO). This decrease was attributed to the high reactivity of the maleic anhydride groups towards the hydroxyl groups in PLA which can give chain extension, branching and even, some crosslinking, all these phenomena having a negative effect on crystallization and formation of imperfect crystals. Quiles-Carrillo et al. [62] also reported a similar effect on PLA by using another reactive compatibilizer derived from soybean oil, namely, acrylated epoxidized soybean oil (AESO).

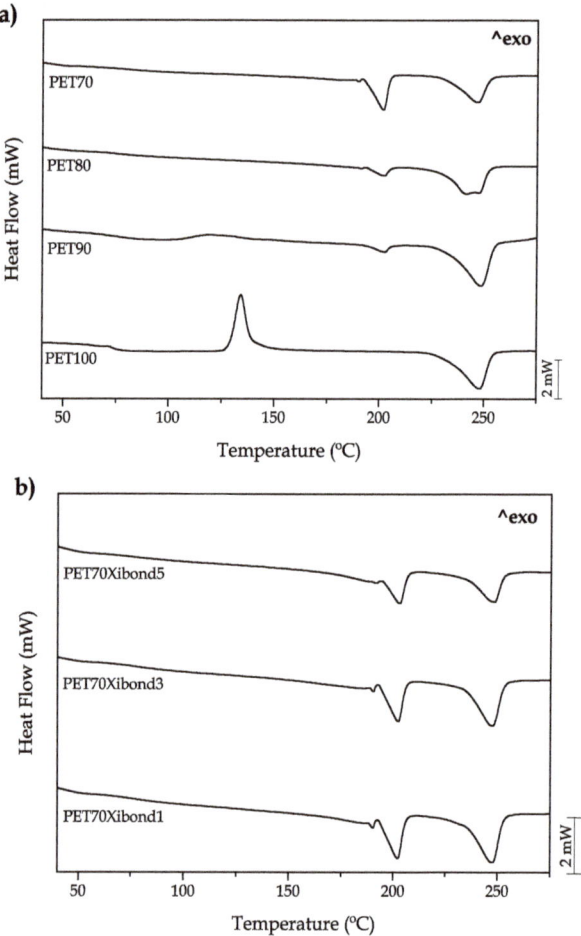

Figure 6. A plot comparative of the differential scanning calorimetry (DSC) thermograms corresponding to binary bioPET/bioPA blends.

The immiscibility of both polymers is also evident from DSC as two melt peaks are obtained with very slight changes in their corresponding peak temperature values. BioPA shows a melt peak located at 202–203 °C with whatever the composition may be. Nevertheless, the crystallinity is also affected by the presence of bioPET as the major component. As expected, the crystallinity of bioPA increases with increasing bioPA content from 9.8% (10 wt %) up to 16.9% (30 wt %) since the presence of higher bioPA loadings promote more intense and independent nucleation and crystal growth processes. Nevertheless, the effect of PS-GMA is the same as in the case of bioPET. The reaction between glycidyl groups in PS-GMA with both bioPET and bioPA polymer chains leads to the formation of imperfect crystals which is responsible for a decrease in the overall crystallinity as seen in Table 4, down to values of 10.2% for the compatibilized blend with 30 wt % bioPA and 5 phr Xibond™ 920. Another interesting finding is that the compatibilizer leads to a slight increase in the T_g of bioPET up to values of 78 °C which is indicating that chain mobility is restricted. Despite this, the determination of T_g by DSC is sometimes complex and inaccurate due to the problems related to the base line and the dilution effect in polymer blends. Similar findings have been reported by D. Garcia-Garcia et al. [63], using reactive extrusion of PHB and PCL with different dicumyl peroxide (DCP) loadings. The reaction of the free radicals generated by DCP can react with both PCL and PHB thus leading to partial compatibilization. These reactions reduce chain mobility as observed by the dynamic mechanical-thermal analysis (DMTA).

Table 4. A summary of the main thermal parameters of binary bioPET/bioPA blends obtained by differential scanning calorimetry (DSC).

Code	bioPET						bioPA		
	T_g (°C)	T_{cc} (°C)	ΔH_{cc} (J g^{-1})	ΔH_m (J g^{-1})	T_m (°C)	χ_c (%)	ΔH_m (J g^{-1})	T_m (°C)	χ_c (%)
PET100	75.2	133.2	27.6	−31.8	248.2	22.7	-	-	-
PET90	75.8	121.9	11.7	−27.5	248.8	21.8	−2.4	202.5	9.8
PET80	75.6	-	-	−25.3	247.6	22.6	−4.9	202.4	10.0
PET70	75.4	-	-	−19.5	246.7	19.9	−12.4	201.9	16.9
PET70Xibond1	78.3	-	-	−19.1	247.9	19.7	−8.8	202.4	12.1
PET70Xibond3	78.6	-	-	−18.4	247.6	19.3	−8.7	202.5	12.2
PET70Xibond5	77.3	-	-	−16.20	248.1	17.3	−7.1	203.1	10.2

Regarding thermal stability (degradation at high temperatures), Table 5 shows a summary of some thermal degradation parameters obtained by thermogravimetry (TGA). Two different characteristic temperatures are gathered in this table, the temperature required for a weight loss of 5% which is representative for the onset degradation ($T_{5\%}$) and the maximum degradation rate temperature (T_{max}) which corresponds to the peak maximum of the first derivative TGA curve (DTG). As can be seen, the $T_{5\%}$ for the neat bioPET is 382.6 °C and the addition of bioPA contributes to delay the onset degradation process as the $T_{5\%}$ characteristic temperature is moved up to 397.4 °C for the uncompatibilized blend containing 30 wt % bioPA. This is because PA1010 has more thermal stability than PET. The effect of the PS-GMA compatibilizer is a slight increase in the onset degradation temperature up to values close to 404 °C with 3 phr Xibond™ 920. Regarding the maximum degradation rate, it is worthy to note a decreasing tendency with increasing bioPA loading on blends. This could be related to the fact that PA1010 is more thermally stable than PET but the degradation rate of PA1010 (change in weight loss with temperature) is higher than PET. For this reason, the T_{max} shows a decreasing tendency. S. Jiang et al. [64] reported that the onset degradation temperature of PA1010 is located at 419.2 °C which is remarkably higher than PET, thus contributing to improved thermal stability.

Table 5. A summary of the thermal degradation of binary bioPET/bioPA blends obtained by thermogravimetry (TGA) analysis.

Code	$T_{5\%}$ (°C)	T_{max} (°C)
PET100	382.6	452.6
PET90	392.8	450.4
PET80	393.2	443.7
PET70	397.4	441.1
PET70Xibond1	399.3	442.6
PET70Xibond3	403.7	446.9
PET70Xibond5	394.7	441.8

Regarding the effect of bioPA and the PS-GMA copolymer on mechanical-dynamical thermal properties, Figure 7 gathers some characteristic curves corresponding to the neat bioPET and the uncompatibilized and compatibilized (5 phr Xibond™ 920) blend with 30 wt % bioPA. Two main effects can be observed on the storage modulus, G'. On the one hand, bioPET is stiffer than its blends with bioPA independently of the PS-GMA compatibilizer. T. Serhatkulu et al. [65] showed this flexibilization phenomenon on PET/PA6 blends. On the other hand, the presence of bioPA minimizes the cold crystallization process as observed in DSC. In fact, some cold crystallization occurs in bioPET/bioPA blends but DSC is not sensitive enough to detect it. However, these slight changes can be observed by DMTA as seen in Figure 7a. Another interesting phenomenon is the shift of the cold crystallization process towards lower temperatures which is in total accordance with the results obtained by DSC. The intensity of the cold crystallization can be observed in Figure 7b as the shoulder located to the right side. The T_g values follow a similar tendency as that observed with DSC but DMTA seems to be more accurate to obtain these parameters. In particular, the T_g for neat bioPET is 79.9 °C while the binary blend with 30 wt % bioPA shows a T_g of 81.1 °C and the compatibilized blend (PET70Xibond5) offers a T_g of 80.5 °C.

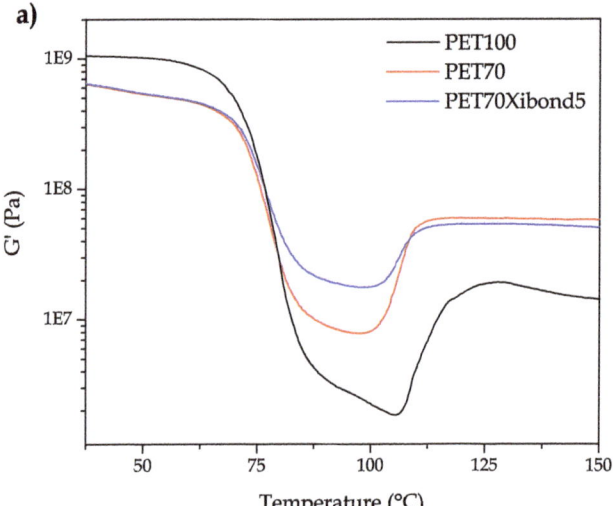

Figure 7. Cont.

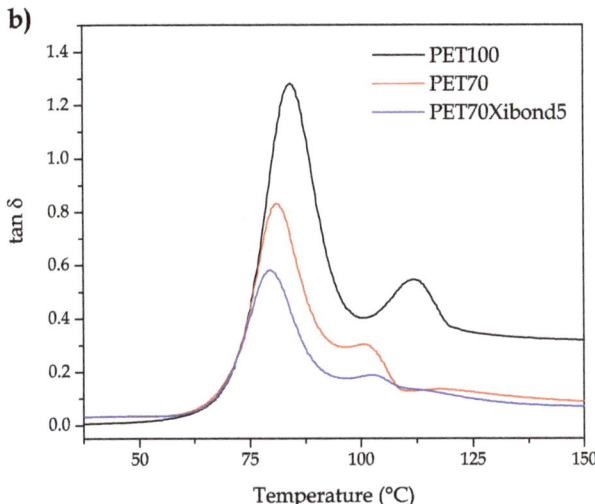

Figure 7. The dynamic mechanical thermal behaviour (DMTA) of binary bioPET/bioPA blends in terms of increasing temperature (**a**) storage modulus, G' and (**b**) dynamic damping factor, tan δ.

In addition to the dynamic mechanical-thermal analysis (DMTA), the dimensional stability has been studied by thermomechanical analysis (TMA). Figure 8 shows the TMA profiles of neat bioPET as well as the uncompatibilized and compatibilized (5 phr Xiboond™ 920) blend containing 30 wt % bioPA. From these TMA curves, it is possible to see the thermal behaviour of these materials. Below 60 °C, all three materials show a linear expansion (see Table 6 for values of the coefficient of linear thermal expansion, CLTE). Below this temperature, the slope is low compared to the slope above 120 °C. The glass transition temperature (T_g) can be observed in the temperature range of 65–80 °C as the onset of a change in the slope. The slope is very high in the rubbery state from 80 °C up to 100 °C. Then, the slope is negative which is indicating increased dimensional stability. This is caused by the cold crystallization process. As seen previously by DSC, the cold crystallization peak is clearly detectable for neat bioPET and it is almost negligible for blends with high bioPA content. These results are in accordance with those shown in Figure 6 as the highest change in the dimensions can be seen for neat bioPET due to the cold crystallization process. Nevertheless, this change is very short for the other developed materials. Finally, above 120 °C, the linear tendency stabilizes, therefore indicating the cold crystallization has finished. Regarding the CLTE values (Table 6), it is worthy to note they follow the same tendency observed for ductile properties. The CLTE for neat bioPET is 152.4 μm m^{-1} K^{-1}, and it increases with increasing bioPA loading up to values of 347.3 μm m^{-1} K^{-1}. The effect of the compatibilizer is that expected since the reaction between the PS-GMA and bioPET and bioPA produces a restriction on chain mobility and this has a positive effect on dimensional stability. Notably, the CLTE value for the blend with 30 wt % bioiPA is compatibilized with 5 phr Xibond™ 920. All these results are in total agreement with the mechanical properties above-mentioned.

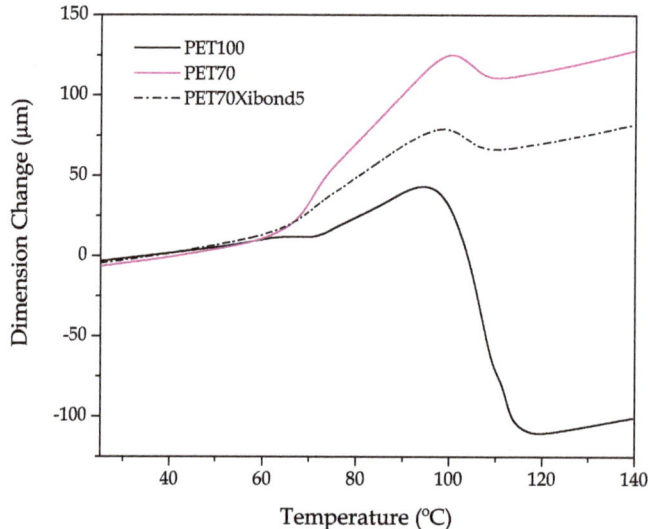

Figure 8. A comparative plot of thermomechanical behaviour of binary bioPET/bioPA blends in terms of increasing temperature.

Table 6. The calculated coefficient of linear thermal expansion (CLTE) of bioPET/bioPA blends obtained by thermomechanical analysis (TMA).

Code	CLTE ($\mu m \cdot m^{-1} \cdot K^{-1}$)*
PET100	152.4 ± 12.2
PET90	162.4 ± 10.4
PET80	262.1 ± 15.1
PET70	347.3 ± 22.9
PET70Xibond1	325.8 ± 45.0
PET70Xibond3	163.2 ± 19.3
PET70Xibond5	172.4 ± 22.9

* The CLTE has been calculated form the slope below 60 °C.

4. Conclusions

This work reports the viability of binary blends of partially bio-based poly(ethylene terephthalate) (bioPET) and fully bio-based poly(amide) 10,10 (bioPA1010) up to 30 wt % bioPA1010. Due to their immiscibility, a poly(styrene-*ran*-glycidyl methacrylate) (PS-GMA) copolymer (Xibond™ 920) is used to provide enhanced interaction. These blends can reach up to 50 wt % bio-based content without compromising other mechanical and thermal properties. The effectiveness of the PS-GMA has been corroborated with an increase in toughness, elongation at yield and tensile strength for a Xibond™ 920 loading of 3 phr. A FESEM study revealed a clear droplet-like structure with a bioPET matrix embedding bioPA-rich spherical (droplets) domains. The exceptional compatibilization effect of Xibond™ 920 in this binary blend is assessed by a remarkable decrease in the droplet diameter changing from almost 4 mm (uncompatibilized blend with 30 wt % bioPA) down to values lower than 1 mm (compatibilized blend with 30 wt % bioPA and 3 phr Xibond™ 920). Regarding the thermal properties, bioPA inhibits cold crystallization and a decrease in the degree of crystallinity of bioPET due to the formation of imperfect crystals. Xibond™ 920 also gives improved dimensional stability to blends thus leading to a new series of binary blends with balanced properties and a clear positive environmental impact since the bio-based content of these blends is close to 50 wt %.

Author Contributions: Conceptualization was devised by L.Q.-C. and R.B.; main experimental procedures were developed by M.J.; methodology, validation, and formal analysis was carried out by M.J., D.L., and S.M.-J.; investigation, resources, data curation, and writing-original draft preparation was performed by R.B. and N.M.

Funding: This research was funded by the Ministerio de Economía, Industria y Competitividad (MICINN) project number MAT2017-84909-C2-2-R. L. Quiles-Carrillo wants to thank GV for his FPI grant (ACIF/2016/182) and MECD for his FPU grant (FPU15/03812). D. Lascano wants to thank UPV for the grant received though the PAID-01-18 program. Microscopy services at UPV are acknowledged for their help in collecting and analyzing FESEM images. Authors thank Polyscope for kindly supplying Xibond™ 920 to carry out this study.

Conflicts of Interest: The authors declare no conflicts of interest.

References

1. Lahtela, V.; Hyvarinen, M.; Karki, T. Composition of Plastic Fractions in Waste Streams: Toward More Efficient Recycling and Utilization. *Polymers* **2019**, *11*, 69. [CrossRef] [PubMed]
2. Avolio, R.; Spina, F.; Gentile, G.; Cocca, M.; Avella, M.; Carfagna, C.; Tealdo, G.; Errico, M.E. Recycling Polyethylene-Rich Plastic Waste from Landfill Reclamation: Toward an Enhanced Landfill-Mining Approach. *Polymers* **2019**, *11*, 208. [CrossRef] [PubMed]
3. Zheng, J.; Suh, S. Strategies to reduce the global carbon footprint of plastics. *Nat. Clim. Chang.* **2019**, *9*, 374. [CrossRef]
4. Castilla-Cortazar, I.; Vidaurre, A.; Mari, B.; Campillo-Fernandez, A.J. Morphology, Crystallinity, and Molecular Weight of Poly(epsilon-caprolactone)/Graphene Oxide Hybrids. *Polymers* **2019**, *11*, 1099. [CrossRef] [PubMed]
5. Puchalski, M.; Szparaga, G.; Biela, T.; Gutowska, A.; Sztajnowski, S.; Krucinska, I. Molecular and Supramolecular Changes in Polybutylene Succinate (PBS) and Polybutylene Succinate Adipate (PBSA) Copolymer during Degradation in Various Environmental Conditions. *Polymers* **2018**, *10*, 251. [CrossRef] [PubMed]
6. Yamane, K.; Sato, H.; Ichikawa, Y.; Sunagawa, K.; Shigaki, Y. Development of an industrial production technology for high-molecular-weight polyglycolic acid. *Polym. J.* **2014**, *46*, 769–775. [CrossRef]
7. Liminana, P.; Garcia-Sanoguera, D.; Quiles-Carrillo, L.; Balart, R.; Montanes, N. Development and characterization of environmentally friendly composites from poly(butylene succinate) (PBS) and almond shell flour with different compatibilizers. *Compos. Part B Eng.* **2018**, *144*, 153–162. [CrossRef]
8. Khalil, F.; Galland, S.; Cottaz, A.; Joly, C.; Degraeve, P. Polybutylene succinate adipate/starch blends: A morphological study for the design of controlled release films. *Carbohydr. Polym.* **2014**, *108*, 272–280. [CrossRef]
9. Garcia-Garcia, D.; Lopez-Martinez, J.; Balart, R.; Stromberg, E.; Moriana, R. Reinforcing capability of cellulose nanocrystals obtained from pine cones in a biodegradable poly(3-hydroxybutyrate)/poly(epsilon-caprolactone) (PHB/PCL) thermoplastic blend. *Eur. Polym. J.* **2018**, *104*, 10–18. [CrossRef]
10. Ferri, J.M.; Garcia-Garcia, D.; Sanchez-Nacher, L.; Fenollar, O.; Balart, R. The effect of maleinized linseed oil (MLO) on mechanical performance of poly(lactic acid)-thermoplastic starch (PLA-TPS) blends. *Carbohydr. Polym.* **2016**, *147*, 60–68. [CrossRef]
11. Domenek, S.; Feuilloley, P.; Gratraud, J.; Morel, M.H.; Guilbert, S. Biodegradability of wheat gluten based bioplastics. *Chemosphere* **2004**, *54*, 551–559. [CrossRef]
12. Song, F.; Tang, D.-L.; Wang, X.-L.; Wang, Y.-Z. Biodegradable Soy Protein Isolate-Based Materials: A Review. *Biomacromolecules* **2011**, *12*, 3369–3380. [CrossRef] [PubMed]
13. Ferrero, B.; Boronat, T.; Moriana, R.; Fenollar, O.; Balart, R. Green Composites Based on Wheat Gluten Matrix and Posidonia Oceanica Waste Fibers as Reinforcements. *Polym. Compos.* **2013**, *34*, 1663–1669. [CrossRef]
14. Xue, Y.; Lofland, S.; Hu, X. Thermal Conductivity of Protein-Based Materials: A Review. *Polymers* **2019**, *11*, 456. [CrossRef] [PubMed]
15. Masood, F.; Yasin, T.; Hameed, A. Polyhydroxyalkanoates—What are the uses? Current challenges and perspectives. *Crit. Rev. Biotechnol.* **2015**, *35*, 514–521. [CrossRef] [PubMed]
16. Wang, Y.; Yin, J.; Chen, G.-Q. Polyhydroxyalkanoates, challenges and opportunities. *Curr. Opin. Biotechnol.* **2014**, *30*, 59–65. [CrossRef]

17. Rydz, J.; Sikorska, W.; Kyulavska, M.; Christova, D. Polyester-Based (Bio)degradable Polymers as Environmentally Friendly Materials for Sustainable Development. *Int. J. Mol. Sci.* **2015**, *16*, 564–596. [CrossRef]
18. Sharma, R.; Ray, A.R. Polyhydroxybutyrate, its copolymers and blends. *J. Macromol. Sci.-Rev. Macromol. Chem. Phys.* **1995**, *35*, 327–359. [CrossRef]
19. Liptow, C.; Tillman, A.-M. A Comparative Life Cycle Assessment Study of Polyethylene Based on Sugarcane and Crude Oil. *J. Ind. Ecol.* **2012**, *16*, 420–435. [CrossRef]
20. Boronat, T.; Fombuena, V.; Garcia-Sanoguera, D.; Sanchez-Nacher, L.; Balart, R. Development of a biocomposite based on green polyethylene biopolymer and eggshell. *Mater. Des.* **2015**, *68*, 177–185. [CrossRef]
21. Ferrero, B.; Fombuena, V.; Fenollar, O.; Boronat, T.; Balart, R. Development of natural fiber-reinforced plastics (NFRP) based on biobased polyethylene and waste fibers from Posidonia oceanica seaweed. *Polym. Compos.* **2015**, *36*, 1378–1385. [CrossRef]
22. Filgueira, D.; Holmen, S.; Melbo, J.K.; Moldes, D.; Echtermeyer, A.T.; Chinga-Carrasco, G. 3D Printable Filaments Made of Biobased Polyethylene Biocomposites. *Polymers* **2018**, *10*, 314. [CrossRef]
23. Winnacker, M.; Rieger, B. Biobased Polyamides: Recent Advances in Basic and Applied Research. *Macromol. Rapid Commun.* **2016**, *37*, 1391–1413. [CrossRef]
24. Jiang, Y.; Loos, K. Enzymatic Synthesis of Biobased Polyesters and Polyamides. *Polymers* **2016**, *8*, 243. [CrossRef]
25. Jasinska, L.; Villani, M.; Wu, J.; van Es, D.; Klop, E.; Rastogi, S.; Koning, C.E. Novel, Fully Biobased Semicrystalline Polyamides. *Macromolecules* **2011**, *44*, 3458–3466. [CrossRef]
26. Ha Thi Hoang, N.; Qi, P.; Rostagno, M.; Feteha, A.; Miller, S.A. The quest for high glass transition temperature bioplastics. *J. Mater. Chem. A* **2018**, *6*, 9298–9331. [CrossRef]
27. Eerhart, A.J.J.E.; Faaij, A.P.C.; Patel, M.K. Replacing fossil based PET with biobased PEF; process analysis, energy and GHG balance. *Energy Environ. Sci.* **2012**, *5*, 6407–6422. [CrossRef]
28. Tachibana, Y.; Kimura, S.; Kasuya, K.-i. Synthesis and Verification of Biobased Terephthalic Acid from Furfural. *Sci. Rep.* **2015**, *5*, 8249. [CrossRef]
29. Neatu, F.; Culica, G.; Florea, M.; Parvulescu, V.I.; Cavani, F. Synthesis of Terephthalic Acid by p-Cymene Oxidation using Oxygen: Toward a More Sustainable Production of Bio-Polyethylene Terephthalate. *Chemsuschem* **2016**, *9*, 3102–3112. [CrossRef]
30. Yasuda, M.; Miyabo, A. Polyamide Derived from Castor Oil. *Sen-I Gakkaishi* **2010**, *66*, P137–P142. [CrossRef]
31. Moran, C.S.; Barthelon, A.; Pearsall, A.; Mittal, V.; Dorgan, J.R. Biorenewable blends of polyamide-4,10 and polyamide-6,10. *J. Appl. Polym. Sci.* **2016**, *133*. [CrossRef]
32. Quiles-Carrillo, L.; Montanes, N.; Boronat, T.; Balart, R.; Torres-Giner, S. Evaluation of the engineering performance of different bio-based aliphatic homopolyamide tubes prepared by profile extrusion. *Polym. Test.* **2017**, *61*, 421–429. [CrossRef]
33. Welle, F.; Bayer, F.; Franz, R. Quantification of the Sorption Behavior of Polyethylene Terephthalate Polymer versus PET/PA Polymer Blends towards Organic Compounds. *Packag. Technol. Sci.* **2012**, *25*, 341–349. [CrossRef]
34. Fabia, J.; Gawlowski, A.; Graczyk, T.; Slusarczyk, C. Changes of crystalline structure of poly(ethylene terephthalate) fibers in flame retardant finishing process. *Polimery* **2014**, *59*, 557–561. [CrossRef]
35. Kuciel, S.; Kuznia, P.; Jakubowska, P. Properties of composites based on polyamide 10.10 reinforced with carbon fibers. *Polimery* **2016**, *61*, 106–112. [CrossRef]
36. Andrzejewski, J.; Szostak, M.; Bak, T.; Trzeciak, M. The influence of processing conditions on the mechanical properties and structure of poly(ethylene terephthalate) self-reinforced composites. *J. Thermoplast. Compos. Mater.* **2016**, *29*, 1194–1209. [CrossRef]
37. Cook, W.D.; Moad, G.; Fox, B.; VanDeipen, G.; Zhang, T.; Cser, F.; McCarthy, L. Morphology-property relationships in ABS/PET blends. 2. Influence of processing conditions on structure and properties. *J. Appl. Polym. Sci.* **1996**, *62*, 1709–1714. [CrossRef]
38. Bartolotta, A.; Di Marco, G.; Farsaci, F.; Lanza, M.; Pieruccini, M. DSC and DMTA study of annealed cold-drawn PET: A three phase model interpretation. *Polymer* **2003**, *44*, 5771–5777. [CrossRef]
39. Chen, Z.; Jenkins, M.J.; Hay, J.N. Annealing of poly (ethylene terephthalate). *Eur. Polym. J.* **2014**, *50*, 235–242. [CrossRef]

40. Chiou, K.C.; Chang, F.C. Reactive compatibilization of polyamide-6 (PA 6)/polybutylene terephthalate (PBT) blends by a multifunctional epoxy resin. *J. Polym. Sci. Part B Polym. Phys.* **2000**, *38*, 23–33. [CrossRef]
41. Samios, C.K.; Kalfoglou, N.K. Compatibilization of poly(ethylene terephthalate)/polyamide-6 alloys: Mechanical, thermal and morphological characterization. *Polymer* **1999**, *40*, 4811–4819. [CrossRef]
42. Huang, Y.Q.; Liu, Y.X.; Zhao, C.H. Morphology and properties of PET/PA-6/E-44 blends. *J. Appl. Polym. Sci.* **1998**, *69*, 1505–1515. [CrossRef]
43. Ferreira, C.T.; da Fonseca, J.B.; Saron, C. Recycling of Wastes from Poly(ethylene tereftalate) (PET) and Polyamide (PA) by Reactive Extrusion for Preparation of Polymeric Blends. *Polim.-Cienc. E Tecnol.* **2011**, *21*, 118–122. [CrossRef]
44. Yan, Y.; Gooneie, A.; Ye, H.; Deng, L.; Qiu, Z.; Reifler, F.A.; Hufenus, R. Morphology and Crystallization of Biobased Polyamide 56 Blended with Polyethylene Terephthalate. *Macromol. Mater. Eng.* **2018**, *303*, 1800214. [CrossRef]
45. Urquijo, J.; Guerrica-Echevarria, G.; Ignacio Eguiazabal, J. Melt processed PLA/PCL blends: Effect of processing method on phase structure, morphology, and mechanical properties. *J. Appl. Polym. Sci.* **2015**, *132*. [CrossRef]
46. Ferri, J.M.; Fenollar, O.; Jorda-Vilaplana, A.; Garcia-Sanoguera, D.; Balart, R. Effect of miscibility on mechanical and thermal properties of poly(lactic acid)/polycaprolactone blends. *Polym. Int.* **2016**, *65*, 453–463. [CrossRef]
47. Xue, B.; He, H.; Zhu, Z.; Li, J.; Huang, Z.; Wang, G.; Chen, M.; Zhan, Z. A Facile Fabrication of High Toughness Poly(lactic Acid) via Reactive Extrusion with Poly(butylene Succinate) and Ethylene-Methyl Acrylate-Glycidyl Methacrylate. *Polymers* **2018**, *10*, 1401. [CrossRef]
48. Garcia-Garcia, D.; Ferri, J.M.; Boronat, T.; Lopez-Martinez, J.; Balart, R. Processing and characterization of binary poly(hydroxybutyrate) (PHB) and poly(caprolactone) (PCL) blends with improved impact properties. *Polym. Bull.* **2016**, *73*, 3333–3350. [CrossRef]
49. Hou, A.-L.; Qu, J.-P. Super-Toughened Poly(lactic Acid) with Poly(epsilon-caprolactone) and Ethylene-Methyl Acrylate-Glycidyl Methacrylate by Reactive Melt Blending. *Polymers* **2019**, *11*, 771. [CrossRef]
50. Jesus Garcia-Campo, M.; Boronat, T.; Quiles-Carrillo, L.; Balart, R.; Montanes, N. Manufacturing and Characterization of Toughened Poly(lactic acid) (PLA) Formulations by Ternary Blends with Biopolyesters. *Polymers* **2018**, *10*, 3. [CrossRef]
51. Torres-Giner, S.; Montanes, N.; Boronat, T.; Quiles-Carrillo, L.; Balart, R. Melt grafting of sepiolite nanoclay onto poly(3-hydroxybutyrate-co-4-hydroxybutyrate) by reactive extrusion with multi-functional epoxy-based styrene-acrylic oligomer. *Eur. Polym. J.* **2016**, *84*, 693–707. [CrossRef]
52. Uribe-Calderon, J.; Diaz-Arriaga, C. The effects of carbon nanotubes, blend composition and glycidyl methacrylate-grafted polypropylene compatibilizer on the morphology, mechanical and electrical properties of polypropylene-polyamide 6 blends. *Polym. Bull.* **2017**, *74*, 1573–1593. [CrossRef]
53. Shin, B.Y.; Ha, M.H.; Han, D.H. Morphological, Rheological, and Mechanical Properties of Polyamide 6/Polypropylene Blends Compatibilized by Electron-Beam Irradiation in the Presence of a Reactive Agent. *Materials* **2016**, *9*, 342. [CrossRef]
54. Li, D.; Song, S.; Li, C.; Cao, C.; Sun, S.; Zhang, H. Compatibilization effect of MMA-co-GMA copolymers on the properties of polyamide 6/Poly(vinylidene fluoride) blends. *J. Polym. Res.* **2015**, *22*, 102. [CrossRef]
55. Lima, M.S.; Matias, A.A.; Costa, J.R.C.; Fonseca, A.C.; Coelho, J.F.J.; Serra, A.C. Glycidyl methacrylate-based copolymers as new compatibilizers for polypropylene/polyethylene terephthalate blends. *J. Polym. Res.* **2019**, *26*, 127. [CrossRef]
56. Pietrasanta, Y.; Robin, J.J.; Torres, N.; Boutevin, B. Reactive compatibilization of HDPE/PET blends by glycidyl methacrylate functionalized polyolefins. *Macromol. Chem. Phys.* **1999**, *200*, 142–149. [CrossRef]
57. McLauchlin, A.R.; Ghita, O.R. Studies on the thermal and mechanical behavior of PLA-PET blends. *J. Appl. Polym. Sci.* **2016**, *133*. [CrossRef]
58. Torres-Huerta, A.M.; Palma-Ramirez, D.; Dominguez-Crespo, M.A.; Del Angel-Lopez, D.; de la Fuente, D. Comparative assessment of miscibility and degradability on PET/PLA and PET/chitosan blends. *Eur. Polym. J.* **2014**, *61*, 285–299. [CrossRef]
59. Carrot, C.; Mbarek, S.; Jaziri, M.; Chalamet, Y.; Raveyre, C.; Prochazka, F. Immiscible blends of PC and PET, current knowledge and new results: Rheological properties. *Macromol. Mater. Eng.* **2007**, *292*, 693–706. [CrossRef]

60. Jazani, O.M.; Rastin, H.; Formela, K.; Hejna, A.; Shahbazi, M.; Farkiani, B.; Saeb, M.R. An investigation on the role of GMA grafting degree on the efficiency of PET/PP-g-GMA reactive blending: Morphology and mechanical properties. *Polym. Bull.* **2017**, *74*, 4483–4497. [CrossRef]
61. Quiles-Carrillo, L.; Montanes, N.; Sammon, C.; Balart, R.; Torres-Giner, S. Compatibilization of highly sustainable polylactide/almond shell flour composites by reactive extrusion with maleinized linseed oil. *Ind. Crops Prod.* **2018**, *111*, 878–888. [CrossRef]
62. Quiles-Carrillo, L.; Duart, S.; Montanes, N.; Torres-Giner, S.; Balart, R. Enhancement of the mechanical and thermal properties of injection-molded polylactide parts by the addition of acrylated epoxidized soybean oil. *Mater. Des.* **2018**, *140*, 54–63. [CrossRef]
63. Garcia-Garcia, D.; Rayon, E.; Carbonell-Verdu, A.; Lopez-Martinez, J.; Balart, R. Improvement of the compatibility between poly(3-hydroxybutyrate) and poly(8-caprolactone) by reactive extrusion with dicumyl peroxide. *Eur. Polym. J.* **2017**, *86*, 41–57. [CrossRef]
64. Jiang, S.; Mi, R.; Yun, R.; Qi, S.; Zhang, X.; Lu, Y.; Matejka, V.; Peikertova, P.; Tokarsky, J. Structure and properties of kaolinite intercalated with potassium acetate and their nanocomposites with polyamide 1010. *J. Thermoplast. Compos. Mater.* **2017**, *30*, 971–985. [CrossRef]
65. Serhatkulu, T.; Erman, B.; Bahar, I.; Fakirov, S.; Evstatiev, M.; Sapundjieva, D. Dynamic-mechanical study of amorphous phases in poly(ethylene terephthalate)/nylon-6 blends. *Polymer* **1995**, *36*, 2371–2377. [CrossRef]

© 2019 by the authors. Licensee MDPI, Basel, Switzerland. This article is an open access article distributed under the terms and conditions of the Creative Commons Attribution (CC BY) license (http://creativecommons.org/licenses/by/4.0/).

MDPI
St. Alban-Anlage 66
4052 Basel
Switzerland
Tel. +41 61 683 77 34
Fax +41 61 302 89 18
www.mdpi.com

Polymers Editorial Office
E-mail: polymers@mdpi.com
www.mdpi.com/journal/polymers

www.ingramcontent.com/pod-product-compliance
Lightning Source LLC
LaVergne TN
LVHW070448100526
838202LV00014B/1687